Micro
CMOS
DESIGN

Circuits and Electrical Engineering Series

Series Editor
Wai-Kai Chen

MicroCMOS Design
Bang-Sup Song

FORTHCOMING

Multiple-Base Number System: Theory and Applications
Vassil Dimitrov, Graham Jullien, and Roberto Muscedere

Micro CMOS DESIGN

BANG-SUP SONG

CRC Press
Taylor & Francis Group
Boca Raton   London   New York

CRC Press is an imprint of the
Taylor & Francis Group, an **informa** business

CRC Press
Taylor & Francis Group
6000 Broken Sound Parkway NW, Suite 300
Boca Raton, FL 33487-2742

First issued in paperback 2017

Version Date: 2011915

ISBN 13: 978-1-138-07236-7 (pbk)
ISBN 13: 978-1-4398-1895-4 (hbk)

Library of Congress Cataloging-in-Publication Data

Song, Bang-Sup.
  MicroCMOS design / Bang-Sup Song.
    p. cm. -- (Circuits & electrical engineering)
  Summary: "Starting at the transistor level, this book covers basic system-level CMOS design concepts applicable to modern SoCs. The text uses practical design examples to illustrate circuit design so that readers can develop an intuitive rather than conventional analytic understanding. System-level knowledge is built upon understanding fundamental concepts of noise, jitter, and frequency and phase noise. This material addresses basic abstract concepts of transistor circuits, as well as advanced topics such as ADCs and PLLs, providing a proper perspective on this advanced SoC design. Other topics include DAC, phase locked loop, frequency synthesizer, clock recovery, and digital assisting"-- Provided by publisher.
  Includes bibliographical references and index.
  ISBN 978-1-4398-1895-4 (hardback)
  1. Metal oxide semiconductors, Complementary. I. Title. II. Series.

TK7871.99.M44S66 2012
621.3815'28--dc23
                                                        2011032460

Contents

Preface

From the early 1970s, the digitization of media has dramatically changed our lives, and shaped the way in which we consume digital information. Because digital systems benefit greatly from advanced complementary metal-oxide semiconductor (CMOS) technologies, they will prevail in the foreseeable cloud-computing and predominantly portable multimedia era through wireless networks. In all digital systems, the processed data are transmitted and received through wireline or wireless channels in the analog form. They are also stored and recovered through magnetic or optical media. Most digital systems are based on massive systems-on-chip (SoCs) with multiple analog/radio-frequency (RF) interfaces. Analog/RF designs in this SoC environment require designers to be familiar with both system- and transistor-level design aspects. As the complexity of SoCs grows, analog/RF systems need to be designed using abstract concepts of the large components, such as operational amplifiers (opamps), analog-to-digital converters (ADCs), and phase-locked loops (PLLs).

This book aims to introduce analog design methodologies with specific emphasis on analog systems that can be integrated into SoCs. The design starts from extracting an abstract concept of both bipolar junction transistor (BJT) and metal-oxide semiconductor (MOS) transistors, and builds larger systems using them. Readers may notice that this book does not focus on the analysis but on the design aspect. Analog design concepts are presented without resorting to derivations of lengthy equations for analysis. More intuitive approaches are taken based on the core design concepts. There are four introductory chapters that emphasize the fundamentals of feedback stability (Chapter 1), the transistor/amplifier concept (Chapter 2), data converter basics (Chapter 4), and PLL basics (Chapter 8). The other five main chapters discuss opamps (Chapter 3), Nyquist-rate converters (Chapter 5), oversampling converters (Chapter 6), high-resolution converters (Chapter 7), and synthesizers and clock recovery (Chapter 9). All feedback systems such as opamps, $\Delta\Sigma$ modulators, and PLLs are presented consistently using common basic circuit concepts and the same parameters for linear small-signal analysis.

This book is mainly written and organized to give proper perspectives on the various designs of data converters and PLLs, which are the two most common analog circuit components in SoCs. The materials covered in the book are suitable for graduate-level students and advanced engineers in the field, though the four introductory chapters are written for engineers with an undergraduate background. Readers are advised to review the four introductory chapters to become familiar with the basic circuit concepts before they move on to the main chapters for practical designs of microCMOS systems.

Bang-Sup Song
La Jolla, California

Acknowledgments

Many thanks go to my graduate students Seung-Hoon Lee, Unku Moon, Tzishiung Shu, Sung-Ung Kwak, Myung-Jun Choe, Alex Bugeja, Woogeun Rhee, Hsinshu Chen, Chun Huat Heng, Seung-Tak Ryu, Sourja Ray, Supisa Lerstaveesin, and Yunshiang Shu, and also to my industry friends M. Tompsett, K. Bacrania, D. Soo, P. Lakers, S. Gillig, T. Cho, D. Kang, Y. Konno, K. Tomioka, Y. Aiba, K. Yamazoe, K. Hamashita, J. Kamiishi, T. Suzuki, S. Takeuchi, K. Koyama, and T. Yoshioka. Without their contributions, this book would not exist.

Special thanks go to Professor Wai-Kai Chen, University of Illinois, Chicago; the staff at Taylor & Francis/CRC Press; and also to Drs. Chong Lee and John Hong, Qualcomm, for their encouragement.

The Author

Bang-Sup Song, Ph.D., received a B.S. from Seoul National University, Korea, in 1973, an M.S. from Korea Advanced Institute of Science in 1975, and a Ph.D. from the University of California–Berkeley in 1983. From 1975 to 1978, he was a member of the research staff at the Agency for Defense Development, Korea. From 1983 to 1986, he was a member of the technical staff at AT&T Bell Laboratories, Murray Hill, New Jersey, and was also a visiting faculty member in the Department of Electrical Engineering, Rutgers University, New Jersey. From 1986 to 1999, Dr. Song was a professor in the Department of Electrical and Computer Engineering and the Coordinated Science Laboratory at the University of Illinois at Urbana. In 1999, Dr. Song joined the faculty of the Department of Electrical and Computer Engineering, University of California, San Diego, where he is endowed with the position of Charles Lee Powell Chair Professor in Wireless Communication.

Dr. Song received a Distinguished Technical Staff Award from AT&T Bell Laboratories in 1986, a Career Development Professor Award from Analog Devices in 1987, and a Xerox Senior Faculty Research Award from the University of Illinois in 1995. His Institute of Electrical and Electronics Engineers (IEEE) activities have been in the capacities of an associate editor and a program committee member for the *IEEE Journal of Solid-State Circuits, IEEE Transactions on Circuits and Systems,* International Solid-State Circuits Conference, and International Symposium on Circuits and Systems. Dr. Song is an IEEE fellow.

1

Amplifier Basics

Amplifiers operate mostly in a linear mode, and their performance is improved by applying feedback. Basic circuit concepts such as complex transfer function, poles, frequency and transient responses, trade-off of gain for bandwidth and linearity, feed-forward zero, and stability criteria of feedback systems are essential to understand, design, and operate amplifier circuits properly.

1.1 Driving-Point and Transfer Functions

Consider any electrical networks with one or two ports as shown in Figure 1.1. We can define the voltage on each port and the current flowing into each terminal. Here, the voltage and current are small signal and frequency dependent.

In the one-port case shown on the left side, the following two ratios of the voltage and current can be defined as

$$Z(s) = \frac{v_i(s)}{i_i(s)}, \quad \text{and} \quad Y(s) = \frac{i_i(s)}{v_i(s)}. \tag{1.1}$$

The former is the driving-point impedance, and the latter is the driving-point admittance. Their units are Ω and $1/\Omega$, respectively. At low frequencies, they are commonly called driving-point resistance and conductance, respectively. That is, if the input and output are referred to the same port, the term *driving point* is used. Similarly, in the two-port case as shown on the right side, the four ratios can be defined as follows:

$$A_v(s) = \frac{v_o(s)}{v_i(s)}, \quad A_i(s) = \frac{i_o(s)}{i_i(s)}, \quad Z(s) = \frac{v_o(s)}{i_i(s)}, \quad \text{and} \quad Y(s) = \frac{i_o(s)}{v_i(s)}, \tag{1.2}$$

where $A_v(s)$ and $A_i(s)$ are the unit-less transfer functions called the voltage and current gains, respectively. The latter two definitions are the same as in Equation (1.1) for the one-port network, but they are now called transimpedance and transadmittance as the input and output ports are referred to two ports. The prefix *trans* relates voltages and currents in two different terminals. Of course, at low frequencies, they are more commonly called transresistance and transconductance, respectively.

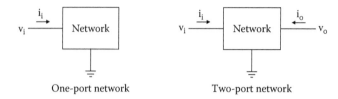

FIGURE 1.1
Driving-point and transfer concepts.

1.2 Frequency Response

In the small-signal steady state, impedances of reactive components such as inductor and capacitor are frequency dependent as sL and 1/sC, where s is the frequency of $j\omega$. The unit of ω is rad/sec. Note that the angular frequency ω is defined as the amount of angle rotation per second, and the ordinary frequency f in the Hertz (Hz) unit is defined as the number of rotations per second. Because one rotation covers an angle of 2π radian, there exists a 2π difference in the relation of $\omega = 2\pi f$. In steady state, all transfer functions of electronic circuits can be represented using a general transfer function $H(s)$, which is a ratio of two polynomials $N(s)$ and $D(s)$ as follows:

$$H(s) = \frac{N(s)}{D(s)} = \frac{b_1 s^m + b_2 s^{m-1} + \cdots + b_m}{a_1 s^n + a_2 s^{n-1} + \cdots + a_n}. \tag{1.3}$$

The solutions of $D(s) = 0$ are poles, and those of $N(s) = 0$ are zeros. Poles and zeros affect both the frequency and transient responses.

If s is replaced by $j\omega$, the steady-state frequency response is obtained. The frequency response is expressed as a complex transfer function $H(j\omega)$, which can be written in a vector form:

$$H(j\omega) = H_R(\omega) + jH_I(\omega) = M(\omega)e^{j\theta(\omega)}. \tag{1.4}$$

For the gain and phase-response plots, the magnitude and phase responses in Equation (1.4) can be rewritten as follows:

$$M(\omega) = |H(j\omega)| = \sqrt{H_R^2(\omega) + H_I^2(\omega)}. \tag{1.5}$$

$$\theta(\omega) = \angle H(j\omega) = \tan^{-1}\frac{H_I(\omega)}{H_R(\omega)}. \tag{1.6}$$

The following is an example of the frequency response defined in Equation (1.4) with three poles and two zeros:

$$H(j\omega) = \frac{a_o\left(1 + \dfrac{j\omega}{\omega_{z1}}\right)\left(1 + \dfrac{j\omega}{\omega_{z2}}\right)}{\left(1 + \dfrac{j\omega}{\omega_{p1}}\right)\left(1 + \dfrac{j\omega}{\omega_{p2}}\right)\left(1 + \dfrac{j\omega}{\omega_{p3}}\right)},$$

(1.7)

where ω_{p1}, ω_{p2}, ω_{p3}, ω_{z1}, and ω_{z2} are pole and zero frequencies. Note that the frequency is normalized to each pole or zero frequency. This factorized transfer function is a convenient form to use for analysis. It enables designers to handle the gain and phase responses separately. Therefore, the most common amplifier design strategy is to set the direct current (DC) gain parameter and to consider the high-frequency effect individually.

At low frequencies where ω/ω_p or ω/ω_z is far smaller than 1, all imaginary terms can be ignored, and the transfer function is very close to the real number a_o with no significant phase shift. Therefore, a_o in the voltage or current gain transfer function is commonly called a small-signal, low-frequency, or DC gain. In general, the small-signal parameter a_o can be any of the small-signal low-frequency parameters such as driving-point resistance, driving-point conductance, voltage gain, current gain, transresistance, or transconductance as defined in Equations (1.1) and (1.2). Now at high frequencies where ω/ω_p or ω/ω_z is no longer negligible compared to 1, the imaginary terms begin to contribute to the transfer function, and its gain and phase vary with frequency. Both the gain and phase responses given by Equations (1.5) and (1.6) can be derived for the example of Equation (1.7) as follows:

$$M(\omega) = \left|\frac{a_o\left(1 + \dfrac{j\omega}{\omega_{z1}}\right)\left(1 + \dfrac{j\omega}{\omega_{z2}}\right)}{\left(1 + \dfrac{j\omega}{\omega_{p1}}\right)\left(1 + \dfrac{j\omega}{\omega_{p2}}\right)\left(1 + \dfrac{j\omega}{\omega_{p3}}\right)}\right| = a_o\sqrt{\frac{\left(1 + \dfrac{\omega^2}{\omega_{z1}^2}\right)\left(1 + \dfrac{\omega^2}{\omega_{z2}^2}\right)}{\left(1 + \dfrac{\omega^2}{\omega_{p1}^2}\right)\left(1 + \dfrac{\omega^2}{\omega_{p2}^2}\right)\left(1 + \dfrac{\omega^2}{\omega_{p3}^2}\right)}}.$$

(1.8)

$$\theta(\omega) = \tan^{-1}\frac{\omega}{\omega_{z1}} + \tan^{-1}\frac{\omega}{\omega_{z2}} - \tan^{-1}\frac{\omega}{\omega_{p1}} - \tan^{-1}\frac{\omega}{\omega_{p2}} - \tan^{-1}\frac{\omega}{\omega_{p3}}.$$

(1.9)

Consider an example of a transfer function with a DC gain of 100 dB (10^5), one zero at 10^7 rad/sec, and three poles at 10, 10^6, and 10^8 rad/sec, respectively. Then, the transfer function of Equation (1.7) is given by

$$H(j\omega) = \frac{10^5\left(1 + \dfrac{j\omega}{10^7}\right)}{\left(1 + \dfrac{j\omega}{10}\right)\left(1 + \dfrac{j\omega}{10^6}\right)\left(1 + \dfrac{j\omega}{10^8}\right)}.$$

(1.10)

Both gain and phase responses can be derived by plugging the real numbers from Equation (1.10) into Equations (1.8) and (1.9).

$$M(\omega) = 10^5 \sqrt{\frac{\left(1 + \dfrac{\omega^2}{10^{14}}\right)}{\left(1 + \dfrac{\omega^2}{10^2}\right)\left(1 + \dfrac{\omega^2}{10^{12}}\right)\left(1 + \dfrac{\omega^2}{10^{16}}\right)}}. \tag{1.11}$$

$$\theta(\omega) = \tan^{-1}\frac{\omega}{10^7} - \tan^{-1}\frac{\omega}{10} - \tan^{-1}\frac{\omega}{10^6} - \tan^{-1}\frac{\omega}{10^8}. \tag{1.12}$$

The standard Bode plots are these two separate gain and phase responses as sketched in Figure 1.2 using the log scale for gain and frequency. The solid straight lines in the plots are just for illustration, and the actual gain and phase responses should make smooth transitions across the pole and zero frequencies as sketched with dotted lines. The overall gain slope varies from −6 dB/oct to −12 dB/oct, and the phase delay is maintained at about −90° at low frequencies but approaches −180° gradually at high frequencies. The zero in the middle makes some irregularities both in the gain and phase slopes.

Note that gain increases and phase leads after passing zero frequencies, while gain decreases and phase lags after passing pole frequencies. That is, the magnitude decreases as a function of $1/\omega$ after each pole, which is equivalent to the gain slope of −6 dB/oct or −20 dB/dec in the Bode gain plot. Here in the unit, the frequency multiples of octave (×2) and decade (×10) are abbreviated as oct and dec, respectively. At a pole frequency, the gain drops by about −3 dB, which is equivalent to the attenuation by $1/\sqrt{2}$. The phase delay is a nonlinear $\tan^{-1}$ function, and at the pole frequency, the phase delay is −45°, which is $-\tan^{-1}1$. The same can be said for each zero with the gain slope of 6 dB/oct or 20 dB/dec,

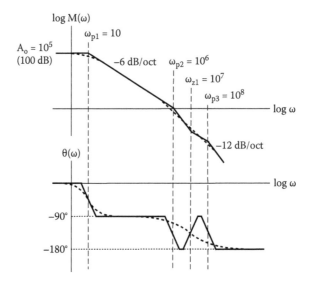

FIGURE 1.2
Bode gain and phase plots.

and the gain of 3 dB and the phase lead of 45° at the zero frequency. These simplified graphical representations of the open-loop Bode gain and phase plots are convenient for stability analysis.

1.3 Stability Criteria

Every circuit node inside a network makes its own pole that is complex, and it character-izes the way the circuit node behaves both in the steady-state and transient conditions. Poles are associated with their unique natural frequencies. Note that poles are only deter-mined by the circuit parameters, regardless of the external sources. If the pole at the ith circuit node is assumed to be p_i, then its transient response is characterized by an expo-nential function. So the transient response of a network can be the linear sum of all natural responses contributed by all of its poles as follows:

$$v_o(t) = k_o + k_1 e^{p_1 t} + k_2 e^{p_2 t} + k_3 e^{p_3 t} + \cdots, \tag{1.13}$$

where p_i is a pole, and k_i is a constant set by the transient condition. Zeros only affect the transient condition, not the natural response. Because p_i is a complex number, it has both real and imaginary parts like $p_i = \alpha + j\omega$. The real part represents the exponential enve-lope, and the imaginary part the frequency component. If poles are negative real poles, the natural response is a pure decaying exponential because the real part of p_i is negative. Otherwise, it is a sinusoid decaying exponentially.

$$e^{p_i t} = e^{(\alpha + j\omega)t} = e^{\alpha t} e^{j\omega t} = e^{\alpha t}(\cos\omega t + j\sin\omega t). \tag{1.14}$$

The real part of Equation (1.14) is the natural response of the complex conjugate poles, and the imaginary part is its quadrature image. Therefore, for any system to be exponentially stable, the real parts of its poles should be negative ($\alpha < 0$) so that any transients may decay exponentially over time.

Figure 1.3 shows three cases with the same negative real pole but with different transfer functions. The circuit on the left side is the Norton's equivalent circuit of the low-pass filter

$$\frac{v_o}{i_i} = \frac{R}{1 + sRC} \qquad \frac{v_o}{v_i} = \frac{1}{1 + sRC} \qquad \frac{v_o}{v_i} = \frac{sRC}{1 + sRC}$$

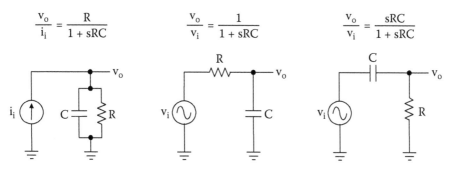

FIGURE 1.3
Single resistance-capacitance (RC) pole cases.

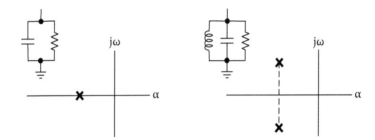

FIGURE 1.4
Location of poles in the complex s-plane.

shown in the middle. The circuit on the right is a high-pass filter. They all have the same time constant of resistance-capacitance (RC), and the negative real pole is at –1/RC. The high-pass filter has a zero at DC due to the capacitor in the signal pass. Note that signal feed-forward through a capacitor always creates a zero. The natural responses of the three are the same because they have the same pole. Because the signal is attenuated by 3 dB at the pole, the pole frequency is also called –3 dB bandwidth or cut-off frequency, as defined below:

$$\omega_{-3dB} = \frac{1}{RC} = \frac{1}{\tau}, \quad f_{-3dB} = \frac{1}{2\pi RC}, \quad \omega_{-3dB} = 2\pi f_{-3dB}, \tag{1.15}$$

where the units for ω_{-3dB}, f_{-3dB}, and τ are rad/sec, Hz, and sec, respectively. Only the time constant τ affects the exponential decay of the natural response, and it is set by the network parameter RC regardless of the signal source.

If there is no feedback applied to any networks that consist of transistors, resistors, and capacitors, their poles are always negative real as shown in Figure 1.4. If they include inductors, poles become complex conjugate but remain in the open left-half complex s-plane. Therefore, all networks without feedback such as open-loop amplifiers are always stable.

Stability is a concern only in feedback amplifiers. If the feedback loop gain is increased, poles may move into the right-half complex s-plane, thereby causing instability. Stability of the feedback network can be analyzed using either the Root-Locus method in the complex s-plane or the Bode gain and phase plots. The latter is far simpler, as the former is difficult to obtain analytically without using computers. Simplified Bode plots as shown in Figure 1.2 are handy for all stability analysis.

1.4 Operational Amplifier (Opamp) in Negative Feedback

An opamp can be considered a device or a circuit block that amplifies a differential input with a very high DC gain. Its input resistance is also very high. Most low-frequency feedback amplifier designs using opamps start with an assumption that opamps are ideal, and only the stability and finite bandwidth effects are considered later. The open-loop transfer function of the opamp is approximated as a single pole roll-off as explained using the Bode plots in Figure 1.5.

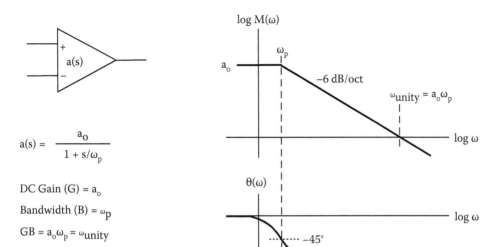

FIGURE 1.5
An opamp and its open-loop gain and phase responses.

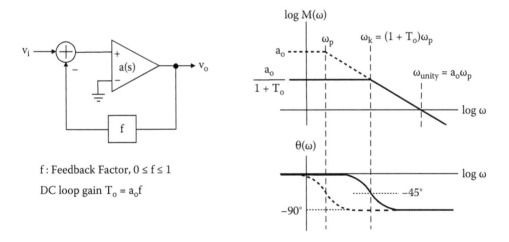

FIGURE 1.6
An opamp-based feedback amplifier.

Assume that opamp has a DC gain of a_o and a single negative real pole at $-\omega_p$, which sets the open-loop −3 dB bandwidth to be ω_p. The phase delay at the pole frequency is about −45°, and the total phase delay after one pole becomes −90°. Therefore, the gain-bandwidth (GB) product is $a_o\omega_p$, which is the frequency where the open-loop gain becomes unity. This open-loop unity-gain frequency is marked as ω_{unity}. Note that opamp is a high-gain amplifier, but its open-loop bandwidth is very narrow. Therefore, the excess gain can be traded for wider bandwidth with negative feedback.

Figure 1.6 shows a closed-loop feedback amplifier using an opamp. In this standard negative feedback example, the output is attenuated by f and subtracted from the input. The forward gain of the opamp is $a(s)$, and the feedback factor is f, which is assumed to be independent of frequency with a value ranging from 0 to 1. If f = 0, there is no feedback, and the amplifier is an open-loop amplifier with the gain of $a(s)$. If f = 1, the whole output is fed back and subtracted from the input. This feedback condition is called the worst-case unity-gain feedback. The loop gain $T(s)$ and the closed-loop gain $H(s)$ can be defined as follows:

$$T(s) = a(s)f.$$

$$H(s) = \frac{v_o(s)}{v_i(s)} = \frac{a(s)}{1+a(s)f} = \frac{a(s)}{1+T(s)}. \qquad (1.16)$$

The following relation between the two low-frequency (DC) parameters, DC loop gain T_o and closed-loop DC gain H_o, clearly explains the feedback effect that the gain decreases as the loop gain increases.

$$T_o = a_o f.$$

$$H_o = \frac{a_o}{1+a_o f} = \frac{a_o}{1+T_o} \approx \frac{a_o}{T_o} = \frac{1}{f}. \qquad (1.17)$$

The negative feedback improves the amplifier performance drastically. Because the open-loop DC gain a_o is lowered by $(1 + T_o)$ after feedback, the closed-loop DC gain H_o is about $1/f$ as given in Equation (1.17) if the DC loop gain T_o is high. This implies that the closed-loop DC gain is no longer affected by the open-loop DC gain a_o, which heavily depends on process, supply voltage, and temperature (PVT) variations. Although the DC gain a_o is reduced after feedback, all other parameters such as linearity, impedance, signal range, bandwidth, and sensitivity to PVT are improved. One of the most notable trade-off effects is that the closed-loop bandwidth ω_k is widened by the same factor $(1 + T_o)$ as sketched using the solid line on the right side of Figure 1.6. Note that ω_k is the frequency where the feedback loop gain is unity. Therefore, ω_k can be called either unity loop-gain frequency or –3 dB bandwidth of the closed-loop gain.

This single-pole negative feedback is absolutely stable with a phase margin of 90° because the total loop phase delay is only –90° at ω_k. However, real opamps have many nondominant high-frequency poles, which contribute some phase shift at ω_k. Therefore, the stability

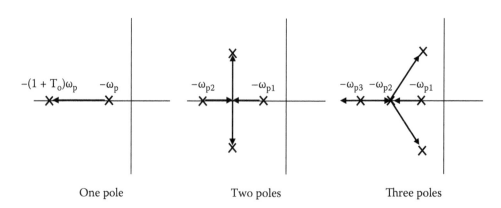

One pole   Two poles   Three poles

FIGURE 1.7
Three cases of Root Locus as T_o increases.

of feedback amplifiers using opamps should be closely looked at. To ensure the stability of a feedback amplifier with opamp, its poles of the transfer function that are the solutions of $1 + T(s) = 0$ should be in the open left-half complex s-plane. Three cases of Root Locus in Figure 1.7 show the poles in the complex s-plane when the loop gain is increased.

In one- or two-pole cases, poles still stay in the open left-half plane, but in the three-pole case, two complex conjugate poles are moving toward the right-half plane as T_o increases. Therefore, three-pole opamps are conditionally stable only with low loop gain and would be unstable for most usable feedback conditions. The situation gets far worse with more poles than three.

1.5 Phase Margin

For an opamp to be stable with negative feedback applied, the total excess loop phase delay should be less than 180° at the unity loop-gain frequency ω_k where the loop gain is unity, or the loop gain should be less than unity at the frequency where the total excess loop phase delay is 180°. This is to prevent the negative feedback from becoming the positive feedback. The Bode gain plot of the open-loop transfer function $a(s)$ can be easily decomposed into two separate plots. One is the feedback loop gain $T(s)$, and the other is the closed-loop gain $H(s)$ as shown in Figure 1.8.

The most important frequency to note is again the unity loop-gain frequency ω_k. The dotted line drawn horizontally in the open-loop gain plot indicates the feedback loop gain of unity. The top portion above the line is the feedback loop gain, and what is left after removing the top portion is the closed-loop gain. If the DC loop gain T_o is taken off from the open-loop DC gain a_o, the closed-loop gain H_o is left. At low frequencies, H_o is about $1/f$. It is lower than the open-loop gain a_o by the DC loop gain $(1 + T_o) \sim T_o = a_o f$. After this gain trade-off, the closed-loop gain H_o now depends only on the passive feedback

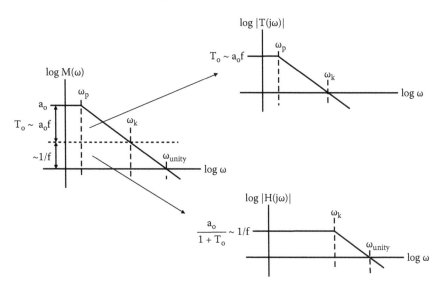

FIGURE 1.8
Bode gain plot with feedback.

network regardless of the open-loop gain a_o. Therefore, the broad-banded closed-loop amplifier with a lower gain H_o becomes more linear. Because the gain slope is –6 dB/oct, the following gain–bandwidth relations still hold:

$$\omega_k = (1+T_o)\omega_p \approx T_o\omega_p = (a_o f)\omega_p = f\omega_{unity}.$$

$$GB = \left(\frac{a_o}{1+T_o}\right)\omega_k = \left(\frac{a_o}{1+T_o}\right)(1+T_o)\omega_p = a_o\omega_p = \omega_{unity}.$$

(1.18)

For frequencies higher than ω_k, the closed-loop gain stays the same as the open-loop gain because the feedback loop gain is less than unity.

The gain or phase margin (GM or PM) is defined as a room for extra loop gain or loop phase until the negative feedback opamp becomes unstable. In order to define the GM, another frequency ω_{180}, which is the frequency where the extra loop phase delay becomes –180°, should be defined. Using ω_k and ω_{180}, the GM and PM are graphically explained in the Bode gain and phase plots of a three-pole feedback opamp in Figure 1.9.

Because each pole contributes a phase delay of 90°, a total of –270° phase delay by three poles can cause a negative feedback opamp to be unstable with high loop gain. Two dotted lines are drawn horizontally in both the gain and phase plots where the loop gain is unity and where the phase shift is –180°, respectively. So, both GM and PM can be estimated as follows:

$$GM = \frac{1}{|T(j\omega_{180})|}, \quad \angle T(j\omega_{180}) = -180^o.$$

$$PM = 180^o - |\angle T(j\omega_k)|, \quad |T(j\omega_k)| = 1.$$

(1.19)

However, it is not necessary to specify both GM and PM, and either one would suffice. In practice, PM is more widely used. For the three-pole one-zero transfer function example

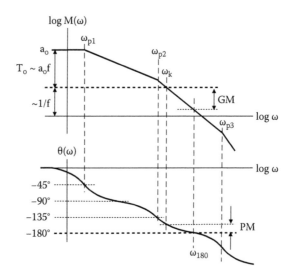

FIGURE 1.9
Open-loop Bode plots of a three-pole opamp.

given in Equation (1.10), PM can be estimated from Equation (1.19) as follows for any given feedback factor f:

$$10^5 f \sqrt{\frac{\left(1+\dfrac{\omega_k^2}{10^{14}}\right)}{\left(1+\dfrac{\omega_k^2}{10^2}\right)\left(1+\dfrac{\omega_k^2}{10^{12}}\right)\left(1+\dfrac{\omega_k^2}{10^{16}}\right)}} = 1. \tag{1.20}$$

$$PM = 180^o + \tan^{-1}\frac{\omega_k}{10^7} - \tan^{-1}\frac{\omega_k}{10} - \tan^{-1}\frac{\omega_k}{10^6} - \tan^{-1}\frac{\omega_k}{10^8}.$$

After finding ω_k from the gain plot, it can be used to estimate the PM in the phase plot.

In most feedback amplifiers, the optimum PM to target is greater than 60°. If PM is smaller than 60°, the frequency response exhibits a slight peaking at the unity loop-gain frequency, and the transient response tends to ring. If PM is 45°, the closed-loop gain at ω_k becomes

$$\left|H\left(j\omega_k\right)\right| = \left|\frac{a\left(j\omega_k\right)}{1+T\left(j\omega_k\right)}\right| = \left|\frac{a\left(j\omega_k\right)}{1+e^{-j135^o}}\right| \approx 1.3\left|a\left(j\omega_k\right)\right|. \tag{1.21}$$

This implies that the frequency response would peak by about 30% at the pass-band edge as sketched in Figure 1.10.

If PM = 0°, the loop gain becomes −1 at ω_k because $T(j\omega_k) = \exp(-j180^\circ)$ in Equation (1.21). Then the closed-loop gain becomes infinite, which means instability. Any noise power at ω_k will grow exponentially. Because $\pm j\omega_k$ are the solutions of the denominator polynomial $D(s) = 0$, they are poles of the closed-loop transfer function $H(s)$. That is, two complex-conjugate poles are at $\pm j\omega_k$ on the imaginary axis. This also implies that the Root Locus of two complex-conjugate poles are right on the imaginary axis. When the unity loop-gain frequency ω_k is close to the second pole, PM is about 45°. Therefore, to achieve a PM greater than 60°, it is necessary to either move the second pole out to frequencies higher than ω_k or add a zero to compensate for the phase delay contributed by the second pole. However, if the pole-zero cancellation occurs at lower frequencies than ω_k, the transient response will still settle slowly with the time constant of the canceled pole. In practice, to avoid peaking in the frequency response and ringing in the transient response, PM greater than 60° should be warranted without any pole-zero cancellation.

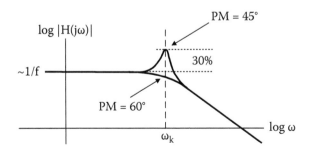

FIGURE 1.10
Phase margin (PM) effect on the frequency response.

1.6 Transient Response

Let us consider a purely exponential transient response settling with a good PM. The open-loop response is still assumed to be one pole roll-off as follows:

$$a(s) = \frac{a_o}{1 + \dfrac{s}{\omega_p}} \approx \frac{a_o \omega_p}{s} = \frac{\omega_{unity}}{s}. \tag{1.22}$$

Then, the closed-loop transfer function is defined as

$$H(s) = \frac{a(s)}{1 + a(s)f} \approx \frac{1}{f} \frac{1}{\left(1 + \dfrac{s}{\omega_k}\right)}. \tag{1.23}$$

Now if the step input with a magnitude V_{step} is applied at $t = 0$, its Laplace transform is

$$V_i(s) = \frac{V_{step}}{s}. \tag{1.24}$$

From Equations (1.23) and (1.24), the output in the complex frequency domain is obtained as

$$V_o(s) = H(s)V_i(s) = \frac{V_{step}}{f} \frac{1}{s\left(1 + \dfrac{s}{\omega_k}\right)}. \tag{1.25}$$

Therefore, by taking its inverse Laplace transform, the transient response is obtained as

$$v_o(t) = \frac{V_{step}}{f}\left(1 - e^{-\frac{t}{\tau}}\right), \quad \tau = \frac{1}{\omega_k} = \frac{1}{f\omega_{unity}}. \tag{1.26}$$

As expected, the natural response to the step input is exponential and has a time constant of $1/\omega_k$, and the output step voltage is V_{step}/f as shown in Figure 1.11.

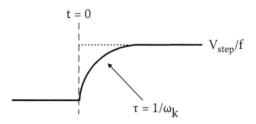

FIGURE 1.11
Transient response of a feedback amplifier.

This most desirable exponential step response is achievable only with negative real poles. However, complex-conjugate poles are generated in any feedback amplifier. In the worst-case unity-gain feedback (f = 1), the step response of the voltage follower can exhibit both exponentially decaying and sinusoidal natural responses as shown in Figure 1.12. Depending on PM, one feature is more prominent than the other.

In real feedback opamps, the zero in the open-loop transfer function or poor PM can cause the transient peaking or ringing, and excessive ringing due to poor PM should be avoided. Assume that a feedback amplifier has a poor PM with two complex-conjugate poles. In the complex-conjugate pole case, the following form of the closed-loop transfer function $H(s)$ is more convenient to use:

$$H(s) = \frac{1}{f} \frac{1}{\left(1 + \frac{1}{Q}\frac{s}{\omega_o} + \frac{s^2}{\omega_o^2}\right)} = \frac{1}{f} \frac{1}{\left(1 + 2\rho\frac{s}{\omega_o} + \frac{s^2}{\omega_o^2}\right)}, \tag{1.27}$$

where Q is the quality factor, and ρ is the damping factor. The complex poles in the s-plane and the overshoot and ringing in the transient output are explained in Figure 1.13.

High Q means that the poles are closer to the imaginary axis. As a result, PM decreases, and the ringing in the transient response lasts longer with a longer time constant. The decaying time constant is inversely proportional to the distance from the imaginary axis. That is, the higher the Q, the longer is the decaying time constant. The ringing

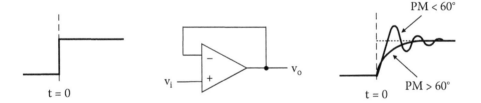

FIGURE 1.12
Transient response of an opamp in a unity-gain feedback.

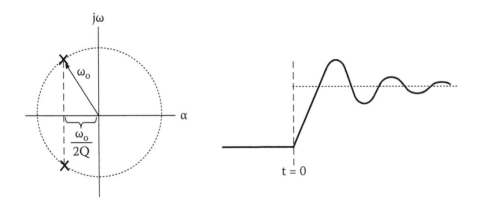

FIGURE 1.13
Conjugate poles with ringing and overshoot in the transient response.

frequency is given as the vertical distance from the real axis in the complex s-plane. Accurate analysis of the ringing is not worth the effort because ringing transient responses are slow and nonlinear and should be avoided by all means when designing feedback amplifiers.

Note that the transient response obtained based on the small-signal linear analysis is only valid when the feedback opamp operates in a small-signal linear mode. With a sudden large step input applied, the opamp input stage can be easily driven into a nonlinear region, and real opamps behave nonlinearly and slew as will be discussed in Chapter 3.

1.7 Feedback Amplifier

The feedback network shown in Figure 1.6 can be implemented using an opamp and a resistive voltage divider as shown in Figure 1.14.

The Bode plots for the feedback loop gain and the closed-loop gain are included in the figure with other important parameters. Again, the stability is determined by the excess phase delay of the loop gain at the unity loop-gain frequency ω_k. PM can be estimated as in Equation (1.20). This configuration has a noninverting gain of $\sim 1/f$, and the input resistance is high as the input looks into the opamp input. Although all opamps reject common-mode

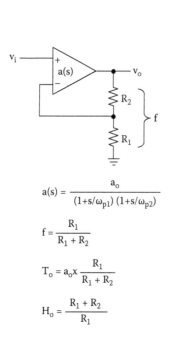

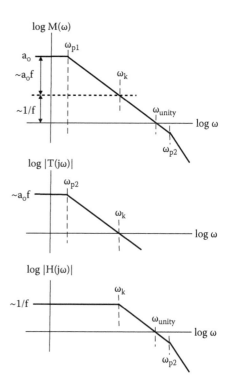

FIGURE 1.14
Opamp feedback with a resistive divider.

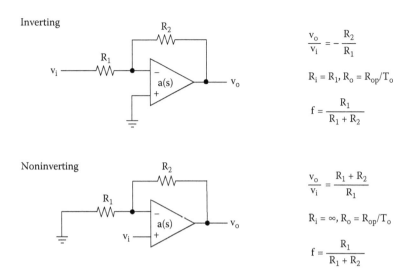

FIGURE 1.15
Inverting and noninverting gain amplifiers.

signals, one drawback of this noninverting amplifier is the input common-mode swing. For good PM greater than 60°, the nondominant pole ω_{p2} should be placed at far higher frequencies than ω_k in this example.

It is also possible to get the inverting gain from the same feedback amplifier. If the input node is grounded to avoid the input common-mode swing, the signal can be applied through the input resistor R_1. Because the opamp input is a virtual ground, the resistor R_1 works as a voltage-to-current converter. The input current passes through the feedback resistance R_2 and generates the negative output. Therefore, the same opamp feedback configuration makes an inverting amplifier. Both inverting and noninverting amplifiers are compared in Figure 1.15.

Although the gains of the two cases are different, both cases are identical in terms of feedback as they use the same resistor-divider feedback network. That is, the feedback factor f stays the same in both cases, as does the unity loop-gain frequency ω_k. Note again that the stability depends on the network parameters regardless of the signal applied. Therefore, both amplifiers have the same PM.

Let's see how the feedback factor f is related to the inverting feedback gain. If we relate the output to the summing node after multiplying the open-loop gain $a(s)$, the following is obtained:

$$v_o = -a(s)\{v_i + (v_o - v_i)f\}. \tag{1.28}$$

Then, by rewriting it in the same closed-loop gain form, we obtain

$$\frac{v_o}{v_i} = \frac{-a(s)(1-f)}{1+a(s)f}. \tag{1.29}$$

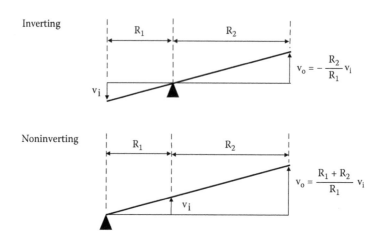

FIGURE 1.16
Inverting and noninverting gain amplifier concepts.

Approximating the DC (low-frequency) gain, f is replaced by $R_1/(R_1 + R_2)$:

$$\frac{v_o}{v_i} = \frac{-a_o(1-f)}{1+a_o f} \approx \frac{-(1-f)}{f} = -\frac{R_2}{R_1}.$$ (1.30)

This negative gain means that the low-frequency signal polarity is inverted because it is equivalent to the 180° phase shift. Although feedback factor is identical, depending on where the input is applied, the gain can be different, but the closed-loop bandwidth stays the same for two cases. The inverting amplifier is more useful in most applications as its input common-mode is grounded.

This concept of inverting and noninverting amplifiers is explained with a simple analogy in Figure 1.16. A straight line can be drawn crossing one reference point marked by a darkened triangle, which is equivalent to the signal ground. As in the example, depending on which part of the line is pushed (input), the other loose end moves by the ratio of the distances from the reference point (output). The direction of movement represents the polarity of the input or output. If two points move in the same direction, the gain is positive. Otherwise, the gain is negative. Because the resistances R_1 and R_2 represent the distance from the reference point, the input and output voltages are obtained by their ratios as shown.

1.8 Feedback Effect

The input and output resistances of the inverting amplifier are modified by the shunt feedback. In general, shunt feedback lowers the impedance level.

Figure 1.17 shows the low-frequency equivalent circuits to find the input and output resistances using the test signals v_x and i_x. Because the same current flows through R_1 and R_2, the following can be written:

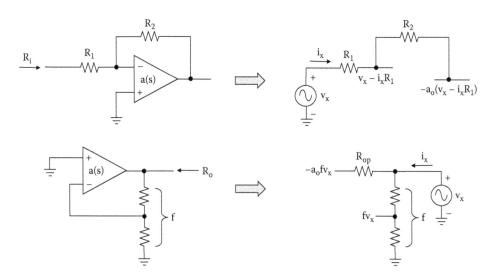

FIGURE 1.17
Input and output resistances of the inverting amplifier.

$$v_x - i_x R_1 + a_o\left(v_x - i_x R_1\right) = i_x R_2. \tag{1.31}$$

Then, the input resistance can be obtained as

$$R_i = \frac{v_x}{i_x} = R_1 + \frac{R_2}{1+a_o} \approx R_1. \tag{1.32}$$

Due to this shunt feedback, the opamp input node becomes the virtual ground with the resistance level of $R_2/(1+a_o) \sim 0$. However, the input resistance of the noninverting amplifier is still very high due to the high input resistance of the opamp.

Similarly from Figure 1.17, the output resistance of the inverting amplifier can be derived. Ignore the current through the feedback network when looking into the output port,

$$i_x = \frac{v_x + a_o f v_x}{R_{op}}. \tag{1.33}$$

Then, the output resistance can be obtained as

$$R_o = \frac{v_x}{i_x} = \frac{R_{op}}{1+a_o f}. \tag{1.34}$$

As shown in Equation (1.34), the output resistance R_{op} of the opamp is reduced by the amount of the loop gain.

1.8.1 Linear Range Improvement

There exist two high-gain amplifiers usually marked as a triangle in circuit diagrams. One is the comparator, and the other is the opamp. The comparator is used as an open-loop high-gain amplifier and even includes a digital latch to amplify a small seed signal to either a high or low digital output. On the other hand, the opamp is always used as a closed-loop feedback amplifier because the negative feedback allows trading gain for bandwidth. Therefore, the stability should be warranted. The worst case is the unity-gain feedback when f = 1 in the noninverting amplifier case. Because the output is subtracted from the input without attenuation in the unity-gain feedback, the amplifier becomes a voltage follower or a unity-gain buffer. That is, if the output is connected to the negative input, the closed-loop gain is about unity.

$$H_o = \frac{a_o}{1 + a_o} \approx 1. \tag{1.35}$$

As in the frequency response, the DC transfer characteristic is also modified when feedback is applied as shown in Figure 1.18. The high/low output swing (V_H/V_L) of the opamp is limited due to the supply rails (V_{DD}/V_{SS}). In the open-loop case (f = 0), the opamp output V_o changes approximately from V_L to V_H for a very small input from $-V_L/a_o$ to V_H/a_o because the DC gain a_o is very high. Furthermore, the input offset voltage, which is defined as an input voltage to make the output zero, is typically on the order of mV or two. As a result, this open-loop configuration is of no use because it is impossible to bias the input due to its high gain, and the bandwidth is too narrow.

With feedback, a wide bandwidth with a realistic gain is attained. As in the example, when the feedback factor f is 0.1, the closed loop gain is ~1/f (~20 dB), and this feedback amplifier now has a wider input range from $-0.1\,V_L$ to $0.1\,V_H$. The closed-loop bandwidth is now widened by the same feedback factor from the narrow open-loop bandwidth ω_p to $0.1\,\omega_{unity}$ ($f\omega_{unity}$). For the unity-gain feedback (f = 1), the input range is widest from $-V_L$ to V_H with the achievable maximum bandwidth of ω_{unity}. However, the closed-loop gain is 0 dB. That is, the feedback effect improves all amplifier parameters. In the unity-gain feedback

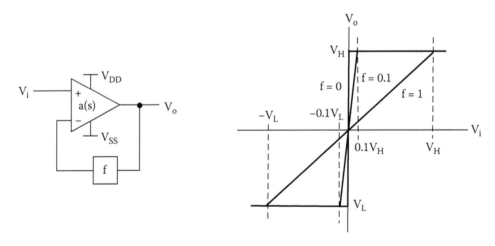

FIGURE 1.18
DC transfer characteristics of the opamp and feedback amplifier.

$(f = 1)$, the open-loop gain is traded for the widest input range, the maximum bandwidth ω_{unity}, and the lowest output resistance.

1.9 Left-Half or Right-Half Plane Zero

Zeros do not affect the stability of the feedback amplifier, but they often let transient responses overshoot due to its phase-leading characteristic. In circuits, there are a couple of cases in which zeros can be created. Most of the time, zeros are created due to the feed-forward of the signal. The most obvious case is when there is a series capacitor in the signal path as explained in Figure 1.3. In this simple high-pass case, the zero is placed at DC. Another case is when a pole in the feedback network becomes a zero. It is commonly assumed that the feedback factor f is a frequency-independent constant, but in real circuits, it can be frequency dependent. That is, if there is a pole in the frequency-dependent feedback factor $f(s)$, it will become a zero of the closed-loop function because $H(s) = a(s)/\{1 + a(s) f(s)\}$. Last, the zero insertion has been the common design tactic at the system level.

In most multistage feedback amplifiers or practical feedback systems such as PLL and $\Delta\Sigma$ modulators, zeros are intentionally introduced to compensate for extra loop phase delay for the stability reason. Assume that the feed-forward signal bypasses an amplifier that has one pole and is directly added to the amplifier output after multiplied by the gain k as shown in Figure 1.19.

Then, depending on the polarity of the gain stage, a zero is created either on the left half or right half of the complex s-plane. Zeros in the left-half lead phase and raise gain as frequency goes higher. They are harmless in most circuits except for the fact that they can cause slow settling in the transient response. However, any feedback systems with zeros on the right-half plane are difficult to stabilize.

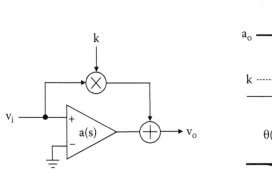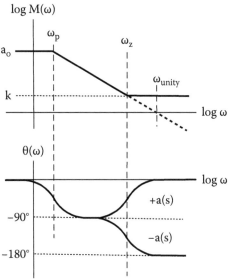

FIGURE 1.19
Creating zero by feed-forwarding.

If the amplifier gain is positive, the transfer function and the zero can be derived as follows because $a_o \gg k$.

$$\frac{v_o(s)}{v_i(s)} = \frac{a_o}{1 + \dfrac{s}{\omega_p}} + k \approx \frac{a_o + k\dfrac{s}{\omega_p}}{1 + \dfrac{s}{\omega_p}}. \tag{1.36}$$

$$z \approx -\frac{a_o \omega_p}{k} = -\frac{\omega_{unity}}{k}. \tag{1.37}$$

It is a negative-half plane zero if the gain is positive ($a_o > 0$). Otherwise, the zero moves to the right-half plane, and the phase lags like a pole. Therefore, the right-half plane zero should be avoided in any feedback system because the delay it causes can easily make the system unstable.

Only one right-half plane zero case exists. A Miller capacitance is added to frequency-compensate two-stage opamps. The unexpected side effect of this pole-splitting Miller compensation is that the input and output of the negative-gain amplifier are shorted together by the capacitance at high frequencies as shown in Figure 1.20.

Assume that the capacitor shunts input and output of the inverting amplifier. Then the transfer function and the right-half plane zero are given by

$$\frac{v_o(s)}{v_i(s)} = \frac{RC\left(s - \dfrac{g_m}{C}\right)}{1 + sRC}, \quad \omega_z = \frac{g_m}{C}. \tag{1.38}$$

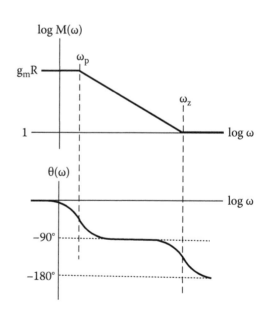

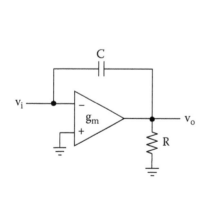

FIGURE 1.20
Right-half plane zero due to Miller capacitance.

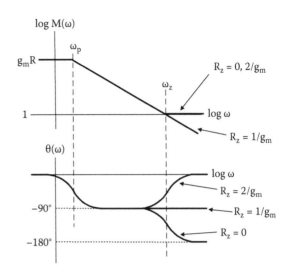

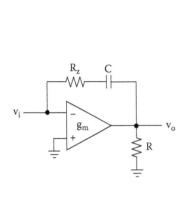

FIGURE 1.21
Right-half plane zero cancellation.

The bandwidth is $\omega_p = 1/RC$, and the right-half plane zero is at the unity-gain bandwidth of $\omega_k = \omega_{unity} = g_m/C$. This implies that if the amplifier gain becomes lower than unity, the signal passes through the capacitance without being inverted.

The simplest way to eliminate the right-half plane zero is to add a resistance R_z in series with the capacitance. Then, Equation (1.38) can be rewritten as follows:

$$\frac{v_o(s)}{v_i(s)} = \frac{RC\left\{s(1 - g_m R_z) - \dfrac{g_m}{C}\right\}}{1 + s(R + R_z)C}, \quad \omega_z = \frac{1}{\left(\dfrac{1}{g_m} - R_z\right)C}. \tag{1.39}$$

This becomes the same equation as Equation (1.38) if $R_z = 0$. Both the pole and right-half plane zero change as the value of R_z increases. If $R_z = 1/g_m$, the zero is completely canceled. If R_z is further increased, the zero moves to the left-half plane. Once in the left-half plane, it helps to lead the phase. With any nonzero value of R_z, the bandwidth becomes $1/(R + R_z)C$. However, because R_z is typically much smaller than R, this narrow-banding effect due to the pole movement is minimal, but the zero movement should be considered. In the three cases of $R_z = 0$, $1/g_m$, and $2/g_m$, the zero is placed at g_m/C, ∞, and $-g_m/C$, respectively, and the Bode gain and phase plots are sketched in Figure 1.21.

To broadband feedback amplifiers, a left-half plane zero is intentionally added to cancel the delay of a pole. The close-by pole-zero pair is called a *doublet*. The doublet is not detrimental to obtaining desired frequency and phase responses, but it affects the transient response. This is because the spectrum of any transient input such as a step or an impulse is spread over a wide range of frequencies. Let's consider a transfer function with a pole w_p and a zero w_z placed close by as follows:

$$\frac{v_o(s)}{v_i(s)} = \frac{1+\dfrac{s}{\omega_z}}{1+\dfrac{s}{\omega_p}} = 1 + \left(\frac{\omega_p - \omega_z}{\omega_z}\right)\frac{\dfrac{s}{\omega_p}}{1+\dfrac{s}{\omega_p}}. \tag{1.40}$$

For example, this transfer function responding to the unit step input of $v_i(s) = 1/s$ can be derived by taking the inverse Laplace transform of $v_o(s)$:

$$v_o(t) = 1 + \left(\frac{\omega_p - \omega_z}{\omega_z}\right)e^{\frac{-t}{\tau}}, \quad \tau = \frac{1}{\omega_p}. \tag{1.41}$$

Note that the transient response has a slow-settling component given by the canceled pole. Its magnitude is proportional to the distance of the zero from the pole. Therefore, when accurate transient settling is required, the doublet should be placed only at higher frequencies than the unity loop-gain frequency.

1.10 Stability of Feedback Amplifiers

As discussed earlier, open-loop opamps are not useful due to their high gain and the difficulty in biasing. Practical amplifiers with proper gain and bandwidth can be made applying feedback. Figure 1.22 shows the four cases of opamps used in feedback.

The resistor feedback in Figure 1.22a is the most common amplifier configuration as discussed earlier, and the feedback factor f affects the loop gain. It is stable if good PM is assured at the unity loop-gain frequency. However, if the feedback resistance is large, an extra pole in the feedback network due to the input capacitance of the opamp can contribute extra phase shift affecting the loop stability. Therefore, it is common to add a bypass capacitance in parallel with the large feedback resistance to improve the PM.

Figure 1.22b is an integrator that is unstable at DC. Integrators are often used to make active filters. However, the global negative feedback at a higher level makes them stable. For example, in active filters, the resonator is stable although it is made of two DC-unstable integrators with positive and negative polarities. The small-signal alternating current (AC) stability can be warranted if the opamp is stable at the unity-gain frequency, because the feedback capacitance shorts the opamp to be in unity-gain feedback at high frequencies. The opamp input capacitance can lower the loop gain and the feedback factor at high frequencies.

The differentiator shown in Figure 1.22c is a unity-gain feedback amplifier at low frequencies, but it works like an open-loop amplifier at high frequencies. The capacitive feedback amplifier shown in Figure 1.22d is the same as the amplifier shown in Figure 1.22a. The difference is that it is unstable at DC because there is no DC path to the opamp input node. In most switched-capacitor circuits, the opamp input is reset periodically.

1.10.1 Frequency-Dependent Feedback Factor

The feedback factor f is usually assumed to be independent of frequency. However, complicated feedback networks are in fact frequency dependent. In the simple examples of the integrator and differentiator, an extra pole is introduced in the feedback network.

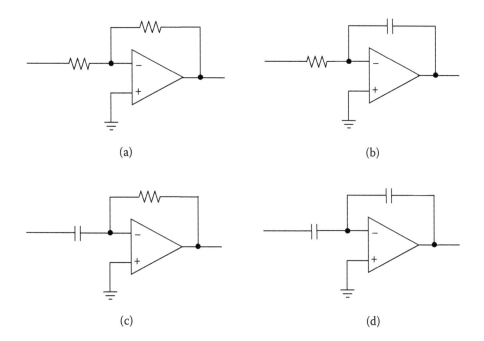

FIGURE 1.22
Opamps in feedback.

Assume that the opamp used in both the integrator and differentiator has a single-pole roll-off with a DC gain of a_o and a bandwidth of ω_p as shown in Figure 1.23. In the integrator case, the following relation can be written:

$$-\left\{v_i + (v_o - v_i) \times \frac{sRC}{(1+sRC)}\right\} \times \frac{a_o}{\left(1+\dfrac{s}{w_p}\right)} = v_o. \tag{1.42}$$

Then the closed-loop transfer function and the feedback factor can be derived as follows:

$$\frac{v_o(s)}{v_i(s)} \approx -\frac{a_o}{(1+a_o)sRC + \dfrac{s^2RC}{\omega_p}} \approx -\frac{\dfrac{1}{sRC}}{1+\dfrac{s}{a_o\omega_p}}. \tag{1.43}$$

$$f = \frac{sRC}{1+sRC}.$$

That is, the feedback factor is a high-pass filter, and above the closed-loop unity-gain frequency 1/RC, f remains constant at 1, which is the unity-gain feedback.

Because the feedback network includes a pole, the closed-loop response of the integrator has two poles as shown. At frequencies lower than 1/RC, Equation (1.43) can be written as

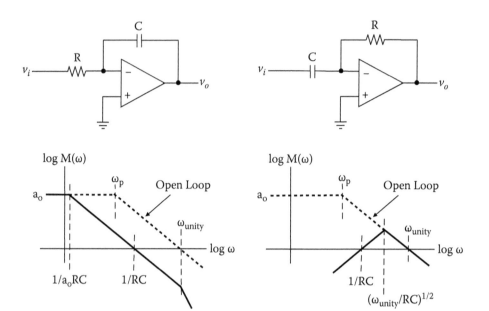

FIGURE 1.23
Frequency responses of the integrator and differentiator.

$$\frac{v_o(s)}{v_i(s)} \approx -\frac{a_o}{1+(1+a_o)sRC} \approx \frac{-1}{sRC}. \tag{1.44}$$

It is the integrator as expected, and the dominant pole is at $1/(1 + a_o)RC$, which is expected considering the Miller capacitance $(1 + a_o)C$ at the input node. At frequencies higher than $1/RC$, Equation (1.43) can be approximated as

$$\frac{v_o(s)}{v_i(s)} \approx -\frac{a_o}{(1+a_o)sRC+\dfrac{s^2RC}{\omega_p}} \approx -\frac{\dfrac{1}{sRC}}{1+\dfrac{s}{a_o\omega_p}}. \tag{1.45}$$

This places the nondominant pole at the open-loop unity-gain frequency $\omega_{unity} = a_o\omega_{po}$ as sketched in Figure 1.23. Therefore, opamps used in the integrator should be compensated up to the unity-gain frequency ω_{unity} with the worst-case feedback factor of $f = 1$.

In the differentiator case also shown in Figure 1.23, the closed-loop transfer function is similarly affected due to the frequency-dependent feedback network, which is now the low-pass filter instead of the high-pass filter in the previous integrator case. Therefore,

$$\frac{v_o(s)}{v_i(s)} = -\frac{a_o sRC}{1+a_o+sRC+\dfrac{s}{\omega_p}+\dfrac{s^2RC}{\omega_p}}. \tag{1.46}$$

$$f = \frac{1}{1+sRC}.$$

At frequencies lower than 1/RC, Equation (1.46) can be approximated as a differentiator.

$$\frac{v_o(s)}{v_i(s)} \approx -\frac{a_o sRC}{1+a_o} \approx -sRC. \tag{1.47}$$

At frequencies higher than 1/RC, Equation (1.46) can be simplified as

$$\frac{v_o(s)}{v_i(s)} \approx -\frac{a_o sRC}{1+a_o+sRC+\dfrac{s^2 RC}{\omega_p}} = -\frac{a_o}{1+\dfrac{1+a_o}{sRC}+\dfrac{s}{\omega_p}}. \tag{1.48}$$

The second term of the denominator shows that the transfer function is a differentiator until the third term gets larger and becomes dominant. As the third term increases, the transfer function follows the open-loop transfer function of the opamp.

The break frequency is obtained by equating the second term and the third term.

$$\omega_{bp} \approx \sqrt{\frac{(1+a_o)\omega_p}{RC}} \approx \sqrt{\frac{\omega_{unity}}{RC}}. \tag{1.49}$$

There are two poles at this frequency, which is the geometric average of two frequencies of 1/RC and ω_{unity} as sketched in Figure 1.23. Therefore, opamps used in the differentiator should be compensated up to this frequency ω_{bp}, where the loop gain is unity.

In general, practical feedback factors are all frequency dependent at high frequencies. However, all feedback systems can be assured to be exponentially stable if the loop phase delay is less than 180° with sufficient PM at the unity loop-gain frequency whether the feedback factor is frequency dependent or not.

2

Amplifier Design

As complementary metal-oxide semiconductor (CMOS) technologies are scaled down to the nanometer range, large analog/radio frequency (RF) mixed-mode systems including most analog functions such as low-noise amplifiers (LNAs), mixers, filters, analog-to-digital converters (ADCs), and even phase-locked loops (PLLs), are being integrated on a single chip together with digital functions. Integrated systems are getting more complicated, and the analog design no longer stays at the transistor level. Although the traditional analog design skill at transistor level is still of paramount interest even in this system-on-chip (SoC) environment, the large-scale SoC trend demands analog circuit designers handle high-level analog components such as voltage-controlled oscillators (VCOs), PLLs, and operational amplifiers (opamps) like building blocks. In this chapter, a simple high-level analog design methodology is developed based on an abstract model of a transistor to help designers cope with the system-level design. Transistors are handled using abstract process parameters based on the Simulation Program with Integrated Circuit Emphasis (SPICE) Level 2 model, and no elaborate empirical higher-level models commonly used in computer simulations are used.

2.1 Abstract Low-Frequency Model of Transistors

An electrical signal defined as a voltage at a circuit node moves from one circuit node to another by going through two conversion processes: voltage-to-current and current-to-voltage conversions. The voltage-to-current conversion is made possible only by active devices such as transistors or vacuum tubes. All active devices are by nature very nonlinear for large signals and are three terminal devices. A bipolar junction transistor (BJT) has an exponential I-V characteristic relating the collector current to its base-emitter voltage, while a CMOS transistor has a square-law I-V function relating the drain current to its gate-source voltage. In fact, the CMOS device has a fourth terminal, which is called the *substrate* or *body* normally tied to a constant voltage or its source terminal.

2.1.1 Large Signals

Two among three terminals take the input voltage V_i, and the third terminal outputs current I_o. The input terminals are marked with an arrow, which indicates the direction of the current flow and also the correct polarity of the bias voltage. Large signal relations between I_o and V_i for BJT and MOS are nonlinear functions as follows:

$$\text{BJT: } I_o = f(V_i) = I_s e^{\frac{V_i}{V_T}}, \quad I_s \sim 1\text{fA}, \quad \text{and} \quad V_T \sim 26 \text{ mV @300°K}. \tag{2.1}$$

$$\text{MOS: } I_o = f(V_i) = \frac{\mu C_{ox}}{2}\frac{W}{L}(V_i - V_{th})^2, \quad \mu_n C_{ox} \sim 200 \text{ μA}/V^2 \quad \text{and} \quad V_{th} \sim 0.4 \text{ V}.$$

V_T is the thermal voltage defined as kT/q, which is about 26 mV at room temperature (300°K). k is the Boltzmann constant (~8.6 × 10^{-5} eV/K), T is the absolute temperature, and q is the charge of an electron (~1.6 × 10^{-19} C). I_s is the reverse saturation current of the base-emitter junction and is directly proportional to the emitter area. It heavily depends on process parameters, and its value doubles at every 8° temperature change. μ_n is the mobility of an N-channel metal-oxide semiconductor (NMOS) transistor, C_{ox} is the gate capacitance per unit area, V_{th} is the threshold voltage, and W/L is the channel width-to-length ratio. Typical N-channel mobility μ_n in most CMOS processes is about 80 ~ 200 cm²/Vsec, and the P-channel mobility μ_p is typically 1/3 of the N-channel mobility. Most MOS processes offer two devices with normal and low threshold voltages and also devices with thick oxides for high-voltage uses.

2.1.2 Small Signals

Assume that an active device is operating at the bias point (I_o,V_i) in its large-signal I-V characteristic as shown in Figure 2.1.

The BJT and MOS transistor are biased into the forward active and saturation regions, respectively. The BJT has the input base-emitter junction forward biased ($V_{BE} > 0$) with the collector-emitter voltage more than the minimum saturation voltage ($V_{CE} > V_{CEsat} ~ 0.2$ V). For the MOS transistor, the same bias condition is that the gate-source voltage is above threshold ($V_{GS} > V_{th}$), and the drain-source voltage has the minimum saturation voltage ($V_{DS} > V_{DSsat} = V_{GS} - V_{th}$). If the bias point is perturbed by the small amounts of i_o and v_i, the following still holds true:

$$I_o + i_o = f(V_i + v_i), \quad \text{and} \quad g_m = \frac{i_o}{v_i} = \frac{dI_o}{dV_i} = f'(V_i). \tag{2.2}$$

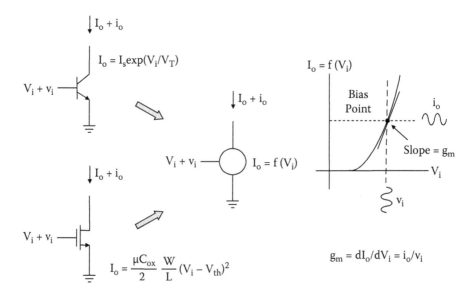

FIGURE 2.1
Generalized linear small-signal model of active devices.

Now these small amounts of fluctuations i_o and v_i of the bias point (I_o, V_i) are defined as small signals. Most active devices are considered linear for small signals even though they are very nonlinear for large signals.

The small-signal linear assumption is valid only in the following conditions:

$$\text{BJT: } I_o + i_o = I_s e^{\frac{V_i + v_i}{V_T}} \approx I_o \left(1 + \frac{v_i}{V_T} \right), \quad \text{if } |v_i| \ll V_T. \tag{2.3}$$

$$\text{MOS: } I_o + i_o = \frac{\mu C_{ox}}{2} \frac{W}{L} (V_i - V_{th} + v_i)^2 \approx I_o \left(1 + \frac{2v_i}{V_i - V_{th}} \right), \quad \text{if } |v_i| \ll (V_i - V_{th}).$$

This implies that BJT transistors are assumed to be linear only for the input signal far smaller than 26 mV at room temperature, while MOS transistors' linear range depends on the overdrive bias voltage $(V_i - V_{th})$. As the signal grows, the former produces infinitely many harmonics, but the latter makes mostly even harmonics. Even harmonics are very common in the square-law devices such as MOS transistors, junction field effect transistors (JFETs), and vacuum tubes. This feature has been advantageously used in making audio amplifiers because even harmonic overtones are very pleasing to the ears. Another reason for using the square-law devices in audio amplifiers is the high-slew rate. As will be discussed later in the transient response (Chapter 3), the slew rate is proportional to the input linear range. BJT has a linear range smaller than V_T of 26 mV, and amplifiers with BJT input devices are slew limited unless they are linearized with a significant amount of local feedback (emitter degeneration).

2.1.3 Transconductance g_m and Output Resistance r_o

For small signals, the transconductance g_m of any active device, which is defined as a ratio of the small-signal output current i_o to the small-signal input voltage v_i, is just a simple derivative of the I-V characteristic at an operating point. Therefore, the g_m of BJT and MOS transistors are defined from Equations (2.2) and (2.3) as

$$\text{BJT: } g_m = \frac{I_o}{V_T} \approx \frac{1}{26\,\Omega}, \quad \text{if } I_o = 1\,\text{mA}. \tag{2.4}$$

$$\text{MOS: } g_m = \sqrt{2 \mu_n C_{ox} \frac{W}{L} I_o} \approx \frac{1}{224\,\Omega}, \quad \text{if } I_o = 1\,\text{mA and } W/L = 50.$$

As noted, the g_m of an NMOS transistor sized with $W/L = 50$ and biased at 1 mA is about one order lower than that of a BJT at the same bias current. The g_m of a P-channel metal-oxide semiconductor (PMOS) transistor is even lower due to its lower mobility.

Ideally, all active devices should work as a transconductance g_m device for small signals, but they all exhibit a finite small-signal output resistance. In BJT, its base width is modulated by the collector-emitter voltage, which is known as the Early effect. In MOS transistor, the same effect appears as the channel length modulation. The small-signal output resistances of the BJT and MOS devices are modeled using the Early voltage V_A and the channel length modulation factor λ as

$$\text{BJT: } r_o = \frac{V_A}{I_o} \approx 100\ k\Omega, \quad \text{if } V_A = 100 \text{ V and } I_o = 1 \text{ mA.} \tag{2.5}$$

$$\text{MOS: } r_o = \frac{1}{\lambda I_o} \approx 10k\ \Omega, \quad \text{if } \lambda = 1/10 \text{ V and } I_o = 1 \text{ mA.}$$

The output resistance of the MOS device is about one order lower than that of the BJT device. It is far worse for modern nanometer short-channel devices. As a result, the maximum direct current (DC) gain $g_m r_o$ achievable using a single MOS amplifier is only in about the 20 ~ 40 dB range, which is 20 ~ 30 dB lower than what is achievable using a BJT amplifier.

2.1.4 Small-Signal Model

One advantage the MOS transistor has is its high input impedance, which is capacitive. At low frequencies, the small-signal input resistance looking into the gate is infinite. On the other hand, the BJT transistor exhibits a low input resistance due to the carrier recombination in the base, which yields the base current. Considering the current gain β, which is the ratio of the collector current to the base current, the input driving-point resistance r_i looking into the base of the BJT transistor is about $\beta/g_m = 2.6$ kΩ if the current gain $\beta = 100$.

The low-frequency small-signal models of BJT and MOS transistors are shown in Figure 2.2. For PNP BJT and PMOS transistors, the polarities of the input voltage and

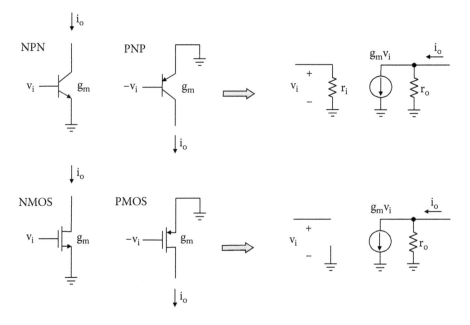

FIGURE 2.2
Low-frequency small-signal model of BJT and MOS transistors.

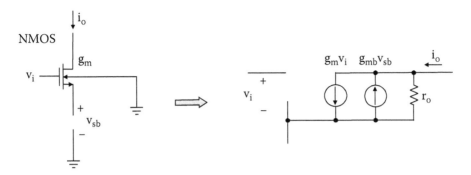

FIGURE 2.3
Low-frequency small-signal model of an NMOS transistor with body effect.

output current are inverted. It is good to remember the simple fact that the output current increases if the input voltage increases regardless of the input polarity. Note that MOS devices have one more terminal, called substrate or body as shown in Figure 2.3.

2.1.5 Body Effect

The body terminal is usually tied to a constant voltage. In most digital or switching circuits, both NMOS and PMOS bodies are always tied to the most negative and positive voltages, respectively. In nonswitching analog circuits, there are a few options. In the common P-substrate N-well process, the NMOS body should be connected to the most negative supply, but the PMOS body in an N-well can be connected either to the most positive supply or to its source. In the twin-tub process, the NMOS body can be also connected to its source.

If the body of the NMOS (PMOS) transistor is tied to the most negative (positive) supply, its source is floating, and the fluctuation of the source-to-body voltage modulates the reverse-biased channel-to-substrate junction. As a result, the threshold voltage of the device is modulated as follows:

$$V_{th} = V_{tho} + \gamma\left(\sqrt{V_{SB} + \phi_{bi}} - \sqrt{\phi_{bi}}\right), \tag{2.6}$$

where V_{SB} is the source-to-body reverse bias voltage, V_{tho} is V_{th} when $V_{SB} = 0$ V, and ϕ_{bi} is the built-in potential between the channel and the substrate in the range of 0.6 ~ 0.7 V. γ is a process dependent parameter typically on the order of 0.4 ~ 0.6 $V^{1/2}$. If the source terminal is floating, and V_{SB} fluctuates by a small amount of v_{sb}, the threshold voltage V_{th} changes accordingly. Therefore, the body transconductance g_{mb} can be defined from Equations (2.1) and (2.6) as

$$g_{mb} = \frac{dI_o}{dV_{SB}} = \frac{i_o}{v_{sb}} \approx -(0.1 \sim 0.2)g_m. \tag{2.7}$$

This implies that the output current decreases if the body voltage increases, and this negative small-signal current is modeled in Figure 2.3 using a current flowing into the opposite direction. The body transconductance g_{mb} heavily depends on the substrate doping density, its profile, and the reverse bias, and is typically in the range of about 10% to 20% of the transconductance g_m.

2.2 Driving-Point Resistances at Low Frequencies

In the previous section, transistors are defined as a three-terminal device. Two of the terminals take the input voltage, and the remaining outputs the current. The arrow indicates the polarity of the input voltage and the direction of the output current flow through the device for both large and small signals. The polarity of the input voltage and the direction of the output current are reversed for PNP and PMOS transistors.

By definition, the driving-point resistance is the ratio of any terminal voltage and the current flowing into the terminal. Small-signal resistances have been commonly derived from small-signal equivalent circuits, and solving small-signal simultaneous equations obtained by applying Kirchoff's law. But a more intuitive way to derive small-signal parameters helps to give a high-level perspective on the transistor circuit design.

Three driving-point resistances looking into three terminals can be defined as shown in Figure 2.4 for BJT and MOS transistors. The first two, one looking into the base and the gate terminals and the other looking into the collector and the drain terminals, are by definition of the parameters. When looking into the emitter or the source terminals, the relationship between the collector current and the base-emitter voltage or the drain current and the gate-source voltage is the transconductance g_m. So the small-signal driving-point resistance is about $1/g_m$. However, considering the output resistance r_o, it becomes the parallel sum of two resistances, $1/g_m$ and r_o. In the BJT case, the collector current is slightly smaller than the emitter current by $\beta/(1+\beta)$, but the approximation by $1/g_m$ is valid because $\beta \gg 1$. Considering the input resistance r_i, the driving-point input resistance R_x looking into the emitter of the BJT is

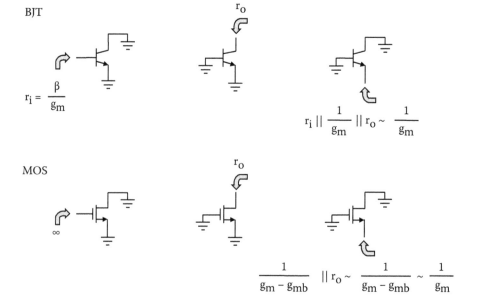

FIGURE 2.4
Driving-point small-signal terminal resistances of transistors.

$$R_x = r_i \left\| \frac{1}{g_m} \right\| r_o \approx \frac{1}{g_m}. \tag{2.8}$$

In the MOS case, there is no input resistance, but one factor to consider is the body effect because the substrate voltage is not constant. Similarly, using Equation (2.7) and assuming that $(g_m - g_{mb})r_o \gg 1$, the driving-point input resistance R_x looking into the source of the MOS transistor is

$$R_x = \frac{1}{g_m - g_{mb}} \left\| r_o \approx \frac{1}{g_m - g_{mb}} \approx (0.8 \sim 0.9)\frac{1}{g_m}. \tag{2.9}$$

The parallel combination of $1/(g_m - g_{mb})$ and r_o can be approximated as $1/(g_m - g_{mb})$, which is about 10% ~ 20% lower than $1/g_m$. That is, if the source is floating, the effective transconductance of a MOS device increases to be $(g_m - g_{mb})$ due to the body effect. However, it is convenient to approximate it just as $1/g_m$ to facilitate the design.

Note that the collector or the drain-side resistance is the highest among the three, and the emitter or the source side resistance is the lowest. The difference is approximately by the amount of the gain factor $g_m r_o$. This gain factor is the most basic parameter in the amplifier design as will be mentioned repeatedly for the rest of this book. The value also depends heavily on the device size, bias, process, supply, and temperature.

In order to define the driving-point resistances for larger circuits made with many transistors, we need to understand how the driving-point resistances are affected if the other terminals are not grounded. That is the case when there are resistances connected to other terminals. Only three cases can be considered. The first is when looking into the base of the BJT or the gate of the MOS transistor as shown in Figure 2.5.

In the BJT case, the input resistance is increased due to the emitter resistance R_E by $(1 + \beta)$. It is often approximated as $r_i(1 + g_m R_E)$, where $(1 + g_m R_E)$ is the feedback loop gain commonly used in the BJT design. Such a local series feedback is called *emitter degeneration*, and the series feedback increases the input and output resistance levels while reducing the transconductance by the same amount of $(1 + g_m R_E)$. However, in the MOS case, the input resistance is infinite regardless of the source-side resistance R_S, because the gate is capacitive. However, as in the BJT case, this source degeneration also reduces the transconductance by the feedback loop gain of $\{1 + (g_m - g_{mb})R_S\}$, where the $(g_m - g_{mb})$ term results because the source is floating. For this reason, in the MOS transistor circuits, the source degeneration is rarely used.

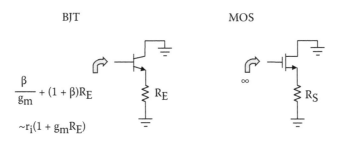

FIGURE 2.5
Resistances looking into the BJT base and MOS transistor gate.

2.3 Resistance Reflection Rules

Let's consider the two remaining cases. The emitter-side resistance of the BJT or the source-side resistance of the MOS transistor is reflected to the collector or the drain side. Similarly, the collector-side or the drain-side resistance is reflected to the emitter or the source side. Consider the small-signal models as shown in Figure 2.6 to find the driving-point resistances of the two cases for MOS transistors.

Assume that the MOS transistor is ideal with a transconductance value of $(g_m - g_{mb})$ ~ g_m, because the source is floating. The small-signal driving-point resistance is defined as the ratio v_x/i_x of the small-signal voltage and current of the test source. First, in the left-hand-side circuit, the current through the MOS device is $-(g_m - g_{mb})i_x R_S$ as explained in Figure 2.3 because the gate-source voltage is $-i_x R_S$. Therefore, the following equation can be written relating i_x to v_x:

$$v_x = \left\{ i_x + \left(g_m - g_{mb} \right) i_x R_S \right\} r_o + i_x R_S, \tag{2.10}$$

because the input voltage v_x is the sum of the voltage drop across r_o and R_S. Then, the driving-point resistance looking into the drain with the source-side resistance R_S is

$$R_x = \frac{v_x}{i_x} = R_S + r_o \left\{ 1 + \left(g_m - g_{mb} \right) R_S \right\} \approx r_o \left\{ 1 + \left(g_m - g_{mb} \right) R_S \right\}$$

$$= \left\{ \frac{1}{\left(g_m - g_{mb} \right)} + R_S \right\} \left(g_m - g_{mb} \right) r_o \approx \left(\frac{1}{g_m} + R_S \right) g_m r_o. \tag{2.11}$$

This can be interpreted as follows. The series feedback increases the transistor output resistance r_o by the loop gain of $\{1 + (g_m - g_{mb})R_S\}$ or the source-side resistance $1/(g_m - g_{mb}) + R_S$ looks larger by the amplifier gain factor of $g_m r_o$ when reflected to the drain side. From Equation (2.11), if the source is grounded ($R_S = 0$), the drain-side resistance is the same output resistance r_o of the MOS transistor as shown in Figure 2.4.

Similarly, in the right-hand side circuit, the current through the MOS device is $-(g_m - g_{mb}) v_x$ because the gate-source voltage is $-v_x$. Therefore, the equation relating i_x to v_x can be derived as

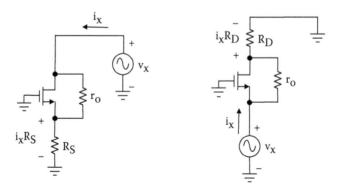

FIGURE 2.6
Small-signal models to find driving-point resistances.

$$v_x = \left\{ i_x - (g_m - g_{mb})v_x \right\} r_o + i_x R_D, \tag{2.12}$$

again because the input voltage v_x is the sum of the voltage drop across r_o and R_D. Therefore, the driving-point resistance looking into the source with the drain-side resistance R_D is

$$R_x = \frac{v_x}{i_x} = \frac{r_o + R_D}{1 + (g_m - g_{mb})r_o} \approx \frac{r_o + R_D}{(g_m - g_{mb})r_o} \approx \frac{r_o + R_D}{g_m r_o}. \tag{2.13}$$

Compared to Equation (2.11), if you look into the source side, the drain-side resistance $(r_o + R_D)$ looks smaller by the same gain factor of $g_m r_o$. This implies that if the drain is grounded ($R_D = 0$), the source-side resistance is simply $1/(g_m - g_{mb})$ as explained in Figure 2.4.

If the body effect is neglected, the transconductance of the MOS transistor can be approximated just as g_m like BJT, and the reflection rules are the same for both the BJT and MOS transistors. Looking into the collector or the drain, the emitter or the source-side resistance looks larger by the gain factor of $g_m r_o$. On the other hand, looking into the emitter or the source, the collector or the drain-side resistance looks smaller by the same gain term of $g_m r_o$. These resistance reflection rules can be generalized as shown in Figure 2.7, where active devices are drawn using an abstract concept given by its transconductance g_m and output resistance r_o.

The arrow indicates the control port and the current direction of the active devices. Resistances in the other ports look either larger or smaller by the same gain factor of $g_m r_o$. These rules are useful in estimating the resistance levels of any circuit node and can facilitate the transistor circuit design.

For BJT, the emitter resistance is effectively shunted by its input resistance r_i due to the finite current gain. Therefore, Equation (2.11) should be modified as follows:

$$R_x = r_i \| R_E + r_o \left\{ 1 + g_m (r_i \| R_E) \right\} \approx r_o \left\{ 1 + g_m (r_i \| R_E) \right\}$$
$$= \left\{ \frac{1}{g_m} + (r_i \| R_E) \right\} (g_m r_o). \tag{2.14}$$

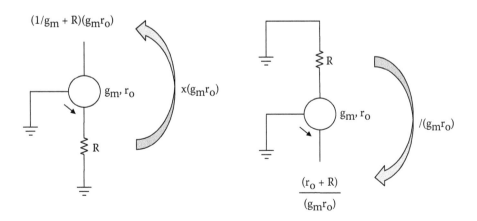

FIGURE 2.7
Resistance reflection rules.

The same conclusion as in the MOS transistor case can be drawn. The emitter-side resistance R_E looks larger by the same gain factor $g_m r_o$. Now the difference in the BJT case is the input resistance r_i, which is connected in parallel with the emitter resistance R_E. So, depending on the value of R_E, we can approximate Equation (2.14) in two cases as follows:

$$R_x \approx r_o \{1 + g_m (r_i \| R_E)\} \approx R_E (g_m r_o), \quad \text{if } r_i \gg R_E.$$
$$\approx r_o (1 + \beta), \quad \text{if } R_E \gg r_i. \tag{2.15}$$

That is, if the emitter degeneration resistor R_E is smaller than the input resistance r_i, the BJT case is no different from the MOS transistor case. However, if R_E is far larger than r_i, the output resistance will be limited by the degeneration by r_i, and the maximum output resistance of the BJT approaches about βr_o as explained in Figure 2.8.

The output resistance of the MOS transistor is similarly limited, but by the leakage of the output node, commonly caused by some nonideal secondary effects such as impact ionization and hot carriers.

The resistance looking into the emitter can be similarly derived for the BJT case. If there is no input resistance r_i, it is given by Equation (2.13). Therefore, r_i can be added in parallel with that.

$$R_x = r_i \| \frac{r_o + R_C}{1 + g_m r_o} \approx \frac{r_o + R_C}{g_m r_o}. \tag{2.16}$$

This approximation is valid if $R_C < \beta r_o$, and the collector-side resistance is reduced by the same gain factor $g_m r_o$ when looking into the emitter. If $R_C > \beta r_o$, it is limited to the level of r_i as sketched in Figure 2.9.

However, in the MOS transistor case, the source-side resistance increases further until the drain-side resistance is limited by the secondary effect as in Figure 2.8.

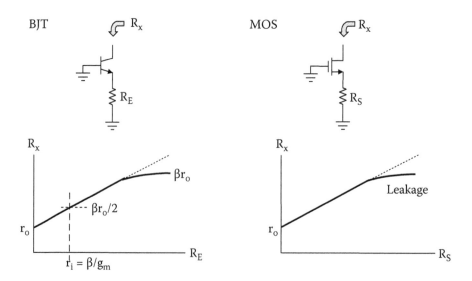

FIGURE 2.8
Resistances looking into the BJT collector and MOS transistor drain.

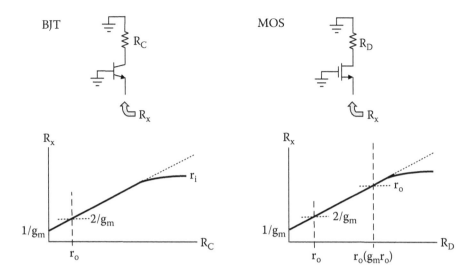

FIGURE 2.9
Resistances looking into the BJT emitter and MOS transistor source.

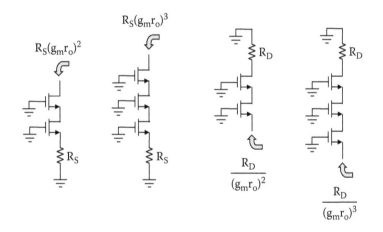

FIGURE 2.10
Examples of the resistance reflection rules.

Therefore, using the results given by the reflection rules, the drain-side and the source-side resistances are reflected as shown in Figure 2.10 by two cascaded transistors with the gain of $g_m r_o$ each if the body effect is neglected.

2.3.1 Local Shunt Feedback

In general, the shunt resistance reduces the input and output resistances by the loop gain while the series resistance increases them. Consider the local shunt feedback case when the resistance is connected between the collector and base of the BJT or between the drain and source of the MOS transistor as shown in Figure 2.11.

The input and output low-frequency driving-point resistances can be defined as shown. Both cases are the same except for the finite input resistance of BJT. Using Kirchoff's

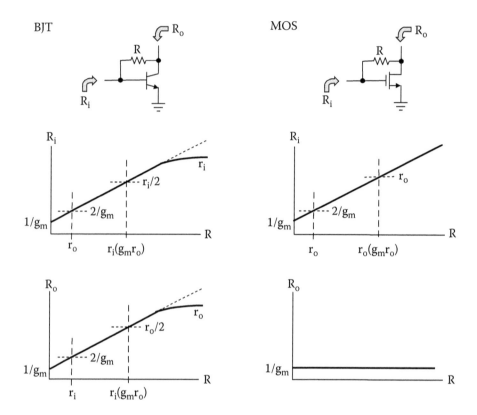

FIGURE 2.11
Resistance reflection rule in the shunt local feedback case.

current law, we can equate the sum of all currents at the two nodes to be zero, and the following two relations can be derived in the MOS case assuming that the input current i_i is applied, and the input and output voltages are v_i and v_o.

$$i_i = \frac{v_i - v_o}{R} = g_m v_i + \frac{v_o}{r_o}. \tag{2.17}$$

By eliminating v_o in two simultaneous equations, the input resistance is obtained.

$$R_i = \frac{v_i}{i_i} = \frac{R + r_o}{1 + g_m r_o}. \tag{2.18}$$

This input resistance is the same as the resistance looking into the source with the drain-side resistance as in Equation (2.13). When R is small, the input resistance is the diode resistance $1/g_m$, and the shunt resistance looks smaller by the gain from the input side. The output resistance R_o is simply $r_o \| (1/g_m)$ regardless of the shunt resistance R, because the MOS gate is open circuited.

Both input and output resistances are sketched in Figure 2.11. If v_i is eliminated, the transresistance transfer function with a typical value of the shunt resistance R ($R \gg 1/g_m$) is defined as follows:

$$\frac{v_o}{i_i} = \frac{\dfrac{1}{g_m} - R}{1 + \dfrac{1}{g_m r_o}} \approx -R. \tag{2.19}$$

This is the common amplifier configuration for the transresistance gain, which is set by the shunt feedback resistance.

The BJT case can be handled by adding the input shunt resistance of r_i to Equation (2.18). That is, the input resistance of the BJT local shunt feedback is

$$R_i = r_i \,||\, \frac{R + r_o}{1 + g_m r_o}. \tag{2.20}$$

The output resistance of the BJT case is made of three components: the transistor output resistance r_o, the input-side resistance $(R + r_i)$, and the transistor with a reduced transconductance $g_m r_i / (R + r_i)$.

$$R_o = r_o \,||\, (R + r_i) \,||\, \frac{R + r_i}{g_m r_i} = r_o \,||\, \frac{R + r_i}{(1 + \beta)}. \tag{2.21}$$

They are also sketched in Figure 2.11. Note that unlike the MOS transistor, the input and output resistances are ultimately limited by r_i and r_o when the shunt resistance is large. Finally, the trans resistance gain of the BJT case with a typical value of the shunt resistance R $(1/g_m \ll R \ll \beta r_o)$ is also given by

$$\frac{v_o}{i_i} = \frac{\dfrac{1}{g_m} - R}{1 + \dfrac{1}{g_m r_o} + \dfrac{1}{\beta} + \dfrac{R}{\beta r_o}} \approx -R. \tag{2.22}$$

The resistance reflection rules are essential in analyzing large circuits without handling lengthy mathematical equations or simulations.

2.4 Three Basic Amplifier Configurations

Transistors have only three terminals. Two of them should be used as input and output ports, and the remaining one terminal is grounded. Depending on which terminal is grounded, three amplifier configurations can be derived. All transistors are basically transconductance g_m devices for small signals. Because the signal is commonly defined in voltage, the output current is converted into a voltage using a resistance. Adding this load resistance, three common-source (CS), common-gate (CG), and common-drain (CD) configurations are sketched for MOS transistors in Figure 2.12. They are equivalent to the common-emitter (CE), common-base (CB), and common-collector (CC) configurations of the BJT amplifiers, respectively.

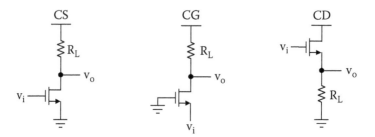

FIGURE 2.12
Three amplifier configurations.

The CS and CG are amplifiers with high gains, but the CD is basically a buffer amplifier transforming the impedance from high to low with no gain. Using low-frequency small-signal driving-point resistances, the low-frequency voltage gain and small-signal input and output driving-point resistances of these three amplifiers can be easily estimated. They can be approximated for the CS and CG amplifiers as follows:

$$CS: A_v = -g_m R_o. \qquad CG: A_v = (g_m - g_{mb})R_o.$$

$$R_i = \infty. \qquad R_i = \frac{r_o + R_L}{1 + (g_m - g_{mb})r_o} \approx \frac{1}{g_m - g_{mb}}, \quad if \ R_L < r_o. \qquad (2.23)$$

$$R_o = r_o \ || \ R_L. \qquad R_o = r_o \ || \ R_L.$$

The CS and CG amplifiers have about the same small-signal gain of $g_m R_o$. In the CG case, the transconductance is a little larger $(g_m - g_{mb})$ due to the body effect. The main difference is the polarity of the gain and the input resistance. The CS is an inverting amplifier with a 180° phase shift, while the CG is a noninverting amplifier. The input resistance of the CG is given by Equation (2.13) because it looks into the source of the MOS transistor. It can be low unless the load resistance R_L is reflected into the input $(R_L < r_o)$. All amplifier gain stages are made with either CS or CG amplifiers.

On the other hand, the CD amplifier, which is commonly called a source follower like the emitter follower of the BJT CC amplifier, has a gain less than unity, but its output resistance can be low because it is driven by the source.

$$CD: A_v = \frac{g_m(r_o \ || \ R_L)}{1 + (g_m - g_{mb})(r_o \ || \ R_L)} \approx \frac{g_m}{g_m - g_{mb}} < 1.$$

$$R_i = \infty. \qquad (2.24)$$

$$R_o = \frac{1}{g_m - g_{mb}} \ || \ r_o \ || \ R_L \approx \frac{1}{g_m - g_{mb}}, \quad if \ R_L < r_o.$$

The small-signal gain in Equation (2.24) can be derived intuitively as follows. The gate-source voltage of the MOS transistor is $(v_i - v_o)$, and the transistor dumps a current of $g_m(v_i - v_o)$ into the total load of $(r_o || R_L)$. In addition, the body voltage v_o adds the current of $g_{mb}v_i$ to the load. Then the output voltage becomes v_o. Formulating this relation in an equation leads to

$$v_o = \left\{ g_m \left(v_i - v_o \right) + g_{mb}v_o \right\} \left(r_o \| R_L \right).$$

(2.25)

From Equation (2.25), the small-signal voltage gain of $A_v = v_o/v_i$ can be derived as in Equation (2.24). This simple manipulation of the circuit parameters helps to analyze and design larger circuits.

2.5 Nine Amplifier Combinations

As discussed in the previous section, there are only three amplifier configurations. Two of them are amplifiers, and one is a buffer. There exist only nine combinations that make transistor circuits as shown in Figure 2.13.

Basically, all electronic circuits are made of these nine combinations. For example, one stage can be CS, CG, and CD, and the following stage can be CS, CG, and CD, too. It is of paramount interest for circuit designers to understand these combinations clearly. From this section on, $(g_m - g_{mb})$ will be approximated by just g_m for simplicity as circuits become more complicated.

2.5.1 CS-CS

This combination makes a two-stage amplifier with two-stage gain of $(g_m R_o)^2$. Two polarity inversions give the output of the same polarity. The input and output resistances are given in Equation (2.23), which are the same as in the CS amplifier.

2.5.2 CS-CG

This combination shown in Figure 2.14 is called a *cascode*. It is most widely used to achieve a high gain and high output resistance. The CS-CG amplifier on the left side is just a two-stage amplifier like the CS-CS. If the first-stage load resistance R_{L1} is removed, the normal cascode configuration is obtained as shown in the middle. The cascode configuration is obtained by just stacking one transistor on the top of another. The cascode has two inputs and is also known as dual-gate MOS transistor in old RF circuits. The RF input is applied to the one input, and the local carrier is applied to the other. Then, this single stage performs

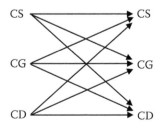

FIGURE 2.13
Nine amplifier combinations.

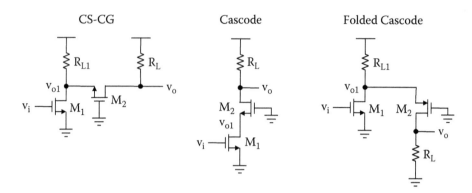

FIGURE 2.14
CS-CG, cascode, and folded-cascode amplifiers.

as a mixer and an amplifier. In most cascodes, one input of the stacked device is grounded to achieve high output resistance as discussed earlier using Equation (2.11). Although cascoding is useful as it increases the output resistance, the signal swing is reduced instead as the output bias point is raised by stacking two devices. One way to restore the signal swing is to fold the cascode stage using another type of device (in this case, PMOS transistor) as shown on the right side. This way, the bias point moves down to make full use of the power supply for large signal swing. All three variations are identical in their operations.

In all three cases, the input resistance is infinite as in the CS amplifier, because it looks into the gate. The output resistance can be estimated using Equation (2.11) by summing the load resistance R_L in parallel with the resistance looking into the drain of M_2 as follows:

$$R_o = R_L \,||\, \left[r_{o1} \,||\, R_{L1} + r_{o2} \left\{ 1 + g_{m2} \left(r_{o1} \,||\, R_{L1} \right) \right\} \right] \approx R_L. \tag{2.26}$$

In the normal cascode case in the middle, R_{L1} is infinite and can be neglected. The driving-point resistance looking into the drain of M_2 is far higher than any load resistance R_L. Similarly, the small-signal gain can be approximated for the two gain stages of v_{o1}/v_i and v_o/v_{o1}.

$$A_v = -g_{m1} \left(r_{o1} \,||\, R_{L1} \,||\, \frac{r_{o2} + R_L}{1 + g_{m2} r_{o2}} \right) \left(g_{m2} - g_{mb2} \right) \left(r_{o2} \,||\, R_L \right) \approx -g_{m1} R_L, \tag{2.27}$$

where Equation (2.23) is used to estimate the gains of the CS and CG amplifiers. In the CS-CG amplifier, the driving-point resistance of the first output node is the parallel sum of three resistances: output resistance of M_1, load resistance R_{L1}, and resistance looking into the source of M_2. The drain-side resistance is reflected into the source side like Equation (2.13).

Judging from Equations (2.26) and (2.27), the cascode amplifier is the same as the CS amplifier except for higher output resistance and higher gain. Two transistors can be used to get one transistor that is more ideal. In integrated circuits, transistors are scaled down and occupy small areas, but short-channel MOS transistors have very low output resistances. Using two transistors to make one transistor with higher output resistance is a good trade-off, although the output swing is somewhat lost due to cascoding. Examples of a single transistor, a cascode, and a triple cascode are shown in Figure 2.15.

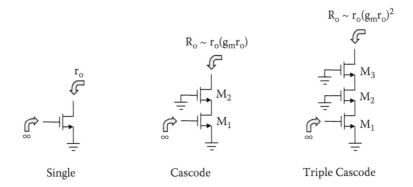

FIGURE 2.15
Single transistor versus cascoded transistors.

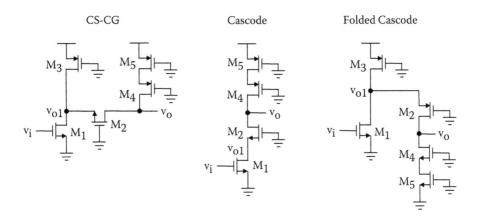

FIGURE 2.16
CS-CG, cascode, and folded-cascode amplifiers with current sources.

They have the same g_m and the same input resistance, but the cascoding increases the output resistance as discussed in the resistance reflection rules of Section 2.3. The output resistance of the cascoded transistor is again derived from Equation (2.11).

$$R_o = r_{o1} + r_{o2}\left(1 + g_{m2}r_{o1}\right) \approx r_{o1}\left(g_{m2}r_{o2}\right). \tag{2.28}$$

If the two transistors are identical and the body effect is neglected, the output resistance r_o is increased by the gain factor of $g_m r_o$. Similarly, the output of the triple cascode will be increased by $(g_m r_o)^2$. More cascoding will raise the output resistance further until it is limited by the leakage at the output port. Therefore, cascoding more than three stages is not common.

Going back to the amplifier configurations, current source loads are more commonly used in the integrated circuits to make high-gain opamps. In such a case, Figure 2.14 is redrawn in Figure 2.16 using current sources as loads. The amplifier output resistance and gain can be approximated depending on the load as in Equations (2.26) and (2.27). In Figure 2.16, the cascode current sources are used as the output load to obtain high gain.

$$R_o = r_{o5}\left(g_{m4}r_{o4}\right)||[r_{o1}\,||\,r_{o3} + r_{o2}\{1 + g_{m2}\left(r_{o1}\,||\,r_{o3}\right)\}].\qquad(2.29)$$

$$A_v = -g_{m1}\left\{r_{o1}\,||\,r_{o3}\,||\,\frac{r_{o2} + r_{o5}\left(g_{m4}r_{o4}\right)}{1 + g_{m2}r_{o2}}\right\}g_{m2}\left\{r_{o2}\,||\,r_{o5}(g_{m4}r_{o4})\right\}.\qquad(2.30)$$

In these equations, $R_{L1} = r_{o3}$, and $R_L \sim r_{o5}(g_{m4}r_{o4})$ are used for simplicity. It appears very complicated, but let's simplify it assuming all transistors have the same parameters, g_m and r_o, and also neglecting the body effect. Then Equations (2.29) and (2.30) can be approximated again as follows:

$$R_o \approx \frac{r_o(g_m r_o)}{3},\qquad A_v \approx -\frac{(g_m r_o)^2}{3}.\qquad(2.31)$$

In the normal cascode case, r_{o3} can be ignored, and Equation (2.31) becomes

$$R_o \approx \frac{r_o(g_m r_o)}{2},\qquad A_v \approx -\frac{(g_m r_o)^2}{2}.\qquad(2.32)$$

This implies that although the cascode configuration is a single-stage amplifier, it achieves a two-stage amplifier gain because its output resistance is raised higher by the one-stage gain factor $g_m r_o$. Therefore, the gain of the cascode amplifier can be derived more intuitively as a single-stage amplifier.

$$A_v = -g_{m1}R_o \approx -g_{m1}\left\{r_{o5}\left(g_{m4}r_{o4}\right)||r_{o1}\left(g_{m2}r_{o2}\right)\right\}.\qquad(2.33)$$

This is the same conclusion we reached for the CS amplifier in Equation (2.23) if an active load is used in place of the resistance load. The only difference is the high output resistance due to the cascoding. Both resistances looking up into the drain of M_4, and the resistances looking down into the drain of M_2 are raised by the same gain factor of $g_m r_o$.

2.5.3 CS-CD

This combination is the CS amplifier followed by a CD buffer amplifier, which is a source follower. Therefore, the total gain is the same as the single CS amplifier gain, but it has a low output resistance of the source follower given by Equation (2.23). The low output resistance can drive the low-impedance load, while the high gain of the CS amplifier is buffered from the load.

2.5.4 CG-CS, CG-CG, CG-CD

These three combinations with the CG amplifier as the first stage are the same as the previous three combinations CS-CS, CS-CG, and CS-CD, respectively. The differences are that there is no polarity inversion in the first CG amplifier, and the input resistance of the CG amplifier is low as given by Equation (2.23).

2.5.5 CD-CS

This combination is not useful in the MOS transistor design. In the BJT design, the emitter follower followed by the common-emitter amplifier is the most useful and is called the Darlington pair. This is because the BJT device has a low input resistance due to the finite current gain. To enhance the input resistance or to increase the current gain, an extra emitter follower is used. Like the previously discussed cascode configuration, two transistors make one transistor with a high current gain. The feature is very desirable in the BJT design. However, MOS input resistance is already high, and there is no need for that. The only case to consider is to shift the DC bias level using the gate source voltage drop. The gain and the output resistance are the same as in the single CS amplifier given by Equation (2.23).

2.5.6 CD-CG

This configuration is a little awkward as shown in Figure 2.17. If two transistors of the same size are biased with the same current, it can be considered as a differential pair driven asymmetrically by one input while the other input is grounded. Then, as will be discussed later in the differential amplifier, the gain is half of the differential-pair gain because the output is taken from one output, but the output resistance looking into the drain of M_2 is doubled as in the differential pair.

If considered as a two-stage amplifier, the source follower is loaded with the input resistance of the second CG stage. Therefore, the gain and the output resistance can be estimated using Equations (2.23) and (2.24) as follows:

$$A_v = \frac{g_{m1}\left(r_{o1} || R_{L1} || \dfrac{r_{o2}+R_L}{1+g_{m2}r_{o2}}\right)}{1+g_{m1}\left(r_{o1} || R_{L1} || \dfrac{r_{o2}+R_L}{1+g_{m2}r_{o2}}\right)} g_{m2}\left(r_{o2} || R_L\right) \approx \frac{\dfrac{g_{m1}}{g_{m2}}}{1+\dfrac{g_{m1}}{g_{m2}}} \times g_{m2}R_L. \qquad (2.34)$$

$$R_o = R_L || r_{o2}\left\{1+g_{m2}\left(r_{o1} || R_{L1} || \dfrac{1}{g_{m1}}\right)\right\} \approx R_L || r_{o2}\left(1+\dfrac{g_{m2}}{g_{m1}}\right). \qquad (2.35)$$

As expected, if $g_{m1} = g_{m2}$, the gain becomes half, and the drain-side resistance is doubled like the differential pair due to the source degeneration.

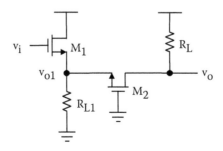

FIGURE 2.17
CD-CG configuration.

2.5.7 CD-CD

This last combination is not useful like the CD-CS combination. In BJT circuits, it is equivalent to the Darlington pair used as an emitter follower. In MOS transistor circuits, it is no different from the single CD source follower. The only case in which to use it is for DC-level shifting.

2.6 Differential Pair

On-chip opamps are commonly made fully differential to take both differential input and output signals. The advantage of the differential signaling is so obvious. As the signal is differential, even the supply noise is rejected together with the common-mode signal. Also due to the symmetry, the differential transfer function is symmetric and produces no even-order harmonics. Furthermore, because the signal magnitude is doubled while the noise power is doubled, it has an advantage of 3 dB in the signal-to-noise ratio.

The differential pair is basically two CS amplifiers biased with a common tail current source I as shown in Figure 2.18. Assume that transistors (M_1, M_2) and resistance loads (R_1, R_2) are matched: $g_{m1} = g_{m2} = g_m$ and $R_1 = R_2 = R$. For the differential small-signal inputs of $\pm(v_i/2)$, the common-source node becomes a virtual ground, and the perfect symmetry is maintained on both sides of the dashed line. As in the CS amplifier, the differential small-signal voltage gain, the input resistance, and the output resistance of the differential pair are defined using Equation (2.23) as

$$A_v = \frac{v_o}{v_i} = \frac{v_o/2}{v_i/2} = -g_{m1}\left(r_{o1} \,||\, R_1\right), \quad -g_{m2}\left(r_{o2} \,||\, R_2\right) \approx -g_m\left(r_o \,||\, R\right).$$

$$R_i = \infty. \quad R_o = \left(r_{o1} \,||\, R_1\right) + \left(r_{o2} \,||\, R_2\right) \approx 2\left(r_o \,||\, R\right). \tag{2.36}$$

Note that the differential-mode gain and the input resistance are the same as the CS amplifier, but the output resistance is doubled because the differential output looks into two CS output nodes.

2.6.1 Common-Mode Rejection

One of the desirable features the differential pair has is that it only amplifies the difference signal while rejecting the common-mode signal. The common-mode signal v_{ic} drives both inputs simultaneously as shown on the right side of Figure 2.18. If the tail current source is ideal, there is no common-mode current driving the output nodes, implying that the common-mode gain is 0. However, if the tail current has a finite output resistance R_{SS}, then the common-mode gain can be defined as follows, neglecting the body effect:

$$A_{cv} = \frac{v_o}{v_{ic}} = -\frac{g_{m1}R_1 - g_{m2}R_2}{1 + \left(g_{m1} + g_{m2}\right)R_{SS}} \approx -\frac{g_m R}{1 + 2g_m R_{SS}}\left(\frac{\Delta g_m}{g_m} + \frac{\Delta R}{R}\right). \tag{2.37}$$

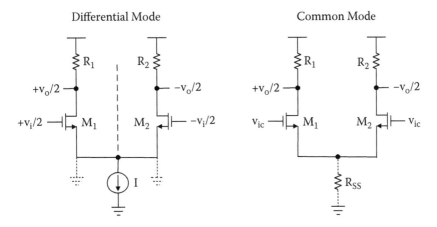

FIGURE 2.18
Differential pair with differential-mode and common-mode inputs.

The denominator term is due to the source degeneration as in Equation (2.24), and it is assumed that the transistors and load resistances are mismatched. The relations of $\Delta g_m = g_{m1} - g_{m2}$ and $g_m = (g_{m1} + g_{m2})/2$ are used in the standard error approximation. The common-mode gain depends on the mismatch of the two branches. If ideally matched, there is no differential signal generated at the output. The useful figure of merit is the common-mode rejection ratio (CMRR) defined using Equations (2.36) and (2.37):

$$CMRR = \frac{A_v}{A_{cv}} = \frac{g_m(r_o \| R)}{g_m R} \frac{1 + 2g_m R_{SS}}{\left(\dfrac{\Delta g_m}{g_m} + \dfrac{\Delta R}{R}\right)} \approx \frac{1 + 2g_m R_{SS}}{\left(\dfrac{\Delta g_m}{g_m} + \dfrac{\Delta R}{R}\right)}. \tag{2.38}$$

This implies that CMRR can be improved by matching as well as by increasing the output resistance of the tail current. The differential pair is the most common amplifier configuration in integrated forms because it can reject common-mode signals coupled from the power supply or bias lines.

2.6.2 Symmetric Transfer Function

The finite tail current constraint modifies the large signal I-V characteristic of the differential pair. That is, the currents in two CS amplifiers are summed to be the tail current I. Using Equation (2.1), the following relations hold true:

$$I_1 = \frac{\mu C_{ox}}{2} \frac{W}{L}\left(V_{GS} - V_{th} + \frac{V_i}{2}\right)^2.$$

$$I_2 = \frac{\mu C_{ox}}{2} \frac{W}{L}\left(V_{GS} - V_{th} - \frac{V_i}{2}\right)^2. \tag{2.39}$$

$$I = I_1 + I_2 = \mu C_{ox}\frac{W}{L}\left\{\left(V_{GS} - V_{th}\right)^2 + \frac{V_i^2}{4}\right\}.$$

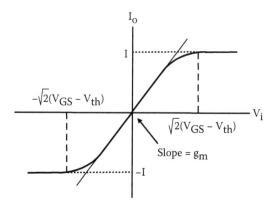

FIGURE 2.19
Transfer function of the differential pair.

Then the differential current output is defined as the difference of the two currents.

$$I_o = I_1 - I_2 = \mu C_{ox} \frac{W}{L} \left(V_{GS} - V_{th} \right) V_i. \tag{2.40}$$

From Equations (2.39) and (2.40), if the overdrive bias term is removed, the large-signal I-V transfer function is derived.

$$I_o = \sqrt{\mu C_{ox} \frac{W}{L} \left(I - \frac{\mu C_{ox}}{4} \frac{W}{L} V_i^2 \right)} \, V_i. \tag{2.41}$$

This large-signal I-V characteristic is symmetric as shown in Figure 2.19.

For the small-signal approximation to be valid, the following condition should be met in Equations (2.39) and (2.41):

$$|v_i| << 2 \left(V_{GS} - V_{th} \right). \tag{2.42}$$

This condition occurs when one conducts I, while the others is off. For small-signal inputs meeting Equation (2.42), the transconductance of the differential pair is defined at $V_i = 0$, and its value is the same as that of the single CS amplifier on each side.

$$g_m = \frac{dI_o}{dV_i}\bigg|_{V_i=0} = \frac{i_o}{v_i} \sim \sqrt{\mu C_{ox} \frac{W}{L} I} = g_{m1}, \; g_{m2}. \tag{2.43}$$

Note that the differential small-signal gain of the differential pair is the same as that of the CS amplifier with common-mode rejection and symmetry.

2.7 Gain Boosting

As discussed in Section 2.5, cascoding (CS-CG) effectively raises the output resistance by the amount of the gain factor of $g_m r_o$. It is logical to assume that the output resistance can

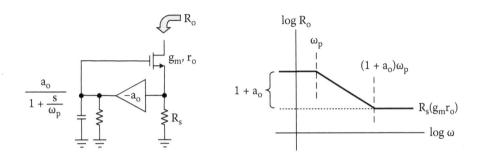

FIGURE 2.20
G_m or gain-boosting concept.

be further increased if the transconductance g_m is boosted further with more feedback. This G_m or gain boosting is conceptually sketched in Figure 2.20 [1,2].

The cascode transistor with small signal parameters of g_m and r_o is gain-boosted using an extra amplifier with a DC gain of a_o and bandwidth of ω_p. Then, the output resistance increases by the amount of the loop gain $(1 + a_o)$ at DC, and decreases as frequency goes higher because the loop gain decreases after the pole frequency of ω_p. Because there is no loop gain beyond the unity loop-gain frequency $(1 + a_o)\omega_p$, it would stay constant at $R_s(g_m r_o)$, which is the output resistance obtained by cascoding without boosting. In fact, the cascoding effect will also start to decrease as frequency approaches the nondominant pole frequency of the cascode node.

From Equation (2.28), we can approximate the output resistance as

$$R_o = R_s + r_o + R_s\left(g_m r_o\right)\left(1 + \frac{a_o}{1 + \dfrac{s}{\omega_p}}\right) \approx R_s\left(g_m r_o\right)\left(1 + a_o\right)\frac{1 + \dfrac{s}{\left(1 + a_o\right)\omega_p}}{1 + \dfrac{s}{\omega_p}}. \tag{2.44}$$

Because the −6 dB/oct slope is flattened after the unity loop-gain frequency, the unity-gain frequency of the gain-boosting loop introduces a zero, and it may introduce a doublet in the overall transfer function. Care must be taken to make sure the local gain-boosting loop has enough bandwidth wider than the unity-loop-gain frequency of the global feedback loop to move the doublet higher than the unity loop-gain frequency.

2.7.1 Doublet Constraints

Figure 2.21 shows how the doublet is created in the G_m-boosted cascode stage. Assume that the output high-impedance node is capacitively loaded by C_L. The dotted line with a −6 dB/oct slope represents the loading capacitance impedance. Also superposed with another dotted line is the output resistance given by Equation (2.44) as shown in Figure 2.20. The total output impedance $Z(\omega)$ is the parallel sum of these two impedances as sketched with the solid line. The impedance curve has two break points. One makes a dominant pole at a low frequency where two impedances are equal. The other is at around the unity loop-gain frequency $(1 + a_o)\omega_p$ of the gain-boosting loop. Therefore, the pole of the local gain-boosting loop always becomes a zero of the transfer function.

There are two constraints in placing this doublet in the transfer function. First, to avoid the slow settling in the transient response, this doublet should always be placed higher than the global unity loop-gain frequency ω_k. Another constraint is the stability of the local gain-boosting loop. For example, if the pole at the cascode node is the nondominant pole ω_{p2}, the unity loop-gain frequency of the gain-boosting loop should be lower than ω_{p2} for PM > 45°. These two constraints set the lower and upper bounds of the unity loop-gain frequency of the boosting amplifier.

$$\omega_k < (1 + a_o)\omega_p < \omega_{p2}. \tag{2.45}$$

As in Equation (2.45), the closed-loop bandwidth of the boosting loop, which is the unity loop-gain frequency of the boosting loop, should be set in the design wider than the global closed-loop bandwidth but lower than the nondominant pole frequency at the cascode node.

2.7.2 Other Gain-Boosting Concepts

Cascoding is the most effective way to achieve high resistance and to enhance the gain of the single-stage amplifier. Other gain-boosting circuit concepts are illustrated in Figure 2.22. On the left side, an extra current given by M_x flows into M_1 while keeping the low current in the load. This effectively increases g_{m1} for higher gain. The same is

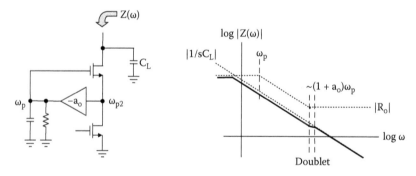

FIGURE 2.21
Output impedance after G_m or gain boosting.

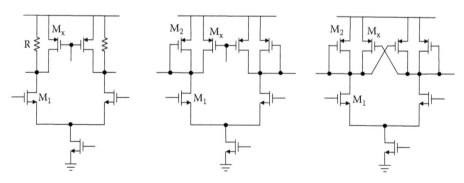

FIGURE 2.22
Gain-boosting examples.

true for the enhancement diode load case in the middle. The small-signal gain is defined as $-g_{m1}/g_{m2}$, and this extra bias current increases g_{m1}. On the right side, rather than using the constant current of M_x, positive feedback is applied through M_x so that a negative resistance can be added to the enhancement diode load. Therefore, the gain is improved as follows:

$$Gain = -\frac{g_{m1}}{g_{m2} - g_{mx}}. \tag{2.46}$$

By making the sizes of M_2 and M_x into a ratio, very high small-signal gain can be obtained using this positive-feedback approach. Either the resistive load or the enhancement diode load needs no common-mode feedback. As a result, this simple gain stage is often used as a low-gain wideband preamplifier for comparators.

2.8 Biasing

Biasing is to set the operating condition of the active device. The current source is a current seed set either by an external resistor or an internally generated voltage reference. It can be made independent of supply and temperature. Input and output common-mode voltages are usually set by local or global feedbacks, but active loads are biased using current mirrors as shown in Figure 2.23.

Routing bias voltages over long distances may make the bias lines sensitive to noise coupling, and it is better to move currents and bypass the gate-source voltages of the current mirrors with large capacitances, as shown. For current mirrors, long-channel devices are used to get high output resistances and to achieve better matching.

The accuracy of the simple current mirror is still sensitive to the drain-source voltage due to the channel length modulation. Cascoding keeps the drain-source voltage constant and improves the output resistance. However, the cascode current mirror requires two gate-source voltage drops $(2V_{GS})$ as shown in Figure 2.24, and the signal swing of the active load is limited only down to $(V_{GS} + V_{DSsat})$ from the supply.

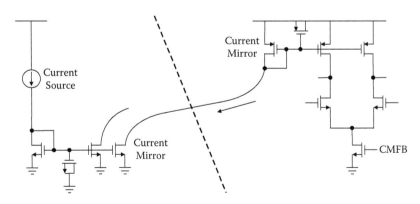

FIGURE 2.23
Bias currents are replicated.

2.8.1 Cascode Biasing for Maximum Swing

To maximize the signal swing, a bias voltage of $(V_{GS} + V_{DSsat})$ needs to be generated to bias the cascode transistor, which will enable the output node swing to be down to $2V_{DSsat}$ from the supply rail as shown in Figure 2.25.

Because the cascode transistor replicates the drain-source voltage to be constant V_{DSsat}, the current mirror can keep the high output resistance with high matching accuracy. To generate the cascode bias of $(V_{GS} + V_{DSsat})$, two transistors M_1 and M_2 are connected to make a diode. These two transistors can be considered as a dual-gate MOS. The bottom transistor M_1 operates in triode, while the top transistor M_2 is in saturation.

Assuming that $V_{th1} = V_{th2}$ and $V_{DSsat1} = V_{DSsat2}$, the ratio of two transistors can be obtained as follows by equating the two currents:

$$\frac{\left(\dfrac{W}{L}\right)_1}{\left(\dfrac{W}{L}\right)_2} = \frac{\left(V_{GS2} - V_{th2}\right)^2}{2\left(V_{GS1} - V_{th1}\right)V_{DSsat1}} \approx \frac{1}{4}. \tag{2.47}$$

By sizing M_1 and M_2 with a ratio of 0.25 or smaller, M_1 can always be operated in triode. However, considering the body effect on the threshold voltage, it is safe to make the ratio slightly lower than 0.25.

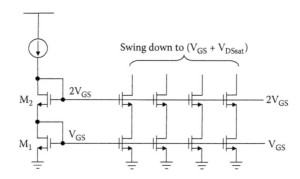

FIGURE 2.24
Cascode current mirror.

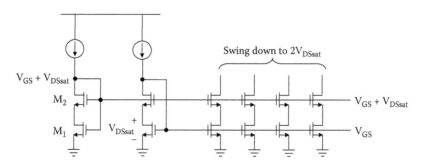

FIGURE 2.25
Cascode current mirror with minimum voltage drop.

Figure 2.26 shows an example of the triple cascode opamp for the maximum signal swing using the ratio of 0.2. All transistors are biased with the drain-source voltages close to the saturation voltage V_{DSsat}.

2.8.2 Current Source Matching

As shown in Figure 2.27, the current matching is measured by the current difference of MOS transistors identically biased with the same gate-source voltages as drawn with a vertical dashed line in the I-V curve. For better matching, transistors should be made with small W/L ratios, and bias currents should be low. Longer channel devices exhibit

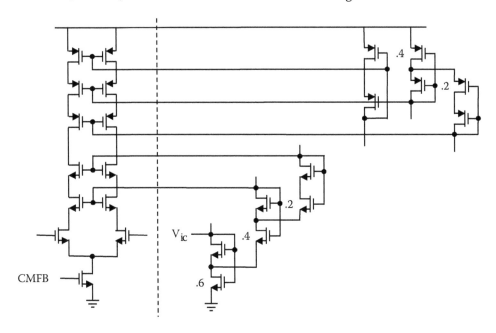

FIGURE 2.26
Triple cascode example with cascode biases.

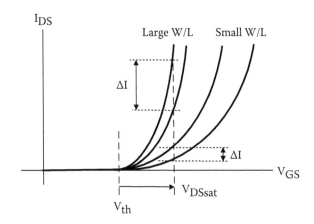

FIGURE 2.27
MOS I-V characteristics for current matching.

better matching characteristics. Therefore, given V_{DSsat}, W/L should be minimized for current source matching, which in turn implies that the bias current should be high. The current source matching with large V_{DSsat} is very important in applications such as current-steering digital-to-analog converters (DACs), but in amplifiers, the signal swing will be greatly reduced if biased with large V_{DSsat}.

2.9 Voltage and Current Sources

All constant bias voltages and currents originate from a voltage or current source. Although an external supply and a reference resistor can be used to generate them, most integrated circuits require on-chip voltage and current references, and either a reference voltage or current should be generated. A few supply independent voltages available in MOS are the gate-source voltage V_{GS}, the gate-source voltage difference ΔV_{GS}, the threshold voltage difference ΔV_{th}, and the bandgap reference voltage.

2.9.1 V_{GS} and ΔV_{GS}-Referenced Current Sources

To generate a supply-independent voltage, self-biasing is required. The self-biased V_{GS} and ΔV_{GS}-referenced current sources are sketched in Figure 2.28. Two current mirrors stacked can reach the bias condition with no current flowing, which is unwanted and should be avoided. A small leakage current or initial current injection using a diode can start up this self-biased circuit. Once the self-biased circuit starts, the voltage developed across the resistor is V_{GS} if M_1-M_2 and M_3-M_4 are matched. The simple current sources can be cascoded to further improve the supply independence and to reduce the channel-length modulation effect.

The temperature coefficient of the V_{GS}-referenced current source also depends on the resistance temperature coefficient (TC) as follows:

$$\frac{1}{I_o}\frac{dI_o}{dT} = \frac{1}{V_{GS}}\frac{dV_{GS}}{dT} - \frac{1}{R}\frac{dR}{dT}. \tag{2.48}$$

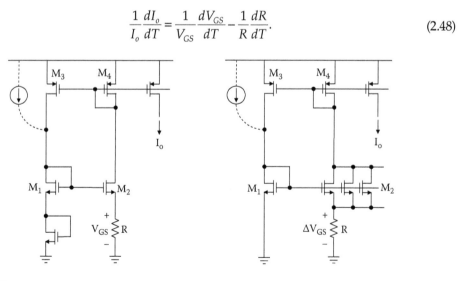

FIGURE 2.28
Self-biased V_{GS} and ΔV_{GS}-referenced current sources.

The V_{GS} voltage has two temperature-dependent terms, and its TC can be predicted approximately as

$$\frac{dV_{GS}}{dT} = \frac{dV_{th}}{dT} + \frac{d}{dT}\left(\sqrt{\frac{I_{DS}}{\mu C_{ox}\,\dfrac{W}{2}\,\dfrac{W}{L}}}\right).$$

(2.49)

The MOS threshold voltage and channel mobility depend heavily on temperature with negative coefficients as follows:

$$\frac{dV_{th}}{dT} \approx -1 \sim 2mV/^{\circ}C, \qquad \mu \propto T^{-3/2} \sim T^{-5/2}.$$

(2.50)

The second ($V_{GS} - V_{th}$) term exhibits a positive TC that can cancel out the negative TC of the threshold voltage. Therefore, V_{GS} can have either a positive or negative TC due to the TC of the overdrive voltage. Depending on the resistance material, the TC of the resistance varies widely within a range from about 100 to 1000 ppm/°C. In the V_{GS}-reference current source, the strong positive TC of ($V_{GS} - V_{th}$) can be partly canceled by the TC of the resistor.

Similarly, ΔV_{GS} can be developed as shown on the right side by making the device ratio of M_2 to M_1 large. Because the threshold voltage term is canceled, ΔV_{GS} is mainly the difference of two overdrive voltages. A very temperature-independent voltage can be generated using ΔV_{th} if two different threshold devices are available [3]. That is, if two same-sized transistors are biased with the same currents, ΔV_{GS} becomes ΔV_{th}, which can be almost independent of temperature.

2.9.2 Bandgap Reference

A constant voltage independent of supply and temperature has been generated using a reverse-biased diode, called a Zener diode. Reverse-biased junction breakdowns occur due to the tunneling effect at lower than 5 V and avalanching at higher voltages. The breakdown exhibits a positive TC of about 2.5 mV/°C, but the low-voltage Zener breakdown exhibits a negative TC. The breakdown voltage heavily depends on the doping densities of the diode. Most integrated circuit (IC) processes provide diodes that break down only at high 6 ~ 7 V. As Zener diodes are noisy and require high voltages, they are not suitable for low-voltage CMOS circuits.

A useful low-voltage temperature-independent reference is the bandgap reference, which relies on the silicon bandgap voltage. The base-emitter voltage V_{BE} of the forward-biased junction diode has an approximate TC of –2 mV/°C as sketched in Figure 2.29. It exhibits a curvature error of a few mV at –55 and 125°C extremes, but overall the slope is linear.

If extrapolated, it crosses about 1.206 V at 0°K, which is the silicon bandgap voltage V_{GO}. At normal room temperature of 300°K, V_{BE} is typically in the range of 0.6 ~ 0.7 V due to its temperature dependence. This temperature dependence can be approximated as the following sum using higher-order temperature-dependent terms [4]:

$$V_{BE}(T) = V_{GO} - k_1 T + k_2 T^2 + k_3 T^3 + \cdots.$$
$$\approx V_{GO} - k_1 T \approx 1.206 - (2mV/^{\circ}C)T.$$

(2.51)

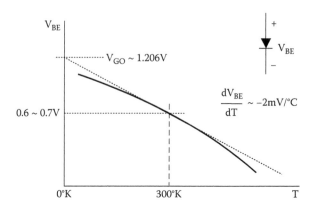

FIGURE 2.29
Temperature dependence of V_{BE}.

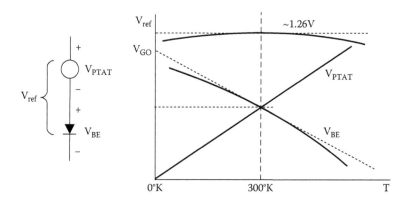

FIGURE 2.30
Temperature dependence of the bandgap reference V_{ref}.

Suppose that we can make a voltage that is proportional to the absolute temperature (PTAT) with a positive TC of 2 mV/°C to compensate for $k_1 T$. Then the reference voltage V_{ref} can be made independent of temperature except for the small curvature variation as shown in Figure 2.30.

$$V_{ref} = V_{BE} + k_1 T$$
$$= V_{GO} + k_2 T^2 + k_3 T^3 + \cdots. \tag{2.52}$$

When the temperature variation of V_{BE} is canceled using a PTAT voltage at room temperature of about 300°K, the bandgap reference voltage V_{ref} is set slightly higher than the silicon bandgap V_{GO}. The actual bandgap reference voltage with a zero TC at a certain temperature varies slightly from process to process, but in typical BJT processes, it is about 1.26 V.

$$\frac{dV_{ref}}{dT} \approx 0 \quad at\ 300^o K, \quad if\ V_{ref} \approx 1.26\ V. \tag{2.53}$$

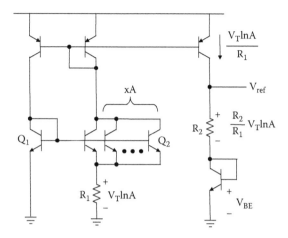

FIGURE 2.31
BJT bandgap reference circuit.

The bandgap reference circuit can be made as shown in Figure 2.31 [5,6]. The PTAT voltage is generated from ΔV_{BE} by sizing two equally biased diodes with a ratio of 1:A. A is the number of multiple parallel devices. Once started, the currents in the two branches are the same assuming the BJT current gain β is large. Then neglecting the Early effect, the voltage across R_1 is simply $V_T\ln A$ because $I_{c1} = I_{c2}$ and $I_{s1} = AI_{s2}$.

$$V_{BE1} - V_{BE2} = V_T \ln A \propto T. \tag{2.54}$$

This current is mirrored into the third branch and generates a voltage across R_2 and the diode voltage V_{BE}. The bandgap reference voltage is the sum of the diode voltage V_{BE} and the PTAT voltage.

$$V_{ref} = V_{BE} + \frac{R_2}{R_1} V_T \ln A \approx 1.26 \ V. \tag{2.55}$$

The PTAT voltage can also be used as a temperature sensor or an electronic thermometer because it is proportional to the absolute temperature. The bandgap reference output can be trimmed to be the magic voltage of about 1.26 V by adjusting either R_1 or R_2.

However, in most bulk CMOS technologies, no good bipolar devices are available. A vertical PNP transistor can be made in CMOS using the N-well base, but the low current gain and large base resistance complicate the CMOS bandgap reference design. Because the substrate becomes the common collector, the PNP transistor can be used only as a diode.

A CMOS version of the bandgap reference circuit that uses only diodes is shown in Figure 2.32. The equally sized transistors M_1 and M_2 generate the PTAT current with self-biasing. However, in addition to the very low current gain β and the large base resistance, the offset of M_1 and M_2 and the current mirror accuracy limit the performance of the CMOS bandgap reference [7].

Another CMOS bandgap reference using opamp is shown in Figure 2.33. The opamp forces two branch currents to be PTAT, and the PTAT voltage across the resistor R_2 is added to the diode voltage to make a bandgap reference voltage. Current mirrors can be

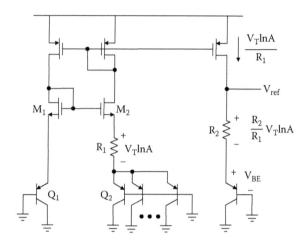

FIGURE 2.32
CMOS bandgap-reference circuit.

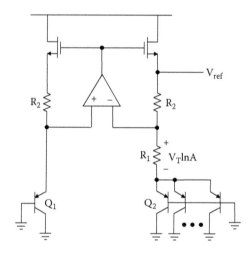

FIGURE 2.33
CMOS low-output resistance bandgap reference circuit.

refined by cascoding for supply independence and better current source matching. When high-gain opamps are in feedback, the closed-loop stability should be considered as in the feedback amplifier.

It is possible to scale the bandgap reference voltage. To generate higher (>1.2 V) or lower (<1.2 V) bandgap references, the diode voltage V_{BE} can be either multiplied or divided, and a proportionately scaled PTAT voltage can be added. BJT bandgap references use accurate film resistors, whose TC is on the order of 100 ppm/°C. Once trimmed, the BJT bandgap reference voltage exhibits an extremely low TC on the order of 10 ~ 20 ppm/°C at 300°K. However, CMOS bandgap references exhibit a TC of 50 ~ 200 ppm/°C due to the poor performance of the bipolar devices and the poor offset and matching of MOS transistors.

To make a temperature-insensitive current source, a bandgap-referenced current source can be made as shown in Figure 2.34. However, the current source I_{ref} is derived from the

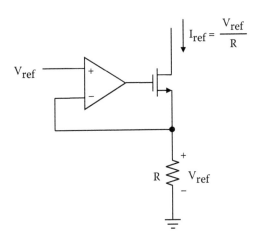

FIGURE 2.34
Temperature-independent current source.

bandgap reference V_{ref} developed across the resistor R, and the TC of the resistor also affects that of the current source. The bandgap reference can be set a little higher so that the extra PTAT voltage can compensate for the TC of the resistor, or an external temperature-stable resistor can be used.

References

1. B. J. Hosticka, "Improvement of the gain of MOS amplifiers," *IEEE J. Solid-State Circuits*, vol. SC-14, pp. 1111–1114, December 1979.
2. K. Bult and G. J. G. M. Geelen, "A fast-settling CMOS opamp for SC circuits with 90-dB DC gain," *IEEE J. Solid-State Circuits*, vol. SC-25, pp. 1379–1384, December 1990.
3. B. S. Song and P. R. Gray, "Threshold voltage temperature drift in ion-implanted MOS transistors," *IEEE J. Solid-State Circuits*, vol. SC-17, pp. 291–298, April 1982.
4. P. R. Gray and R. G. Meyer, *Analysis and Design of Analog Integrated Circuits*, New York: Wiley, pp. 256, 1977.
5. R. J. Widlar, "New developments in IC voltage regulators," *IEEE J. Solid-State Circuits*, vol. SC-6, pp. 2–7, February 1971.
6. A. P. Brokaw, "A simple three-terminal bandgap reference," *IEEE J. Solid-State Circuits*, vol. SC-9, pp. 288–393, December 1974.
7. B. S. Song and P. R. Gray, "A precision curvature-compensated CMOS bandgap reference," *IEEE J. Solid-State Circuits*, vol. SC-18, pp. 634–643, December 1983.

3

Operational Amplifier (Opamp)

Off-the-shelf opamps have two input terminals for differential input and a single-ended output terminal. Their direct current (DC) gain is very high, but bandwidth is very narrow. Opamps are always used with feedback applied. Although they can be designed with three or more stages to achieve high DC gain, only one or two gain stages are commonly used for stability reasons. In integrated circuits, opamps are made fully differential to make them immune to noises from the supply or bias lines. Therefore, common-mode feedback circuits are needed to operate them. However, most integrated complementary metal-oxide semiconductor (CMOS) opamps have no output buffers because most of them drive only internal capacitive loads.

3.1 Small-Signal Model of the Operational Amplifier

A simple CMOS version of the standard two-stage opamp is shown in Figure 3.1. Like the off-the-shelf opamp, the output is single ended. As this opamp has two gain stages with two high-impedance nodes marked as 1 and 2. The first stage of the opamp is a differential pair with a current mirror load driving the first node 1. The second stage is a single-ended common-source (CS) amplifier. In bipolar junction transistor (BJT) opamps, a Darlington pair is typically used as a second common-emitter (CE) amplifier so that it may not load node 1. However, in CMOS opamps just a CS amplifier will suffice. Because the high-impedance node 2 can drive only capacitive loads, an output buffer is required if the load is resistive.

Assume the small-signal transconductances of two gain stages are G_{m1} and G_{m2}, respectively. Also, R_1 and R_2 and C_1 and C_2 are total driving-point resistances and capacitances at two high-impedance nodes 1 and 2, respectively. As the signal goes through two high-impedance nodes 1 and 2, it is delayed at each node by its own time constant. That is, only the small-signal currents i_1 and i_2 can drive two high-impedance nodes 1 and 2, respectively. The small-signal operation inside the opamp is explained using two equivalent circuits in Figure 3.2.

The input v_i is converted into the current ($i_1 = G_{m1}v_i$) in the first differential pair, and the current i_1 drives the first high-impedance node 1, and is converted back to the first-node voltage v_1. This first-node voltage v_1 is converted again into the current ($i_2 = G_{m2}v_1$) in the second stage, and the current i_2 drives the second high-impedance node 2 to generate the second-node voltage v_2.

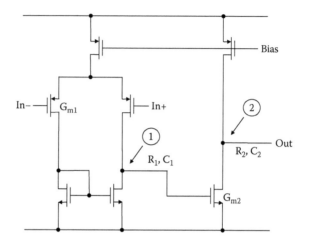

FIGURE 3.1
Standard two-stage opamp.

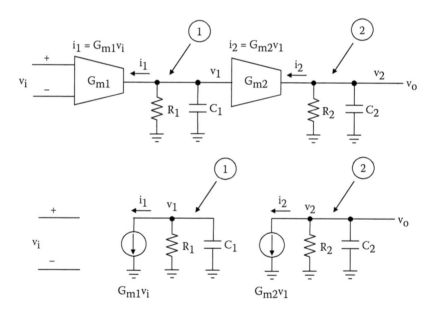

FIGURE 3.2
Small-signal equivalences of the two-stage opamp.

At low frequencies, the impedances of the capacitances C_1 and C_2 are high, and they can be considered as open circuits. Then, the low-frequency small-signal parameters such as DC gain (G) and input and output resistances can be written as follows:

$$G = G_{m1}R_1G_{m2}R_2,$$

$$R_i = \infty,$$

$$R_o = R_2.$$

(3.1)

Unlike two-stage BJT opamps with very high gain over 100 dB, metal-oxide semiconductor (MOS) opamps have typically low gains of 50 ~ 80 dB depending on the process and power consumption.

Now if two capacitances C_1 and C_2 are considered, two high-impedance nodes generate two negative real poles at $-\omega_{p1}$ and $-\omega_{p2}$ as follows:

$$B = \omega_{p1} = \frac{1}{R_1 C_1}, \qquad \omega_{p2} = \frac{1}{R_2 C_2}. \tag{3.2}$$

The dominant pole frequency becomes the open-loop bandwidth (B). These two poles at $-\omega_{p1}$ and $-\omega_{p2}$ result from total parasitic capacitances C_1 and C_2. Considering the DC gain and poles given by Equations (3.1) and (3.2), the open-loop frequency response can be sketched using Bode plots as shown in Figure 3.3.

If the unity loop-gain frequency is set to be the second pole frequency ($\omega_k = \omega_{p2}$), the phase margin (PM) becomes 45°. The maximum amount of the feedback loop gain with a PM of 45° is only about ω_{p2}/ω_{p1} as explained in Figure 3.3. The high closed-loop gain of about 1/f within the narrow bandwidth is not useful in practice, and the opamp frequency response should be narrow-banded. For the worst-case maximum feedback down to the unity gain, it is necessary to separate two poles by the amount of DC gain to have a PM greater than 45°. That is,

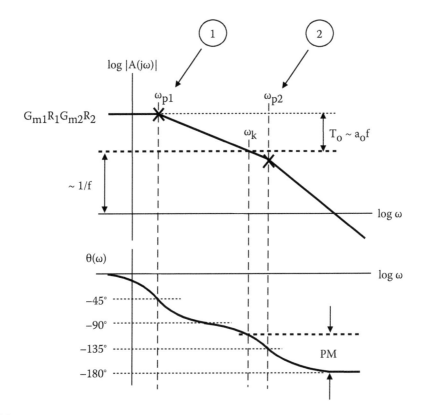

FIGURE 3.3
Bode plots of the two-stage opamp.

$$\omega_{p1} \leq \frac{\omega_{p2}}{a_o}. \tag{3.3}$$

Because the nondominant pole is limited by many design factors such as process and power consumption, the dominant pole should be moved to a lower frequency, and this narrow-banding process is known as frequency compensation. Most feedback amplifiers using opamps need a PM greater than 60°. To achieve that, it is necessary to push the second pole ω_{p2} higher than the unity loop-gain frequency ω_k.

3.2 Opamp Frequency Compensation

All opamps should be narrow banded or frequency compensated so that they can have a one-pole roll-off frequency response up to the unity loop-gain frequency ω_k. There are two ways to compensate opamps. One is a C-shunt method, and the other is a pole-splitting Miller compensation. The former narrow-bands single-stage opamps by adding a shunt capacitor so that the dominant pole can move to lower frequencies. On the other hand, the latter does the same to two-stage opamps but by splitting two poles widely using the Miller effect. Both use compensation capacitors to lower the dominant pole ω_{p1} by increasing the total effective capacitance at node 1. After compensated, ω_{p1} and ω_k are separated at least by the amount of the loop gain T_o for PM > 45°.

3.2.1 Shunt Compensation

With an extra shunt capacitor C_s added as shown in Figure 3.4, the total capacitance at node 1 is increased to be $(C_1 + C_s)$, and the dominant pole moves to a lower frequency.

$$\omega_{p1} = \frac{1}{R_1 (C_1 + C_s)}. \tag{3.4}$$

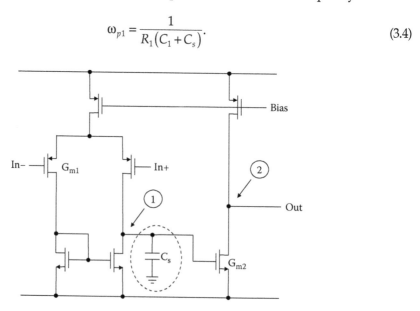

FIGURE 3.4
Frequency compensation by adding a shunt capacitor C_s.

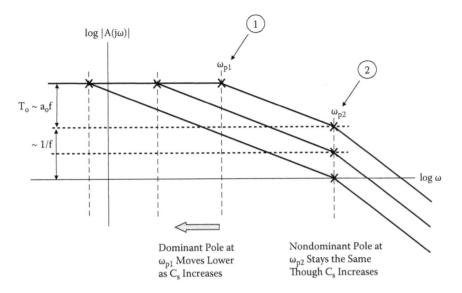

FIGURE 3.5
Shunt compensation for different loop gains.

When this method is used to compensate two-stage opamps, the compensation capacitor C_s becomes too large to integrate because the nondominant pole ω_{p2} at high-impedance node 2 is also very low.

Figure 3.5 shows the effect of the shunt compensation on the frequency response. From Equations (3.3) and (3.4), C_s should be sized to meet the following condition to have a PM greater than 45°:

$$\frac{1}{R_1\left(C_1+C_s\right)} \leq \frac{\omega_{p2}}{T_o} \approx \frac{\omega_{p2}}{a_o f}. \tag{3.5}$$

In single-stage opamps, the loading capacitor C_L works as a shunt capacitor to the high-impedance output node. Therefore, no additional C_s is needed if the nondominant pole ω_{p2} is higher than the unity loop-gain frequency ω_k, which is given by G_m/C_L.

3.2.2 Pole-Splitting Miller Compensation

Another more efficient alternative approach is a pole-splitting Miller compensation using shunt feedback. The shunt feedback always lowers driving-point input/output impedances. That is, if a capacitor is connected in a shunt feedback loop across a negative-gain opamp as shown in Figure 3.6, the effective input capacitance looks larger by the amount of feedback loop gain $(1 + a_o)$.

The Miller effect is easily understood because the voltage across the capacitor is now $(1 + a_o)$ v_i considering the opamp gain. That is, the capacitor looks, in effect, larger by the gain factor of $(1 + a_o)$. For the same reason, the shunt-feedback resistance is $(1 + a_o)$ times smaller when looking into the input, as also shown in Figure 3.6, which is the transimpedance amplifier principle.

If a compensation capacitor C_c is connected between the two high-impedance nodes 1 and 2 as shown in Figure 3.7, two poles ω_{p1} and ω_{p2} at nodes 1 and 2 are separated widely. The dominant pole at node 1 moves down due to the Miller effect.

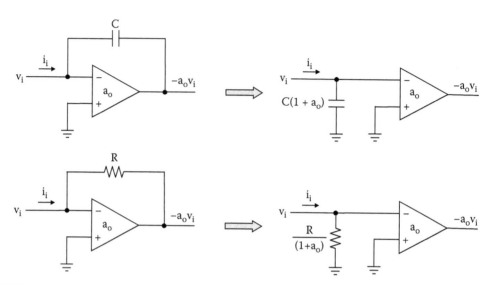

FIGURE 3.6
Effect of shunt feedback.

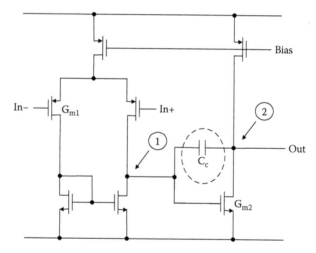

FIGURE 3.7
Frequency compensation by pole-splitting Miller capacitance.

Because the second stage's gain is $G_{m2}R_2$, the compensation capacitor C_c looks larger by $(1 + G_{m2}R_2)$ times, and the dominant pole ω_{p1} at node 1 moves to far lower frequencies as estimated below:

$$B = \omega_{p1} = \frac{1}{R_1\{C_1 + C_c(1+G_{m2}R_2)\}} \approx \frac{1}{G_{m2}R_1R_2C_c}. \tag{3.6}$$

In integrated CMOS opamps, it falls in the range of a few hundred Hz to low kHz, and a small compensation capacitor C_c up to 10 ~ 20 pF can be integrated on a chip.

Figure 3.8 explains the pole-splitting effect. The Miller effect alone cannot explain the nondominant pole though it predicts the dominant pole accurately. At low frequencies

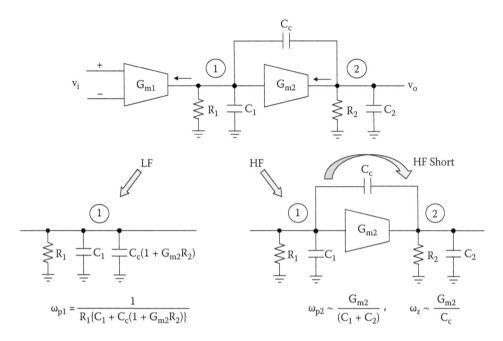

FIGURE 3.8
Pole-splitting effect of Miller compensation.

(LF), the signal at node 1 is inverted and amplified to reach node 2. However, as the second-stage gain decreases at high frequencies, the signal does not go through the inverting G_{m2} stage and become amplified. At high frequencies (HF), two nodes are shorted by the large compensation capacitor C_c and become one node. If any G_m cell's input and output terminals are shorted, it becomes a diode with a small-signal driving point resistance of $1/G_m$. Therefore, the nondominant second pole created at the shorted nodes 1 and 2 can be approximated using their total resistance and capacitance.

$$\omega_{p2} = \frac{1}{\left(R_1 \| \frac{1}{G_{m2}} \| R_2\right)(C_1 + C_2)} \approx \frac{G_{m2}}{C_1 + C_2} \approx \frac{G_{m2}}{C_L}, \tag{3.7}$$

where the total parasitic capacitance ($C_1 + C_2$) at nodes 1 and 2 can be approximated as C_L because it is much smaller than the typical loading capacitance C_L and the compensation capacitance C_c in most opamp designs.

Also note that this signal feed-forward through the inverting stage creates a right-half-plane zero as discussed in Chapter 1.

$$\omega_z = \frac{G_{m2}}{C_c}. \tag{3.8}$$

What to note about this right-half-plane zero is that its phase lags like a pole.

Figure 3.9 shows that two poles split as the compensation capacitance C_c increases. As approximated in Equations (3.6) and (3.7), the dominant pole ω_{p1} moves lower while

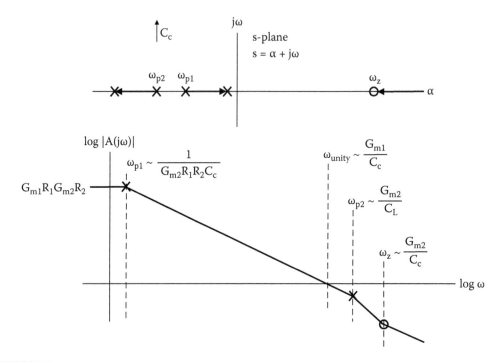

FIGURE 3.9
Pole-splitting Miller effect.

the nondominant pole ω_{p2} moves higher to several hundreds of MHz range because the load capacitance C_L is driven by $1/G_{m2}$. A zero moves in from infinite to the right-half plane as the value of C_c increases as given by Equation (3.8). After Miller compensation, the open-loop bandwidth becomes narrow, and the gain drops at the rate of −6 dB/oct or −20 dB/dec until it reaches the unity-gain frequency ω_{unity}, which is the gain-bandwidth product (*GB*).

$$\omega_{unity} = GB \approx G_{m1}R_1G_{m2}R_2 \times \frac{1}{G_{m2}R_1R_2C_c} = \frac{G_{m1}}{C_c}. \qquad (3.9)$$

   To ensure stability down to the worst unity-gain feedback, the nondominant pole should be placed at higher than ω_{unity}, and the pole-splitting Miller compensation achieves this goal using a small on-chip capacitor. Once the compensation is done for the worst-case feedback up to f = 1, stability is warranted for all values of f, but the down side is that if the feedback factor f is small (f ≪ 1), the closed-loop bandwidth becomes very narrow. Therefore, it is desirable to set C_c to maximize the bandwidth ω_k for any given f values. The opamp frequency response is modified as shown in Figure 3.10 depending on the size of the compensation capacitor C_c. For low feedback factors, C_c can be sized so that ω_{p2} can be higher than ω_k.

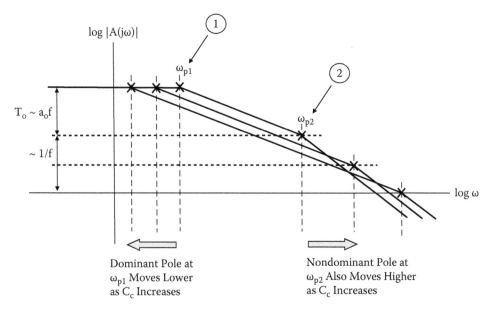

FIGURE 3.10
Miller compensation for different loop gains.

3.3 Phase Margin of Two-Stage Miller-Compensated Opamps

The Bode gain plot of a two-stage Miller-compensated opamp resembles that of an ideal opamp with one pole roll-off. After compensated with PM > 45°, only one pole is allowed below the unity loop-gain frequency ω_k. Therefore, the frequency response of the Miller-compensated two-stage opamp can be handled with only a few circuit parameters such as DC gain (G), dominant pole or bandwidth ($B = \omega_{p1}$), unity-gain frequency or gain-bandwidth product ($GB = \omega_{unity}$), nondominant pole (ω_{p2}), and right-half-plane zero (ω_z) extracted from the previous intuitive analysis.

The unity loop-gain frequency is related to the open-loop unity-gain frequency by the feedback factor f.

$$\omega_k \approx f\omega_{unity}. \tag{3.10}$$

Therefore, the PM with a feedback factor of f can be approximated as

$$PM(\omega_k) = 90^o - \tan^{-1}\frac{\omega_k}{\omega_{p2}} - \tan^{-1}\frac{\omega_k}{\omega_z}. \tag{3.11}$$

Note again that the phase due to the right-half-plane zero lags like a pole. Considering C_L and C_c and using Equations (3.7), (3.8), and (3.9), the PM given by Equation (3.11) can be rewritten as

$$PM(\omega_k) = 90^o - \tan^{-1} f\left(\frac{G_{m1}}{G_{m2}}\right)\left(\frac{C_L}{C_c}\right) - \tan^{-1} f\left(\frac{G_{m1}}{G_{m2}}\right). \qquad (3.12)$$

As expected, PM is worst when f = 1 in the unity-gain feedback. To reduce the right-half-plane zero effect, most BJT opamps are designed with the second-stage G_{m2} much greater than the first-stage G_{m1}.

Because the transconductance is a linear function of the collector current, it is easy to achieve a ratio of $G_{m2}/G_{m1} \gg 20$ in BJT opamps. Then, the right-half-plane zero degrades PM by less than 3°. However, it is very difficult to make such a large transconductance ratio in CMOS opamps. A ratio of 4 is typical in CMOS design, and PM suffers by as much as $\tan^{-1}(1/4) = 14°$. If $C_c = C_L/2$ and $G_{m1} = G_{m2}/4$, PM with f = 1 is about

$$PM(\omega_{unity}) = 90^o - \tan^{-1}\frac{1}{2} - \tan^{-1}\frac{1}{4} \approx 90^o - 26.5^o - 14^o \approx 49.5^o. \qquad (3.13)$$

This implies that the right-half-plane zero contributes a significant amount of extra phase delay. To achieve a usable PM greater than 60°, the right-half plane zero should be canceled or eliminated in CMOS opamps.

To maximize the closed-loop bandwidth for any given PM, the value of the Miller compensation capacitor C_c can be chosen using Equations (3.10) and (3.12) as

$$C_c \geq fC_L\left(\frac{G_{m1}}{G_{m2}}\right)\cot(90^o - PM). \qquad (3.14)$$

For example, the second pole ω_{p2} should be higher than ω_k by cot(90° − PM), which is about 1.73 ω_k for PM > 60°. If C_c is larger than this value, two poles split farther apart, and the closed-loop bandwidth ω_k becomes narrower. Because the size of C_c greatly affects the transient response, opamps should not be overcompensated with C_c larger than given by Equation (3.14).

Figure 3.11 shows both open-loop and closed-loop frequency responses for three examples when compensating two-stage opamps using the pole-splitting Miller effect. Assume that the feedback factor f is set to be, for example, 0.1 for 20 dB noninverting gain. Without compensation, two poles ω_{p1} and ω_{p2} are at low frequencies, and the closed-loop bandwidth ω_k can be wider than the nondominant second pole frequency ω_{p2} but with PM smaller than 45°. However, considering extra high-frequency stray poles, the feedback amplifier can be unstable. If overcompensated, PM improves to be almost 90°, but the closed-loop bandwidth is narrowed.

3.4 Right-Half Plane Zero Cancellation in Two-Stage Opamps

After passing the right-half plane zero, the gain increases with a 6 dB/oct slope like other zeros, but the phase lags by 90° like poles. It is not easy to keep a high ratio of G_{m2}/G_{m1} in CMOS opamps. There are three ways to eliminate the right-half plane zero.

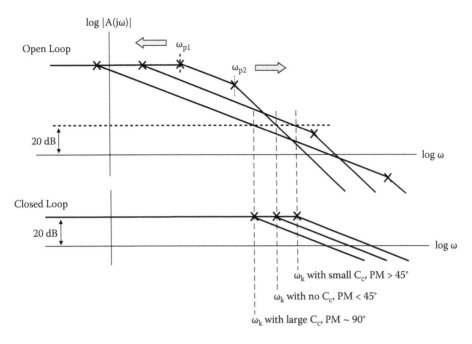

FIGURE 3.11
Bandwidth versus Miller capacitor.

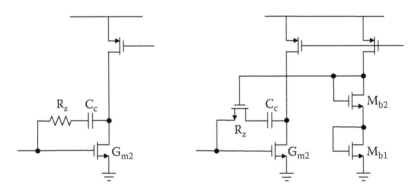

FIGURE 3.12
Zero cancellation with a series resistance.

3.4.1 Inserting a Series Resistance

One easy way is to insert a resistance R_z in series with C_c shown in Figure 3.12. As discussed in Chapter 1, if $R_z = 1/G_{m2}$, the right-half-plane zero moves to infinite, and if $R_z > 1/G_{m2}$, it gets back into the left-half plane. This zero in the left-half plane can be advantageously used to make up for the phase lag. A fixed resistance can be chosen to match $1/G_{m2}$. Due to the process and temperature variations, the resistance value R_z does not track the value of $1/G_{m2}$, but this zero cancellation does not require an accurate matching as long as the right-half-plane zero can be moved far higher than the unity loop-gain frequency to avoid the doublet.

The resistance can be replaced by a MOS transistor operating in triode as shown on the right side of Figure 3.12. To make the resistance R_z keep track of $1/G_{m2}$, the bias circuit generates a proper gate-source voltage V_{GS} of the MOS transistor in triode. The V_{GS} of M_{b1} replicates that of the G_{m2} transistor, while the V_{GS} of M_{b2} replicates that of the R_z transistor. Although M_{b2} operates in saturation while the R_z transistor is in triode, their threshold voltages are assumed to track each other as the source bias voltage is about the same. Because two transistors M_{b1} and M_{b2} in the bias string can be made into ratios, R_z and $1/G_{m2}$ can be derived as follows:

$$R_z = \frac{1}{\mu C_{ox} \left(\dfrac{W}{L}\right)_{R_z} \left(V_{GS} - V_{th}\right)_{R_z}}$$

$$\frac{1}{G_{m2}} = \frac{1}{\mu C_{ox} \left(\dfrac{W}{L}\right)_{G_{m2}} \left(V_{GS} - V_{th}\right)_{G_{m2}}}.$$ (3.15)

Because the gate-source overdrive voltage of the G_{m2} transistor is replicated using that of M_{b1} in the replica bias circuit, the device size of the R_z transistor can be chosen by equating $R_z = 1/G_{m2}$.

$$\left(\frac{W}{L}\right)_{R_z} = \left(\frac{W}{L}\right)_{G_{m2}} \sqrt{\frac{\left(\dfrac{W}{L}\right)_{M_{b2}}}{\left(\dfrac{W}{L}\right)_{M_{b1}}}}.$$ (3.16)

This replica bias circuit will keep the proper matching between R_z and $1/G_{m2}$ for any variations over process, voltage, and temperature.

3.4.2 Using Source Follower for Feedback

Another way is to use a source follower for feedback as shown in Figure 3.13, which eliminates the feed-forward path that creates the right-half-plane zero. The Miller effect is not affected by this feedback, but the feed-forward signal is stopped at the output of the source follower due to its low driving-point output impedance.

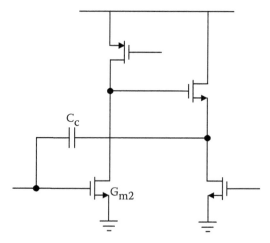

FIGURE 3.13
Using source follower for feedback.

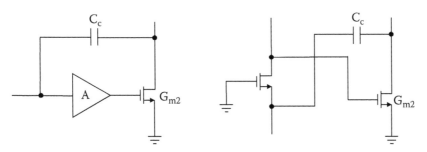

FIGURE 3.14
G_m boosting with common-gate (CG) amplifier.

This arrangement has two main drawbacks. One is extra power consumption and chip area for the source follower, and the other is the left-half-plane zero introduced by the pole of the feedback network. The source follower output resistance is about $1/g_m$, which drives the total loading capacitance C_p. Then, because the pole at the source follower output is $\omega_p = g_m/C_p$, it becomes the left-half plane zero.

3.4.3 Boosting G_m with Extra Gain Stage

In Equation (3.8), a right-half-plane zero is created when an inverting G_m stage is bypassed by a capacitor. A G_m-boosting concept is also used to increase the output resistance of the cascoded transistor.

For example, an extra stage with a gain of A can be added to a G_m stage as shown in Figure 3.14, and the effective G_{m2} will increase by A to be AG_{m2}. From Equations (1.38) and (3.8), the right-half-plane zero is now at an A-times higher frequency AG_{m2}/C.

This extra gain stage can be implemented using a CG amplifier. Because the first stage of the opamp needs no large output swing, its output is usually cascoded for high gain. Therefore, the Miller capacitor can be fed into the source of the cascode stage without adding an extra gain stage. This local two-stage feedback has a different implication in the transient response. Unless the CG amplifier is ideal, its pole makes a zero inside the global feedback loop and can cause high-frequency peaking or ringing in transients.

3.5 Transient Response of Opamp in Feedback

In Chapter 1, the transient response of opamps in feedback is discussed based on the assumption that the amplifier is operating in small-signal linear mode. Under such an assumption, the transient response is the sum of all exponentially decaying terms. Either zeros or complex-conjugate poles make the transient response peak or ring like a sinusoidal decaying exponentially. However, responses to any arbitrary waveforms are difficult to analyze analytically. Most transient responses can be understood using either a sinusoidal or step function. The transient step response is mainly dominated by pole frequencies and PM. That is, the linear model of transistor is still valid, and the steady-state analysis like frequency response can provide valuable information on the transient step response.

3.5.1 Slew Rate

However, in reality, the actual circuits used in opamps are far from being linear, in particular, for large signals. With a large input signal suddenly applied to the opamp input, the opamp first stage becomes nonlinear and behaves quite differently from the one predicted by the linear analysis.

The nonlinear transient behavior called slewing is explained in Figure 3.15. The maximum current flowing into or out of the first high-impedance node of the two-stage opamp shown in Figure 3.7 is limited by the tail current I_{bias}. The overdrive voltage of the differential pair input limits the linear range of the input. That is, if the input differential voltage is larger than $\sqrt{2}(V_{GS} - V_{th})$, the maximum output current is I_{bias}. In such a case, the second stage with the Miller shunt feedback behaves as an integrator, and the output voltage increases or decreases linearly as shown.

Figure 3.16 shows the slew and settling aspect of the transient step response. Assume that an opamp has a unity-gain bandwidth of 10 MHz and a good PM in a unity gain feedback. If a sudden step of 5 V is applied to the input, the transient response expected from the linear analysis predicts the settling of the output within about 50 ns as sketched. So the settling time constant τ is about 16 ns. However, in practice, it will take far longer. For example, if the input device is biased with 50 µA and the compensation capacitor is 5 pF, it takes about an additional 500 ns (almost 10 times longer) for the output to settle. The output changes almost linearly with a ramp rate of about 10 V/µs. This is called the slew rate, which results from the nonlinear operation of the input differential pair.

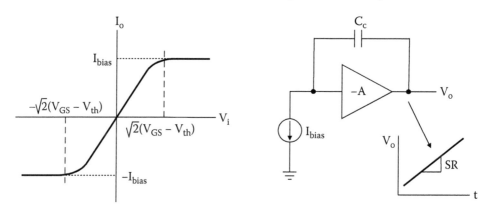

FIGURE 3.15
I-V characteristic of the input differential pair and the following integrator.

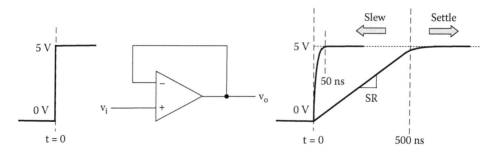

FIGURE 3.16
Slew and settling in the transient response.

For large input, the differential pair is fully switched on and off, and the maximum current out of the differential pair is limited by the bias tail current I_{bias}. Therefore, the slew rate of the Miller-compensated two-stage opamp is defined as

$$Slew\ Rate\ (SR) = \frac{I_{bias}}{C_c}. \tag{3.17}$$

The time it takes for the output to approach the final value for the step input depends on the magnitude of the signal. When the step is small, the output is just settling with an exponential time constant τ. Depending on accuracy, the settling time varies. For example, it takes 6.9τ (~7τ) to achieve 0.1% settling accuracy because the exponential term decays as $\exp(-t/\tau)$.

$$e^{-6.9} = 0.001 = 0.1\%. \tag{3.18}$$

When the step is large, the total settling time is the slewing time plus the exponential settling time. It can be approximated as follows for the final 0.1% settling accuracy:

$$Total\ Settling\ Time \approx \frac{V_{final} - V_{initial}}{SR} + 7\tau. \tag{3.19}$$

3.5.2 Maximum Power Bandwidth

The slew rate affects the sinusoidal response differently. When a sinusoidal signal is generated with a slew-limited opamp, the sine wave will be distorted if the output changes faster than the slew rate as explained in Figure 3.17. To avoid slewing, the maximum change of the signal should be less than the slew rate. For sinusoidal waveforms, the maximum signal change occurs at the zero-crossing points. To generate a sinusoidal of $V_o = A_o\sin\omega t$ with the peak magnitude of A_o, the following condition should be met:

$$\left.\frac{dV_o}{dt}\right|_{t=0} = A_o\omega\cos\omega t\big|_{t=0} = A_o\omega < SR. \tag{3.20}$$

Therefore, the maximum power bandwidth is defined as a sine wave frequency that can be generated with a magnitude of A_o as

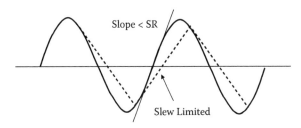

FIGURE 3.17
Slew-limited operational amplifier (opamp) distorts sinusoidal output.

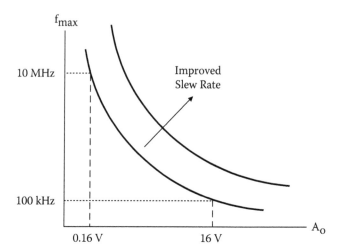

FIGURE 3.18
Maximum power bandwidth of slew-limited opamps.

$$f_{max} = \frac{SR}{2\pi A_o}. \tag{3.21}$$

Typical slew rates of two-stage CMOS opamps are in the range of about 10 ~ 100 V/µs. For example, with a slew rate of 10 V/µs, the peak magnitude and the maximum power bandwidth are related as shown in Figure 3.18. If the slew rate is improved, the power bandwidth is widened with larger peak magnitude.

Note that slew rate greatly affects both the step and sinusoidal responses, and high slew rate is desirable. There are ways to improve slew rate. We can derive the relation between the slew rate and the unity-gain bandwidth for standard two-stage CMOS opamps shown in Figure 3.7.

$$SR = \frac{I_{bias}}{C_c} = \frac{I_{bias}}{G_{m1}}\omega_{unity} = \sqrt{\frac{I_{bias}}{\mu C_{ox}\dfrac{W}{L}}} \times \omega_{unity}. \tag{3.22}$$

For high slew rate, the ratio of I_{bias}/G_{m1} or the linear range should be increased. Also, increasing the bias current I_{bias} or decreasing the compensation capacitor C_c is an effective way to improve the slew rate. For feedback amplifiers, ω_k is a small-signal bandwidth, and ω_{max} is a large-signal bandwidth. Thus, it is necessary to compensate opamps with minimum compensation capacitances to warrant the maximum bandwidth. The standby bias current limits the slew rates of all class-A amplifiers. Therefore, using class-AB amplifiers is another way to improve the slew rate, because they can source or sink far more output currents than the standby bias current.

In general, BJT has a small linear range limited by V_T (26 mV) per each device, and a MOS or junction field effect transistor (JFET) device has a wide linear range depending on the bias condition, for example, 0.5 ~ 1 V or even higher. Therefore, commercial JFET input opamps exhibit at least 20 times higher slew rate than BJT opamps. The slew rate of typical BJT opamps ranges from 1 to 10 V/µs, while that of JFET input opamps ranges from

100 to 1000 V/µs. So, high slew rate amplifiers using JFET, MOS, and even vacuum tubes are preferred in fast transient purposes, such as in audio amplifiers, as they can reproduce sudden music transients faithfully.

3.6 Opamp Design Examples

Opamps are designed with numerous constraints, such as DC gain, bandwidth, supply voltage, power consumption, slew rate, noise, offset, output signal swing, input common-mode range, and output load driving capability. Therefore, it is unlikely that any one opamp can meet all the requirements. Most precision analog circuits require a typical DC loop gain of 70 ~ 80 dB. Because modern scaled CMOS provides a low maximum gain $g_m r_o$ of about 20 ~ 30 dB, three gain stages are usually needed.

Most on-chip opamps are fully differential and designed in class A depending on the load capacitance and power requirement. For class-A designs, the DC gain and signal swing requirements set how many gain stages are needed. Class-AB opamps can drive large loads with very low standby currents, but the bandwidth is still set by the standby current. They also benefit from high slew rate when driving low impedance at high supply voltages. However, class-AB opamps in scaled CMOS exhibit no merits in high-speed applications, because they rarely slew at low supply voltages.

3.6.1 Telescopic Triple Cascode Opamp

For class-A design, the DC gain and signal swing requirements set how many gain stages are needed. The simplest three-stage opamp can be made using a single-stage triple cascode structure as shown in Figure 3.19, where all driving-point resistances are shown assuming all devices have the same small-signal parameters such as g_m and r_o for simplicity. As analyzed in Chapter 2, this one-stage opamp gives a DC gain of

$$\frac{v_o}{v_i} = -g_m\left\{r_o\left(g_m r_o\right)^2 \| r_o\left(g_m r_o\right)^2\right\} = -\left(g_m r_o\right)^3/2. \tag{3.23}$$

Furthermore, the alternating current (AC) and transient performance of the cascode opamp is superb. Figure 3.20 sketches the Bode gain plot intuitively derived from the impedance of the cascode nodes. Assume three nodes are loaded by the loading capacitance C_L and the parasitic capacitances C_p, as shown. The loading capacitance is usually far larger than the device parasitics. Therefore, at high frequencies, the loading capacitance C_L starts to short the output node to ground first because it also has the highest driving point resistance. As a result, the single-stage opamp needs no frequency compensation. The dominant pole is placed at

$$\omega_{p1} \approx \frac{1}{\left\{r_o\left(g_m r_o\right)^2/2\right\}C_L}. \tag{3.24}$$

If the output impedance becomes low due to this shorting, the driving-point resistance looking up from the source side of the cascode transistor becomes $1/g_m$ as marked. That is, the impedance $Z(j\omega)$ looking up from the cascode node has a zero approximately at

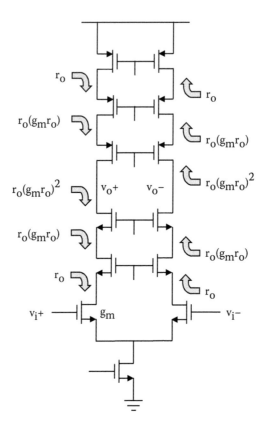

FIGURE 3.19
Single-stage triple-cascode opamp.

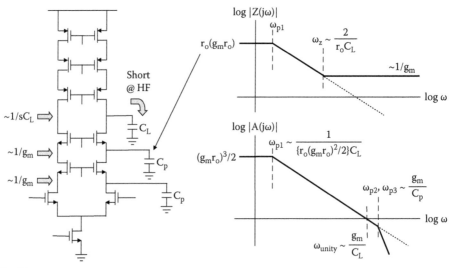

FIGURE 3.20
Frequency response of a single-stage triple-cascode opamp.

$$\omega_z = r_o\left(g_m r_o\right) \times \frac{1}{\left\{r_o\left(g_m r_o\right)^2/2\right\}C_L} \times g_m \approx \frac{2}{r_o C_L}. \tag{3.25}$$

This break frequency is where the impedance $1/sC_L$ of the load capacitance becomes about half the device output resistance r_o.

Also as explained in Figure 2.9, note that the drain-side low resistance is reflected to the source side after divided by the gain factor $g_m r_o$. Therefore, the high-frequency nondominant poles at two cascode nodes are pushed to much higher frequencies than the unity-gain frequency, because they are driven by the same low-resistance $1/g_m$, but the parasitic capacitances are much smaller than the loading capacitance C_L. That is,

$$\omega_{p2}, \omega_{p3} \approx \frac{g_m}{C_p} \quad \gg \quad \omega_{unity} \approx \frac{g_m}{C_L}. \tag{3.26}$$

This single-pole role-off is very close to the ideal opamp frequency response. As a result, it is the fastest-settling opamp with good PM. Compared to the nondominant pole of the two-stage opamp, these nondominant poles are very high-frequency parasitic poles.

This single-stage opamp consumes low power due to the current reuse and achieves wide bandwidth. In fact, you can cascode one more stage to get the gain of four stages, but the down side is the difficulty in biasing the input common mode and the limited signal swing. If each transistor is biased with a minimum overdrive voltage of V_{DSsat}, the input common mode should be biased at $(V_{GS} + V_{DSsat})$ higher than the negative supply. Furthermore, the peak signal swing is limited to $(V_{DD} - 7V_{DSsat})$ because seven devices are stacked between the supplies. If supply voltages are scaled down, this limited signal swing becomes a serious disadvantage.

3.6.2 Folded-Cascode Opamp

One way to restore the input common-mode bias to the middle of the supply rails is to fold the cascode stage as shown in Figure 3.21.

Both the N-channel metal-oxide semiconductor (NMOS) and P-channel metal-oxide semiconductor (PMOS) triple-cascode input stages can be folded using the PMOS and NMOS differential pairs, respectively, as discussed in Chapter 2. Once folded, the input common-mode biasing is no longer critical, as it is free from the input common-mode constraint, and it becomes possible to use cascode opamps in continuous time.

Assume that all transistors have the same bias currents, same g_m, same r_o, and same parasitics as in the triple-cascode case. Then the obvious disadvantage of the cascode folding is that it consumes twice as much power as one additional branch for the differential pair is added. Because two currents are summed, the size of the current source should be doubled as marked, and its output resistance is lowered to be half. Then compared to the standard cascode case, the output resistance of the folded-cascode output decreases by 1/3, and the parasitic at the cascode node is also doubled. Therefore, the gain of the folded-cascode opamp is slightly lower than that of the unfolded cascode opamp, while the gain-bandwidth product or the unity-gain frequency ω_{unity} stays about the same. However, one of the nondominant pole frequencies is lowered due to the increased parasitic at the cascaded node. Ignoring these slight differences, the folded-cascode topology can be considered as a single stage because the nondominant pole is still set by the parasitic capacitance.

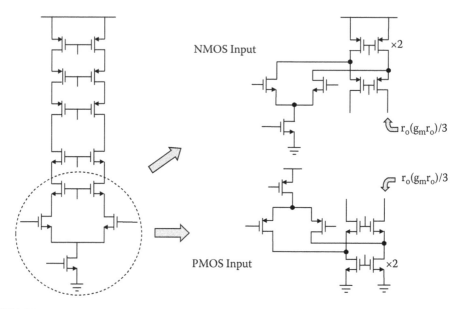

FIGURE 3.21
Folded-cascode opamp.

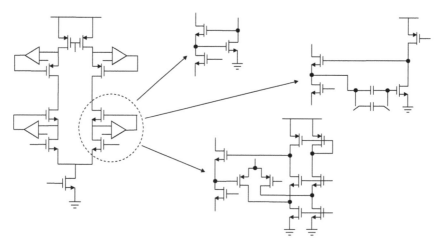

FIGURE 3.22
Gain-boosted cascode opamp.

3.6.3 Gain-Boosted Cascode Opamp

The gain boosting by feedback discussed in Chapter 2 is also effective in achieving high gain. Figure 3.22 shows three ways of implementing the gain-boosting amplifier.

The first is just an inverter. Although it is effective, it needs to be biased with a fixed V_{GS} and reduces the signal swing. The second is to use a capacitive level shifting to bias the inverter input with V_{GS}, while the cascode node can be biased separately. This will increase the signal swing. Because it uses only NMOS transistors for boosting, the parasitic capacitance is small, and the amplifier is fast while consuming low power. However, the drawback is the dynamic level shifting works only for switched-capacitor circuits. The third example

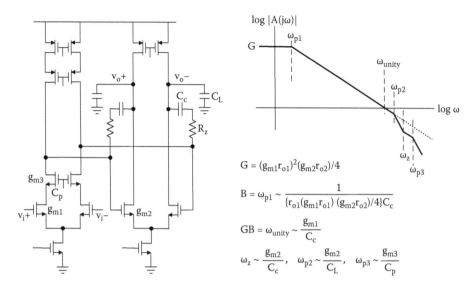

$$G = (g_{m1}r_{o1})^2(g_{m2}r_{o2})/4$$

$$B = \omega_{p1} \sim \frac{1}{\{r_{o1}(g_{m1}r_{o1})\ (g_{m2}r_{o2})/4\}C_c}$$

$$GB = \omega_{unity} \sim \frac{g_{m1}}{C_c}$$

$$\omega_z \sim \frac{g_{m2}}{C_c}, \quad \omega_{p2} \sim \frac{g_{m2}}{C_L}, \quad \omega_{p3} \sim \frac{g_{m3}}{C_p}$$

FIGURE 3.23
Miller-compensated two-stage opamp.

is to use a folded-cascode opamp as a boosting amplifier. NMOS and PMOS cascodes need PMOS and NMOS inputs, respectively. This gives the freedom to choose the input common-mode bias but consumes extra power and introduces extra pole in the gain-boosting loop. There are also many different ways to make inverting amplifiers for gain boosting, but they all consume power, and the extra pole of the boosting amplifier becomes a zero. The unity-gain bandwidth of the gain-boosting loop should be controlled to be higher than the unity loop-gain frequency of the main loop but lower than the nondominant pole.

3.6.4 Two-Stage Opamp

If the output swing is still limited due to cascoding, the only way to achieve large output swing is to add one more stage of the CS amplifier to make a standard two-stage opamp as discussed earlier. Figure 3.23 shows the same two-stage opamp with the first stage cascoded to get a higher gain than the one in Figure 3.1. A differential pair replaces the second CS amplifier stage for fully differential operation. Also shown are small-signal parameters of the opamp, assuming all devices have the same g_m and r_o. The right-half plane zero is created when the series resistance R_z is not inserted. The right-half plane zero moves out as R_z value increases.

Another way to push the right-half plane zero is to boost the second-stage G_m as explained in Figure 3.24, where the Miller capacitance is returned to the cascode node instead of the first-stage output. However, the CG amplifier pole produces an additional zero at high frequencies, which causes slight peaking in transient responses.

The third pole is generated at the cascode node as in the triple cascode opamp. Note that the second pole is generated because the load capacitance C_L is driven by $1/g_{m2}$. So compared to the cascode opamp, the second pole moves in. Therefore, the transconductance of the second gain stage is of paramount interest in achieving good PM in two-stage opamps. In this case, the peak output swing is $(V_{DD} - 3V_{DSsat})$, and the input common-mode voltage is still $(V_{GS} + V_{DSsat})$ above the negative supply rail. The second-stage input common-mode

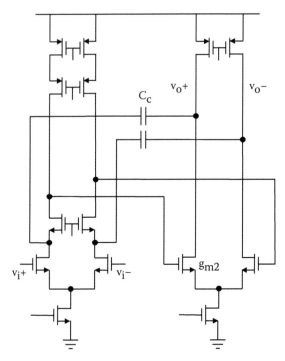

FIGURE 3.24
Another Miller-compensated two-stage opamp.

voltage should also be ($V_{GS} + V_{DSsat}$), and the first stage should operate with this output common-mode voltage. Therefore, the following condition should be met:

$$V_{GS} + V_{DSsat} > 3V_{DSsat}. \tag{3.27}$$

That is, V_{DSsat} should be smaller than the threshold voltage of the second-stage input device. Note that the gain of the first cascode stage can be boosted further to achieve a gain of four stages.

3.7 Common-Mode Feedback

Unlike the single-ended opamps, the ouput common-mode (CM) voltage of the fully differential opamps is not defined due to the mismatch between two different current sources. One current source pulls up the high-impedance output node, while the other pulls it down. Therefore, the output nodes of the fully differential opamp should be stabilized applying the CM feedback. Switched-capacitor CM feedback (CMFB) applied to the triple cascode opamp is explained in Figure 3.25 [2].

As expected, the CM voltage is not defined without capacitor C_1. If C_1 is charged with a constant DC, the output CM change is directly fed to the gate of the tail current source so that the tail current can match the sum of the two PMOS load currents. Because the gate of the tail current is capacitive, a DC path to the gate should be provided. This can be done

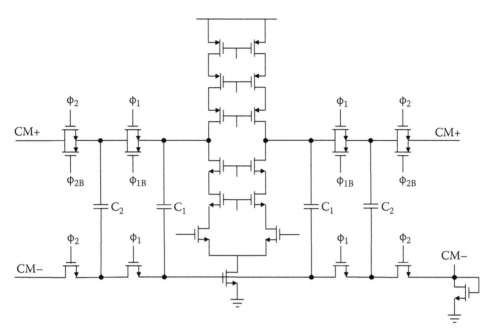

FIGURE 3.25
Capacitive common-mode feedback (CMFB) for triple-cascode opamp.

by periodically switching another capacitor C_2 back and forth using two nonoverlapping clocks ϕ_1 and ϕ_2. Because the charge loss due to the leakage from the source or drain area of the switch is small, C_2 can be made far smaller than C_1. However, if C_2 is too small, it takes too many cycles to fill up C_1 at the beginning. Note that differential signals are averaged by two equal C_1 and do not affect the CM feedback. Unfortunately, the differential output node is loaded by C_1, which is directly added to the total loading capacitance. If C_1 is made small to avoid the loading effect, the CM feedback gain is reduced $2C_1/(2C_1 + C_i)$ due to the capacitive divider formed with the input capacitance C_i of the tail current source.

In the CM feedback shown in Figure 3.25, both CM and differential-mode (DM) loops share the same gain stage, but the CM gain is as high as the DM gain because the output resistance is higher when looking down the NMOS side cascoded by one more stage. The DM loop sees C_1 as a load, while the CM loop sees two series capacitances $2C_1$ and C_i as a load. If the load capacitance is much larger than these, both CM and DM loops have about the same unity-gain bandwidths and, therefore, settle with about the same time constants.

3.7.1 Common-Mode Loop Requirements

Due to the high common-mode rejection ratio (CMRR) in most differential opamps, the CM settling is not as stringent as the DM settling. For example, in most transient switched-capacitor amplifiers, the CM signal swing is small. Assume that it settles with a time constant of τ. Due to CMRR, the CM signal does not need to settle as accurately as the DM signal, and the CM loop does not need to be as broad-banded as the DM loop. If CMRR is 40 dB, the following settling time can be saved:

$$t = \tau \times \ln CMRR \approx 4.6\tau. \tag{3.28}$$

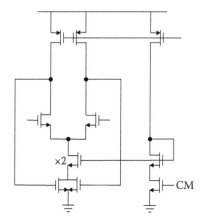
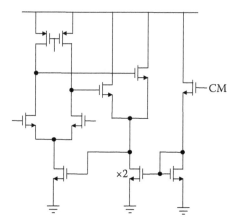

FIGURE 3.26
CMFB with triode inverters and source followers.

For example, this implies that the DM loop needs to settle for nine time constants while the CM loop needs to settle only for 4.4 time constants. That is, the CM loop bandwidth can be only half of the DM loop bandwidth. Considering that the CM signal does not vary much once set, the CM loop bandwidth requirement is far less stringent than the DM loop bandwidth requirement, and even narrow bandwidths of 1/3 to 1/5 of the DM bandwidth are still acceptable. For the same reason, the loop gain of the CM loop does not need to be made unnecessarily high.

3.7.2 Continuous-Time Common-Mode Feedback

Although the capacitive CMFB is useful, it only works for discrete-time switched-capacitor circuits. For continuous-time circuits, CMFB needs to be implemented in continuous time. Shown on the left side of Figure 3.26, two triode-biased transistors sense the output CM voltage and apply negative feedback directly [1]. Compared to the capacitive feedback, both CM loop gain and bandwidth are lower, and the offset will be higher because V_{GS} of a triode-biased transistor is large.

On the right side, CMFB is applied using source followers. Both source followers should be biased with large overdrive voltages. The output swings are asymmetrically sensed due to the source follower. The high swing is easily pulled up, but the low swing is uncontrolled as the source follower is turned off. Therefore, the CM voltages tend to go lower as the signal swing gets larger.

This CM shift can be avoided if the differential pair is used for CMFB as shown on the left side of Figure 3.27. However, the differential pair has a high gain and adds too much gain to the CM feedback loop. To avoid the frequency compensation, the feedback loop gain can be reduced using the diode load. Like the source follower CMFB, the differential swing is still limited by the overdrive voltage of the differential pair.

One way to avoid biasing transistors with large overdrive voltages is to use a resistive divider as shown on the right side. The drawbacks of this arrangement are the resistor loading to high-impedance nodes and the extra pole generated with the input capacitance of the CMFB amplifier. The former reduces the gain of the main amplifier, and the latter contributes extra phase shift in the CMFB loop. The resistors can be bypassed at high frequencies using parallel capacitors, but the extra bypassing capacitors also load the main

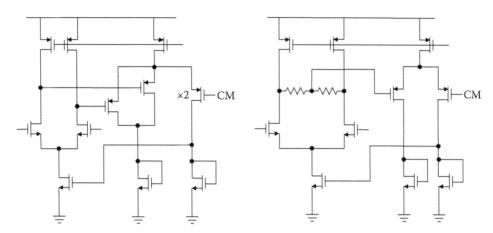

FIGURE 3.27
CMFB with differential pair and resistive feedback.

amplifier. This CMFB is the most exact CMFB and is conceptually similar to the capacitive CMFB. The resistor divider is quite insensitive to the DM signal.

3.8 Offset Cancellation

The constant offset of the opamp does not affect most system functions, and all digital systems can easily null the system offsets digitally. However, most low-frequency applications suffer greatly from the offset and the low-frequency 1/f noise. The low-frequency spectrum at DC can be either chopped (modulated to higher frequencies) or sampled and canceled in switched-capacitor circuits. The chopper concept using a clock frequency higher than the signal bandwidth is shown in Figure 3.28.

The input is chopped twice using the same binary pulse of 1 and –1, while the offset and the low-frequency 1/f noise are chopped once. The binary pulse of 1 and –1 has no DC component, and frequency components are only odd harmonics of the clock tone. Therefore, the low-frequency spectrum is modulated out of the signal band.

In most switched-capacitor circuits, the offset and the low-frequency 1/f noise are directly sampled on the input capacitors and subtracted in the subsequent half-clock period ($T/2$) as shown in Figure 3.29. In the closed-loop offset sampling, the opamp is connected in a unity-gain feedback using two switches to sample the input at the bottom plates of the capacitors. That is, the offset transfer function becomes $(1 - z^{-1/2})$.

$$\text{Re}\left(1 - e^{-j\frac{\omega T}{2}}\right) = 1 - \cos\frac{\omega T}{2}. \tag{3.29}$$

This offset and 1/f noise cancellation is effective only when the 1/f noise corner frequency is far lower than the sampling frequency, $\omega T \ll 1$. Although offset is canceled, the mismatch between the two feedback switches leaves the charge injection and feed-through error on

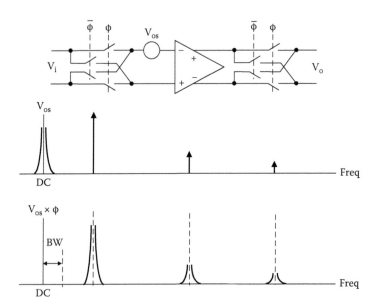

FIGURE 3.28
Chopping low-frequency spectrum.

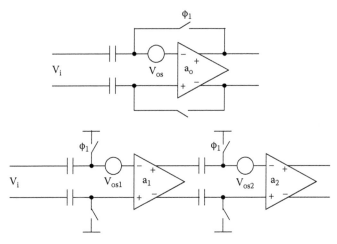

FIGURE 3.29
Closed-loop and open-loop offset sampling.

the sampling capacitors, which results in a residual offset. Furthermore, this closed-loop sampling requires that the opamp bandwidth be very wide.

Alternatively, switched-capacitor circuits commonly employ bottom-plate sampling. In this open-loop sampling, the opamp input is grounded, and the offset is amplified. The amplified offset is then sampled on the input of the second-stage opamp. Therefore, the effective input-referred offset would be

$$V_{os} = V_{ft1} + \frac{V_{ft2} + V_{os2}}{a_1},$$ (3.30)

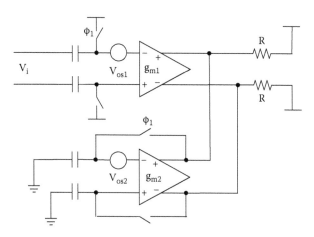

FIGURE 3.30
Closed-loop sampling of the amplified offset.

where V_{ft1} and V_{ft2} are feed-through voltages of the two sampling stages, respectively. Although Equation (3.30) is written in a mathematical form, all offsets and feed-through voltages are random variables, and only their root-mean-square (RMS) values are meaningful. The problem with the open-loop offset sampling is that the first stage should drive the capacitive load in open loop. So the high gain of a_1 is difficult to achieve, though high gain is desired to reduce the offset of the second stage. To facilitate the offset sampling, a parallel amplifier can be used as shown in Figure 3.30.

Assume the amplifier is made of a transconductance stage followed by the resistive load, and two g_m stages are used. One is to amplify the offset, and the other is to sample the amplified offset. Note that unlike Figure 3.28, the offset sampling capacitor is driven using the closed-loop amplifier made of g_{m2}. Therefore, the closed-loop offset sampling can be faster. The closed-loop g_{m2} makes a diode, and it becomes a load to the input g_{m1}. Considering the loading effect, the input-referred offset becomes

$$V_{os} = \frac{V_{os1}}{1 + g_{m2}R} + \frac{V_{os2}}{1 + g_{m1}R}.$$

(3.31)

Similarly, the feed-through voltages are referred to the input similarly as

$$V_{os} = V_{ft1} + \frac{g_{m2}}{g_{m1}} V_{ft2}.$$

(3.32)

Therefore, a compromise should be made when canceling the feed-through voltage and the second-stage offset simultaneously.

The switch feed-through voltage resulting from the mismatch between two sampling switches contributes significantly to the offset. Therefore, low-frequency systems sensitive to offset use the double-sampling scheme to take the difference of two sampled values. As shown in Figure 3.31, two stages are sampling at two different time points. The first stage samples first, and the second stage waits until the first stage amplifies the offset and the

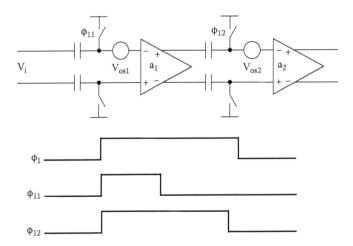

FIGURE 3.31
Correlated-double sampling.

feed-through voltages. Therefore, the double sampling can get rid of the first-stage switch feed-through error. When this double sampling scheme is applied to image sensors, the first sampling is to get the black or reference level, and the second sampling gets the image sensor output relative to the previously sampled reference level to take the difference.

3.9 Opamp Input Capacitance

In most high-frequency switched-capacitor designs such as filters and analog-to-digital converters (ADCs), the input sampling capacitor gets smaller while the input device gets larger to achieve wider bandwidth and fast settling. The opamp input capacitance affects the feedback factor f, which results in low loop gain and narrow bandwidth. Therefore, it is of paramount interest to reduce the opamp input capacitance. The opamp input is made of a differential pair, and its input capacitance is mainly the gate-source capacitance C_{gs} plus the Miller capacitance, which is the gate-drain overlap capacitance C_{gd} multiplied by the gain.

Figure 3.32 shows three common examples of the input differential pair with low-input capacitance. The first example is to add an equal amount of the negative capacitance by connecting the same overlap capacitance to the positive gain side. This is done using a half-sized transistor that is normally biased off to cancel the Miller capacitance. This positive feedback approach can slow the opamp settling as it takes time for the positive feedback to work. The second example is to use the source follower input [3]. The source follower input capacitance is mostly the overlap capacitance C_{gd}, and the gate capacitance C_{gs} does not count because there is no Miller effect. However, the g_m of the source follower should drive the input capacitance of the input differential pair. This pole should be pushed higher than the unity loop-gain frequency, and it consumes extra power. Because extra power is consumed, the source follower stage can be replaced by a stage with a low gain of 3 or 4 in the third example. Compared to the source follower, the input capacitance can be slightly higher due to the Miller effect, and the diode-load g_m should be designed high enough to drive the differential pair input capacitance.

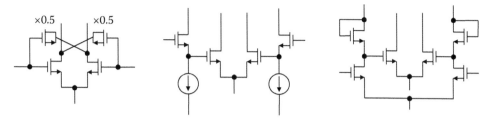

FIGURE 3.32
Three ways to reduce the opamp input capacitance.

3.10 Opamp Offset

The offset of the differential pair is due to the asymmetry of the differential pair. As shown in Figure 3.33, the offset is the DC input voltage to make the output to be zero. If there is an offset of the next stage, it is referred to the input after being divided by the first-stage gain. Therefore, usually the offset is limited by the asymmetry of the first stage. Offsets are either random or systematic. The former is simply due to the random fluctuation of the device geometries and process parameters, but the latter is when the asymmetry is built into the design.

The offset results mainly from the mismatches of the M_1-M_2 and M_7-M_8 pairs, which are supposed to have identical geometries and device parameters. The mismatches of the M_3-M_4 and M_5-M_6 pairs do not affect the input referred offset, because their mismatches only affect the high impedance nodes. That is, the offset currents of the M_1-M_2 and M_7-M_8 pairs make the offset current. Therefore, the input-referred random offset of the differential pair shown in Figure 3.33 is obtained after summing the power of two random offsets and taking a square root of that.

$$V_{os} = \sqrt{\left(V_{os1-2}\right)^2 + \left(\frac{g_{m7-8}}{g_{m1-2}}\right)^2 \left(V_{os7-8}\right)^2} \approx V_{os1-2}, \qquad (3.33)$$

where V_{os1-2} and V_{os7-8} are the offsets of M_1-M_2 and M_7-M_8 pairs, and g_{m1-2} and g_{m7-8} are the transconductances of M_1-M_2 and M_7-M_8 pairs, respectively. That is, the low-offset opamp design sets the g_m of the M_1-M_2 differential pair to be greater than that of the M_7-M_8 current source pair so that the offset can be limited by the matching of the input differential pair as in Equation (3.33).

Note that offsets have the same transfer functions as the signal, but due to their randomness, their power transfer functions should be used. After summing all the offset powers, taking the square root of the power sum will yield the RMS value of the offset. It is convenient to refer all the offsets to the input. Therefore, offsets of the later stages become smaller when referred to the input due to the gains.

One factor to consider for low offset is the overdrive voltage. As shown in Figure 3.34, the offset is the difference in the input voltages to make the output currents the same as drawn with a horizontal dotted line. The overdrive voltage should be minimized for low V_{os}. That is, the offset is limited by the mismatch of the threshold voltage if the overdrive voltage is small. So it is better for devices to have large ratios of W/L or low bias currents.

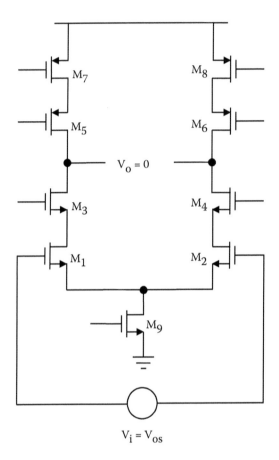

FIGURE 3.33
Cascoded differential pair with offset condition.

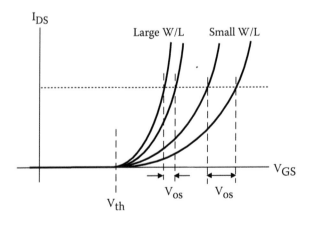

FIGURE 3.34
Metal-oxide semiconductor (MOS) I-V characteristics for low offset.

Longer channel devices exhibit better matching characteristics. Note that this contradicts the conclusion drawn in Chapter 2 in matching current sources.

3.11 Opamp Noise

Another nonideal random parameter designers should handle is noise. Any resistor has a thermal noise (or Johnson noise). Its power is proportional to the resistance value. The voltage noise power spectral density of the resistance R is

$$\frac{v^2}{\Delta f} = 4kTR, \tag{3.34}$$

where k is the Boltzmann constant, and T is the absolute temperature. Therefore, if $R = 1k\Omega$ at room temperature, its voltage noise density is approximately $16 \times 10^{-18} \, V^2/Hz$ or $4 \, nV/Hz^{1/2}$. This series thermal noise can be represented using a parallel current noise using the Norton's equivalence:

$$\frac{i^2}{\Delta f} = \frac{v^2}{\Delta f} \times \frac{1}{R^2} = 4kT\frac{1}{R}. \tag{3.35}$$

This current noise density is about $16 \times 10^{-24} \, A^2/Hz$ or $4 \, pA/Hz^{1/2}$. The thermal noise is a white noise because it is spread uniformly over the GHz range as sketched on the left side of Figure 3.35. Like the random offset, the noise can be handled like the signal using the same transfer functions. However, because the noise is random, noise powers (not voltages or currents) are added, and the square root of the sum becomes the RMS value.

MOS is a surface device with a resistive channel, and it exhibits two types of noises. One is the thermal noise of the resistive channel, and the other is the flicker noise of the surface device. The latter is also called 1/f noise, because the noise power is concentrated at DC and decays gradually as frequency goes higher. When saturated, the effective channel conductance is about $(2/3)g_m$. Therefore, using Equation (3.35), the drain-source current noise of the MOS device becomes

$$\frac{i^2}{\Delta f} = 4kT\frac{2}{3}g_m + \frac{K_{1/f}}{f}, \tag{3.36}$$

where $K_{1/f}$ is a constant depending on the process. If the output current noise is referred to the input, the equivalent input voltage noise becomes

$$\frac{v^2}{\Delta f} = \frac{i^2}{\Delta f} \times \frac{1}{g_m^2} = 4kT\frac{2}{3}g_m + \frac{K_{1/f}}{g_m^2 f}. \tag{3.37}$$

Two noise sources are sketched on the right side of Figure 3.35. Note that the frequency where the 1/f noise level decreases to the thermal noise floor. It is called 1/f corner frequency, and it depends on the process and temperature. The 1/f noise is critical in several applications, such as in audio ADCs, image sensors, low-noise amplifiers (LNAs),

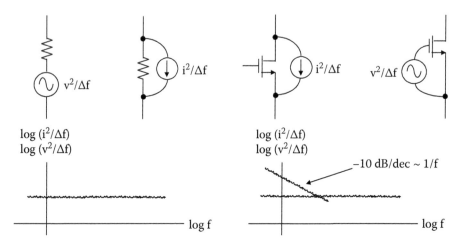

FIGURE 3.35
Thermal noise and 1/f noise spectral densities of a resistor and transistor.

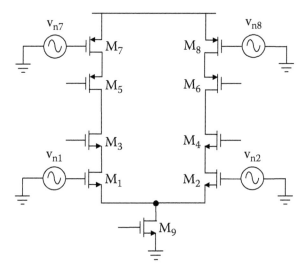

FIGURE 3.36
Cascode differential pair with noise sources.

mixers, and radio frequency (RF) voltage-controlled oscillators (VCOs). It mostly affects the low-frequency performance in low-noise systems with wide dynamic ranges, but in RF VCOs, the low-frequency 1/f noise is directly modulated up to the RF frequency due to the switching activity in the VCO, and it affects its phase noise. Depending on applications, the typical 1/f corner frequency ranges from a few 100s of Hz in audio to 10s of MHz in RF LNA. Note that the low-frequency 1/f noise can be canceled during the offset sampling together with the offset.

Similarly to the offset referring to the input side, all noise sources can be referred to the input side for convenience. As shown in Figure 3.36, all transistors are noisy, but only noises of the four transistors M_1, M_2, M_7, and M_8 are dominant. The noise of M_9 just drives the common node and becomes the common-mode noise, and the current noises of M_3, M_4, M_5, and M_6 drive the high-impedance node as mentioned in the offset case. As in Equation (3.33), the input-referred RMS noise is

$$v_n = \sqrt{(v_{n1})^2 + (v_{n2})^2 + \left(\frac{g_{m7-8}}{g_{m1-2}}\right)^2 \left\{(v_{n7})^2 + (v_{n8})^2\right\}} \approx \sqrt{2} v_{n1,2}. \tag{3.38}$$

The same conclusion can be drawn as in the offset case. For low noise, the g_m of the input differential pair should be designed to be higher than that of the load. Then the opamp noise is limited by the noise of the differential pair, whose noise is also set by the input devices. Therefore, the design of the input differential pair is the most important in achieving both low offset and noise.

3.12 Opamp Common-Mode Rejection

As discussed in Chapter 2, the finite resistance of the tail current source and the g_m and load matching of the differential pair limit the CMRR of the differential pair.

If the tail current of the differential pair is just a current source, a very high CMRR is achieved at DC as shown in Figure 3.37. However, it drops quickly at high frequencies because any parasitic capacitance C_{ss} at the common node starts to short the summing node. This corner frequency $1/R_{ss}C_{ss}$ is very low because the current source output resistance is usually designed to be very high. In particular, for RF applications, no high CMRR is achievable although the differential pair is used. Therefore, a resonant circuit can be used so that the parasitic capacitance can be tuned out using an inductor as shown on the right side to achieve a high CMRR at the resonant frequency.

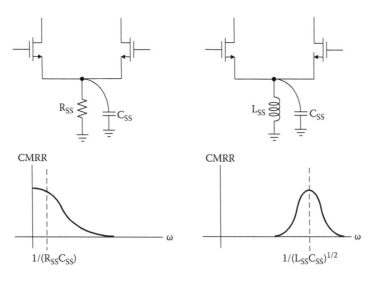

FIGURE 3.37
Tail current impedance effect on common-mode rejection.

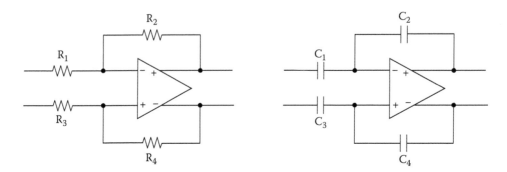

FIGURE 3.38
Opamp in feedback.

However, in most opamp applications that require high CMRR, the opamp is used in feedback as shown in Figure 3.38. If opamps are in feedback, CMRR is often limited by the matching of the external feedback components, because the gain on the positive side is different from that on the negative side. That is, if the ratio R_2/R_1 (or C_1/C_2) is not matched to those of R_4/R_3 (or C_3/C_4), the input common-mode signal leaks to the output as a differential signal. Although the opamp CMRR is infinite, the CMRR of the opamp in feedback is limited by the passive component matching regardless of the opamp CMRR.

$$CMRR \approx \frac{R}{\Delta R}, \frac{C}{\Delta C}. \tag{3.39}$$

This implies that CMRR is ultimately limited either by the matching accuracy of the opamp input differential pair or by the matching of the feedback network, whichever is lower. Furthermore, high CMRR is effective only at low frequencies.

References

1. T. Choi, R. T. Kaneshiro, R. W. Brodersen, P. R. Gray, W. B. Jett, and M. Wilcox, "High-frequency CMOS switched-capacitor filters for communications application," *IEEE J. Solid State Circuits*, vol. SC-18, pp. 652–664, December 1983.
2. R. Castello and P. R. Gray, "A high-performance micropower switched-capacitor filter," *IEEE J. Solid State Circuits*, vol. SC-20, pp. 1122–1132, December 1985.
3. W. C. Song, H. W. Choi, S. U. Kwak, and B. S. Song, "A 10-b, 20-M sample/s low-power CMOS ADC," *IEEE J. Solid State Circuits*, vol. SC-30, pp. 514–521, May 1995.

4

Data Converter Basics

All signals in the real world are analog in nature, and their waveforms are continuous in time. Because most signal processing is done numerically in discrete time, devices that convert an analog waveform into a stream of sampled discrete digital numbers or the digital number stream back into an analog waveform are needed. The former is called analog-to-digital converter (ADC or A/D converter), and the latter is called digital-to-analog converter (DAC or D/A converter).

4.1 Analog-to-Digital Converter Basics

Digitizing signals is a prerequisite in the implementation of all digital processing systems. There can be many different types of ADCs that vary widely in their resolutions and conversion rates, and key parameters related to their performance need to be closely looked at.

4.1.1 Aliasing by Sampling

The sample and hold (S/H) is an integral part of discrete-time signal processing. The digitization or quantization process is shown in Figure 4.1. To meet the Nyquist sampling requirement, the input signal should be band-limited within half the sampling frequency of $f_s/2$. Sampling is the multiplication of the input signal with the impulse sequence, whose frequency response also repeats every f_s. Because the multiplication in the time domain is the convolution in the frequency domain, the spectrum of the sampled-data discrete-time waveform repeatedly occurs at any multiples of f_s as sketched. Therefore, if the input signal is not band limited, the spectrum at the multiples of f_s will be aliased into the signal band. For example, if any unwanted signal exists at $3f_s$ as shown in Figure 4.2, it is aliased into all the signal bands after sampled.

Therefore, the purpose of the anti-aliasing filter placed before the S/H is obvious, and the signal band limited by $f_s/2$ is called the Nyquist band. However, the aliasing effect can be used intentionally to down-sample band-pass signals, as shown in Figure 4.3.

It is to sample the signal band, which exists at frequencies higher than the sampling frequency. However, even in such a band-pass down-sampling case, the signal band should still be band limited within the Nyquist bandwidth. The maximum Nyquist bandwidth for the band-pass signal is the same as $f_s/2$.

4.1.2 Quantization Noise

Resolution is a term used to describe a minimum amount of voltage or current that ADC/DAC can resolve. Resolution is usually quoted with the number of binary bits representing the analog input. The fundamental limit is the quantization step due to the finite number

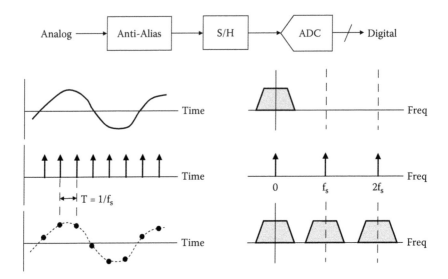

FIGURE 4.1
An ADC system and its discrete-time spectrum.

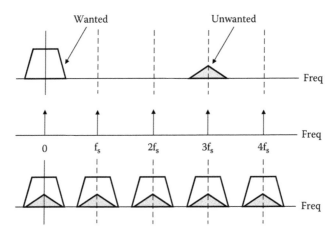

FIGURE 4.2
Aliasing if not band limited.

of bits used. In an N-bit ADC, the minimum input analog voltage $V_{ref}/2^N$ can be resolved with a full-scale input range of V_{ref}. That is, a total of 2^N digital codes are available to represent a continuous analog input as explained in Figure 4.4.

Similarly, in an N-bit DAC, 2^N input digital codes can generate 2^N distinct analog output levels within V_{ref}. This minimum resolvable step Δ can be defined as

$$\Delta = \frac{V_{ref}}{2^N}. \tag{4.1}$$

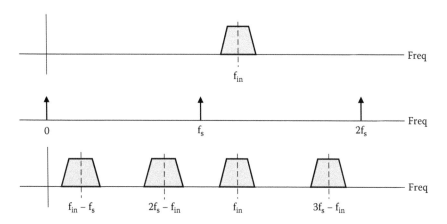

FIGURE 4.3
Down-sampling band-pass signals.

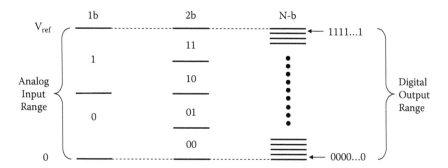

FIGURE 4.4
Quantization and digital representations.

The finer is the step Δ, the higher is the resolution. The step also gets smaller as N increases. The analog-to-digital conversion process is to map the sampled input to one digital representation as follows:

$$D = \left[\frac{V_{in}}{\Delta}\right] = \left[\frac{V_{in}}{V_{ref}} \times 2^N\right], \tag{4.2}$$

where the brackets [ ] represent an operation of rounding to a lower integer.

The binary number system is the most common, but the flash ADC also produces the thermometer-coded or 1-out-of-2^N coded digital output. The Gray coding is the same as the binary, but only one bit changes from one code to the next and is often used in flash ADCs to reduce the sparkle noise. Examples of four different digital codes are shown in Figure 4.5. When represented in a binary number like 11010…10, the first digit 1 is the most significant bit (MSB), and the last digit 0 is the least significant bit (LSB).

The continuous analog input is represented using only a finite number of digital numbers. Therefore, the transfer function of the ADC is made of uniform steps, and the quantization error varies for the input magnitude as shown in Figure 4.6.

Binary	Thermometer	1 out of 2^N	Gray
11	0111	1000	10
10	0011	0100	11
01	0001	0010	01
00	0000	0001	00

FIGURE 4.5
Digital representations.

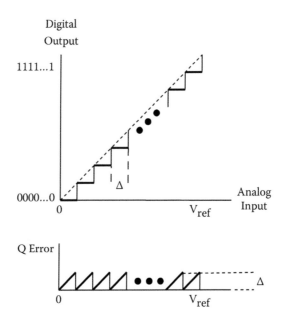

FIGURE 4.6
ADC quantization error.

Therefore, if the input varies, the time-domain error occurring during the quantization process appears as a white noise in the frequency domain uniformly spread over the Nyquist band. Assuming that the magnitude of the quantization error is independent of the input and random between $-\Delta/2$ and $\Delta/2$, the mean squared quantization noise power is estimated as follows:

$$\overline{q^2} = \frac{1}{\Delta} \int_{-\Delta/2}^{\Delta/2} x^2 dx = \frac{\Delta^2}{12}. \qquad (4.3)$$

This is the total quantization noise power integrated up to half the sampling frequency.

4.1.3 Signal-to-Noise Ratio

The spectral density of the quantization noise power depends on the sampling frequency because the total noise power given by Equation (4.3) is uniformly spread over the Nyquist band as shown in Figure 4.7.

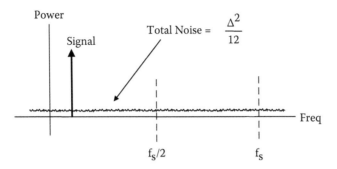

FIGURE 4.7
Quantization noise spread over the Nyquist bandwidth.

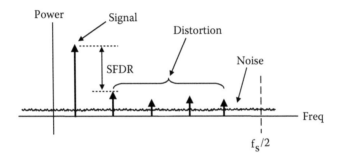

FIGURE 4.8
Nonideal quantized spectrum.

Signal-to-noise ratio (SNR) is defined as a ratio of the signal power to the uncorrelated in-band noise power. At the Nyquist sampling rate, the SNR of an ideal N-b ADC is therefore approximated from Equations (4.1) and (4.3) as follows:

$$SNR = \frac{\left(\dfrac{V_{ref}}{2\sqrt{2}}\right)^2}{\dfrac{\Delta^2}{12}} = \frac{3}{2}2^{2N}. \qquad (4.4)$$

In practice, it is more convenient to use the dB unit for SNR. That is,

$$10\log SNR \approx (6N + 1.8)\,dB. \qquad (4.5)$$

For example, an ideal 16b ADC has an SNR of about 97.8 dB.

The ADC resolution can be described in terms of SNR, but SNR accounts only for the uncorrelated noise. Limited by many nonideal factors, the ADC output spectrum even with a pure sinusoidal input includes harmonic and spurious tones in addition to the quantization noise as explained in Figure 4.8.

Real ADC performance is better represented by the signal-to-noise and distortion ratio (SNDR), which is the ratio of the signal power to the total in-band noise power including all harmonic and spurious tones.

$$SNDR = \frac{S}{N+D}. \qquad (4.6)$$

SNDR is always lower than SNR. Another definition widely used in radio frequency (RF) receiver systems is the spurious-free dynamic range (SFDR), which is defined as the ratio of the carrier to the highest spurious tone.

$$SFDR = \frac{Max.\,Signal}{Max.\,Spurious\,Tone}. \qquad (4.7)$$

SFDR is important in RF receivers because the desired signal is quantized together with neighboring strong blocker channels for digital signal processing. SFDR can be much higher than SNDR only if the quantization process is linear.

A slightly different term is often used in audio applications in place of SNR. The dynamic range (DR) is defined as the power ratio of the maximum signal to the minimum signal. The minimum signal is defined as the smallest signal when SNDR is 0 dB; the maximum signal is the full-scale signal without gross distortion.

$$DR = \frac{Max.\,Signal}{Min.\,Signal\,@\,SNDR = 0dB}. \qquad (4.8)$$

In practice, the ADC performance is limited not only by the quantization noise but also by many other nonideal factors such as noises from other circuit components, power supply coupling, noisy substrate, timing jitter, settling, and nonlinearity. Therefore, SNR is always lower than DR, because the ADC noise floor can be elevated with a large signal present. An alternative definition of resolution is the effective number of bits (ENOB), which is derived from the definition of SNDR.

$$ENOB \approx \frac{SNDR - 1.8}{6} \, (Bits). \qquad (4.9)$$

Usually, ENOB is defined at the Nyquist sampling rate when the signal frequency is half the sampling frequency. The ENOB represents the true and most rigorous definition of the ADC resolution including all nonideal factors.

4.1.4 Differential and Integral Nonlinearity

The input/output ranges of an ideal N-bit ADC are equally divided into 2^N small units, and one least significant bit (LSB) in the digital code corresponds to an analog voltage unit of $V_{ref}/2^N$. Therefore, the static ADC performance is defined by differential nonlinearity (DNL) and integral nonlinearity (INL). DNL is a measure of deviation of an actual ADC step from the ideal LSB step, and INL is a measure of deviation of the ADC output from the ideal straight line drawn between two end points of the transfer characteristic as shown in Figure 4.9.

Both DNL and INL are measured using the unit of one LSB. In practice, the largest positive and negative numbers are usually quoted. These DNL and INL definitions are as follows:

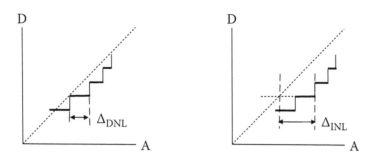

FIGURE 4.9
Differential nonlinearity (DNL) and integral nonlinearity (INL) definitions.

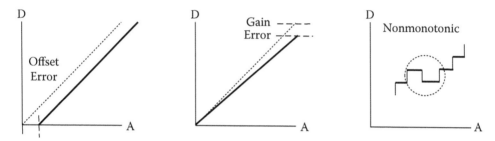

FIGURE 4.10
Offset, gain, and nonmonotonicity.

$$DNL = \frac{\Delta_{DNL} - \Delta}{\Delta}(LSB), \quad INL = \frac{\Delta_{INL} - \Delta}{\Delta}(LSB). \tag{4.10}$$

However, several different definitions of INL may result depending on how two end points are defined. In actual systems, two end points are not exactly 0 and V_{ref}, and INL can be better defined using a straight-line linearity concept rather than the end-point linearity. A straight line can be defined as a line connecting two end points of the actual transfer function, or as a theoretical line adjusted for the best fit. The former is called end-point linearity; the latter is called best-straight-line linearity.

Three nonideal ADC transfer functions are explained in Figure 4.10. The nonideal reference point causes an offset error; the nonideal full-scale range gives rise to gain error. In most applications, these offset and gain errors do not matter, but the low-level linearity measured by missing codes and nonmonotonicity matter a lot. Ideally, the digital output code should increase over the full range as the analog input increases. Otherwise, the transfer function becomes nonmonotonic.

Nonmonotonicity makes certain digital codes occur more frequently and results in large DNL error greater than 1 LSB. Missing codes are the other extreme. The negative DNL should be smaller than −1 LSB to avoid missing codes. If DNL = −1 LSB, $\Delta_{DNL} = 0$ by definition. That is, the step of zero width does not exist, and a digital code at this step is missing.

4.1.5 DNL-Related Low-Level Distortion

ADC with large INL suffers greatly from poor SNDR and SFDR due to the harmonically related tones created when the signal is large. However, if ADC is nonmonotonic and

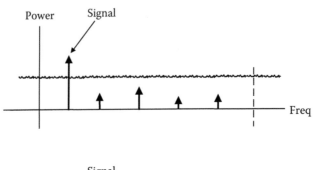

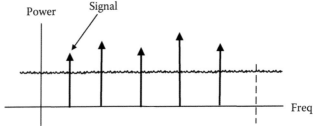

FIGURE 4.11
Low-level distortion: INL and DNL limited.

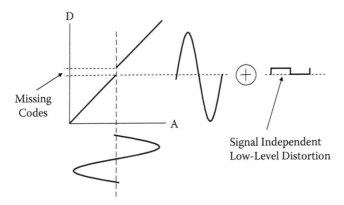

FIGURE 4.12
Missing codes at the operating point.

has many missing codes, DNL-related noises appear in the output spectrum even at very low signal levels. Such high distortion levels with small signals are not allowed in most ADC applications, such as in digital audio, video, control, and intermediate-frequency (IF) quantizer. DNL should be either removed or controlled to be smaller than 1 LSB.

Figure 4.11 shows the low-level linearity of the output spectrum. The top spectrum is the INL-limited case, and the bottom one is DNL limited. In the former INL-limited case, if the signal is small, the harmonic tones also become small. However, in the DNL-limited case, the harmonic tones stay constant although the signal is very small.

Assume that there are DNL errors such as missing codes or nonmonotonicity errors at the operating point, but otherwise, the transfer function is linear as shown in Figure 4.12. The problem with this small DNL error is that it creates the signal-dependent distortion every time the small signal trips over the discontinuity. In order to achieve low DNL in high-resolution ADCs, randomization techniques such as dithering are mandatory to smooth out the DNL error in the ADC transfer function by spreading it over neighboring codes.

4.1.6 Nyquist-Rate Sampling versus Oversampling

Both high-resolution ADCs and DACs operating at the low end of the spectrum, such as for digital audio, voice, and instrumentation, have been successfully implemented using oversampling techniques. As shown in Figure 4.13, the same quantization noise power of $(\Delta^2/12)$ spreads over a wider Nyquist band when oversampled, and the noise power inside the signal band f_{BW} is reduced by the oversampling ratio M.

Although Nyquist-rate ADCs can achieve comparable resolution, they are more sensitive to nonideal factors such as process, component matching, and even environmental changes. Oversampling can achieve higher resolution by trading speed for accuracy, and reduce the effect of the quantization noise and clock jitter. Operated in $\Delta\Sigma$ modulators, the quantization noise of the oversampling ADC can be suppressed even further. However, only Nyquist-rate sampling or subsampling ADCs can meet the high sampling rate requirements with medium resolution as in the wireless receivers and high-speed digital communication channels.

As discussed earlier, the sampling process moves the Nyquist-band spectrum to all multiples of the sampling frequency. Figure 4.14 shows the operation of the Nyquist-rate ADC with the Nyquist sampling rate f_N, which is twice the bandwidth. In this case, the signal bandwidth f_{BW} is half the sampling frequency $f_N/2$. To avoid the aliasing of the spectrum centered at multiples of f_N into the signal band, the analog anti-aliasing low-pass filter (LPF) is required to filter out any spectrum over the Nyquist band before sampling as marked with the dotted line.

On the other hand, if oversampled by M times and decimated as shown in Figure 4.15, the signal band is widely separated from the aliasing band, and the anti-aliasing requirement is far less stringent than the Nyquist sampling case. Whether or not using $\Delta\Sigma$ modulators, the quantization noise is spread, and the in-band quantization noise decreases by the oversampling ratio M. Therefore, many inaccurate decisions made at higher sampling rates can be averaged out to yield a more accurate decision at the Nyquist sampling rate. Because resolution and SNDR are not limited by process variations or by matching, such converters are widely used as high-resolution ADCs for digital audio and instrumentation.

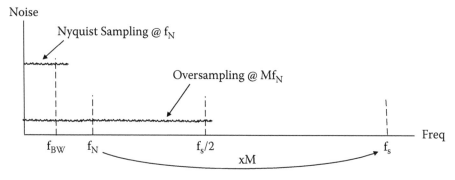

FIGURE 4.13
Quantization noise spectrum when oversampled.

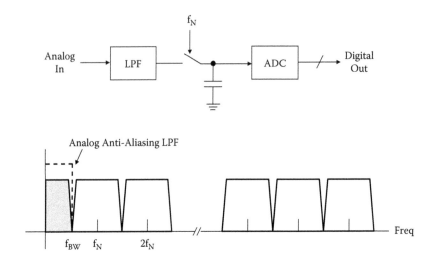

FIGURE 4.14
Anti-aliasing requirement for Nyquist-rate ADC.

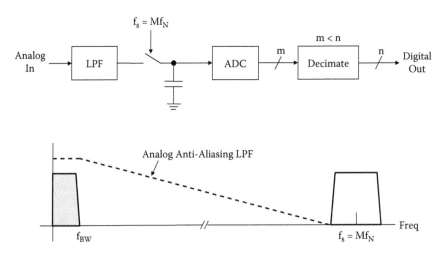

FIGURE 4.15
Anti-aliasing requirement for oversampling ADC.

4.2 Sample and Hold

The function of sample and hold (S/H) is to freeze the analog input signal for discrete-time sampled-data processing. It is the basic functional block for all switched-capacitor filters and data converters. The analog signal can be stored as an amount of charge on capacitor or as an amount of flux on inductor, but sampling on the capacitor is common. The track and hold (T/H) concept is slightly different from the S/H concept.

In all discrete-time sampled-data processing, T/H is not required because only the value at the sampling time matters. The simplest S/H is made of a switch and a capacitor, and its equivalence using a MOS switch is shown in Figure 4.16. If it is an ideal switch, it

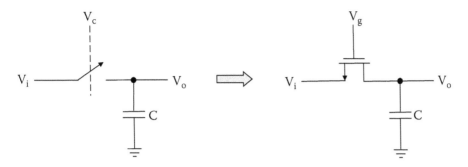

FIGURE 4.16
Ideal switch versus MOS switch.

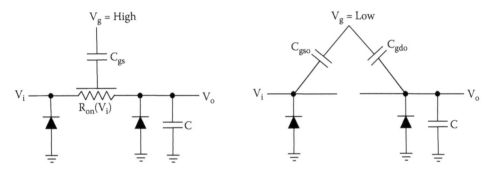

FIGURE 4.17
MOS switch model when on and off.

is short when the control voltage V_c is high but open when low. However, the MOS switch is far from being ideal. Its on-resistance R_{on} is a function of many process parameters and heavily depends on the input V_i, the gate voltage V_g, and the device threshold V_{th}.

Using the simple MOS model, the switch on-resistance becomes

$$R_{on} \approx \frac{1}{\mu C_{ox} \dfrac{W}{L}(V_g - V_i - V_{th})}. \tag{4.11}$$

When the gate voltage is low, it is close to the open circuit. In addition to this on-resistance, the control gate capacitance is connected to the source and drain, and the source and drain areas have nonlinear reverse-biased diode capacitances. All these affect the accuracy of the S/H for large high-frequency inputs.

The switch model including nonidealities is shown in Figure 4.17. C_{gs} is the gate-channel capacitance when the switch is on, and C_{gso} and C_{gdo} are the gate-source and gate-drain overlap capacitances, respectively, when the switch is off. The reverse-biased junction capacitance C_j is the major nonlinearity source, in particular at high frequencies.

$$C_j \approx \frac{C_{jo}}{\left(1 + \dfrac{V_{bias}}{\phi_{bi}}\right)^m}, \tag{4.12}$$

where C_{jo} is the junction capacitance with no junction bias, V_{bias} is the reverse junction bias voltage, and ϕ_{bi} is the junction built-in potential of about $0.6 \sim 0.7$ V. The exponent m is typically in the range of $0.3 \sim 0.5$. C_{jo} is a process parameter on the order of $1 \sim 2$ fF/μ^2. The MOS threshold voltage is heavily nonlinear and limits the performance of the S/H at all frequencies.

4.2.1 Charge Injection and Clock Feed-Through

When the sampling switch is suddenly turned off, a part of the channel charge stored under the gate of the MOS switch is injected into the sampling capacitor, and the clock is still fed through the overlapping capacitance even after the switch is turned off, as shown in Figure 4.18.

The channel charge is split and injected into the source and capacitor sides depending on the transient impedances looking into both sides. However, if it is assumed that the charge is evenly divided, the sampled voltage decreases due to the charge injection by

$$\Delta V_{ci} \approx -\frac{C_{ox}WL}{2C}(V_{DD} - V_i - V_{th}). \tag{4.13}$$

When the gate voltage V_g goes below $(V_i + V_{th})$, the channel is open. However, if the gate voltage continues to fall, the output voltage is pulled down further by the capacitive coupling through the overlap capacitance C_{gdo} by

$$\Delta V_{ft} \approx -\frac{C_{gdo}}{C + C_{gdo}}(V_i + V_{th}). \tag{4.14}$$

Because these charge injection and the feed-through errors occur simultaneously during sampling, it is difficult to handle them separately, but the total charge injection and feed-through error is the sum of them. The total sum can be called either charge injection or feed-through error. Both charge injection and feed-through errors are nonlinear due to their dependence on the threshold voltage.

The sampling network has a small-signal bandwidth limited by the product of the switch on-resistance and the sampling capacitor.

$$BW = \frac{1}{2\pi R_{on}C} \approx \frac{\mu C_{ox}\dfrac{W}{L}(V_{DD} - V_i - V_{th})}{2\pi C}. \tag{4.15}$$

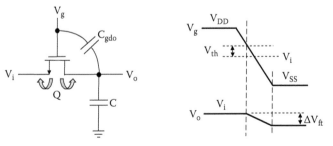

FIGURE 4.18
Charge injection and clock feed-through.

If $C = 1$ pF, $\mu C_{ox} = 200$ $\mu A/V^2$, $C_{ox} = 5$ fF/μ^2, $W/L = 10$ $\mu/0.2$ μ, and the overdrive voltage $(V_{DD} - V_i - V_{th}) = 1$ V, the channel charge injection ΔV_{ci} amounts to about -5 mV approximated using Equation (4.13). If the overlap capacitance is 600 pF/m and $(V_i + V_{th})$ is 2 V, the clock feed-through ΔV_{ft} is about -1.2 mV from Equation (4.14). The switch on-resistance R_{on} is about 100 Ω when estimated using Equation (4.11). Then, from Equation (4.15), the input sampling bandwidth amounts to 1.6 GHz. That is, the simple S/H samples an input with an accuracy of a few mV within a GHz bandwidth.

4.2.2 Nonlinearity of Sampling Switch

In addition to the charge injection and feed-through error just discussed, the nonlinearity of the sampling switch greatly affects the sampling accuracy. The sampling network is far from being ideal due to its signal dependency as shown in Figure 4.19.

The actual voltage to be sampled on the capacitor is the voltage at the time when the sampling switch is turned off. This implies that the voltage to be sampled is modified by the sampling R_{on}-C network. The switch on-resistance R_{on} is a nonlinear function of V_i as given by Equation (4.11), and the voltage drop across R_{on} depends on V_i and frequencies. For example, consider that a $1 \sim 2$ V signal is sampled with $V_{DD} = 3$ V. If the switch size is $W/L = 10$ $\mu/0.2$ μ, $V_{th} = 0.5$ V, and $\mu C_{ox} = 200$ $\mu A/V^2$, the on-resistance R_{on} varies nonlinearly to be 66.7, 100, and 200 Ω when V_i is 1, 1.5, and 2 V, respectively. The switch on-resistance and on-conductance versus the input are sketched in Figure 4.20. Because the body effect increases the threshold voltage V_{th}, the on-resistance is higher while the on-conductance is lower. Due to this nonlinearity of R_{on}, both the voltage drop across the switch and the sampled voltage on the capacitor would be nonlinear.

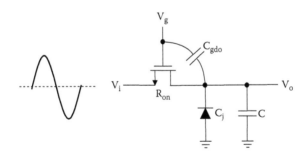

FIGURE 4.19
Nonlinearity in sampling.

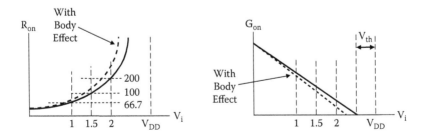

FIGURE 4.20
Switch on-resistance and on-conductance.

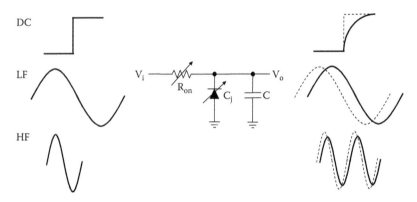

FIGURE 4.21
Effect of nonlinear sampling.

At high frequencies, the junction capacitance of the sampling switch also distorts the signal as the junction capacitance is signal dependent. That is, the current through the nonlinear junction is also nonlinear. If $R_{on} \ll 1/\omega C$, the sampling error would be negligible. For example, if $R_{on} = 100\,\Omega$ and $C = 1$ pF, $1/\omega C$ is 16 MΩ for the 10 kHz input, but 16 kΩ for the 10 MHz input. As a result, R_{on} and C tend to get small, and the charge injection and feed-through error would also increase. Furthermore, the C_j nonlinearity gets worse as the nonlinear junction diode area of the switch gets larger.

The input and output relationship is shown in Figure 4.21 for three different inputs of DC, low- and high-frequency (LF and HF) inputs. For the DC input, no nonlinearity errors occur as long as the R_{on}-C circuit is allowed to settle long enough. However, the AC inputs will suffer from the group delay variation, the gain reduction, and the nonlinearity of R_{on} and C_j. The switch and capacitor nonlinearity error is troublesome only in the S/H, but the charge injection and feed-through error applies to all cases.

4.2.3 Bottom-Plate Sampling

Sampling on the capacitor top plate is affected by the clock falling and aperture delay. Most nonlinearity error sources in the top-plate S/H can be eliminated if the bottom-plate sampling scheme is used as shown in Figure 4.22.

Assume that a switch is added to the top plate of the sampling capacitor. Because the input AC signal is divided in the sampling network made of R_{on}, C, and R_{top} as illustrated, the signal to be sampled on the capacitor depends heavily on both frequency and input. In most switched-capacitor circuits, the top plate of the capacitor is usually connected to the high-impedance operational amplifier (opamp) input node. The top plate is initialized to the ground or the opamp input common-mode level using a small N-channel metal-oxide semiconductor (NMOS) switch, and the bottom plate of the capacitor is switched to the signal source using both large CMOS (NMOS and P-channel metal-oxide semiconductor [PMOS]) switches. If the top-plate initialization switch is turned off earlier than the bottom-plate switch, the total charge injection and clock feed-through error stays constant, and even the nonlinear junction capacitor (C_j) effect of the sampling switch disappears. The charge injection and feed-through error becomes independent of the input V_i because the constant input is sampled repeatedly.

The only problem that remains in the bottom-plate sampling is the nonlinearity of the bottom-plate switch. R_{on} is a nonlinear function of V_i, and its nonlinearity is the major

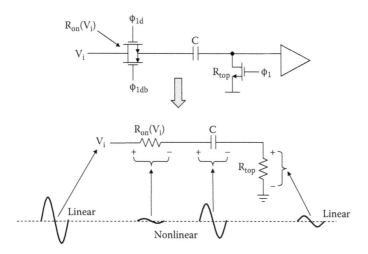

FIGURE 4.22
Bottom-plate sampling.

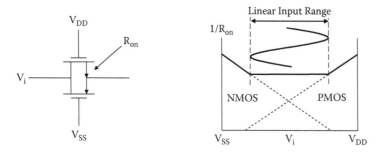

FIGURE 4.23
On-resistance of the CMOS switch.

error source in the bottom-plate sampling. If the voltage across this on-resistance is non-linear, the sampled voltage on the capacitor will be nonlinear. From Equation (4.11), the on-resistance of the CMOS switch is approximately

$$R_{on} \approx \frac{1}{\mu_n C_{ox}\left(\dfrac{W}{L}\right)_n (V_{DD} - V_i - V_{thn}) + \mu_p C_{ox}\left(\dfrac{W}{L}\right)_p (-V_{SS} + V_i + V_{thp})}. \qquad (4.16)$$

If the NMOS and PMOS gain terms of $\mu_n C_{ox}(W/L)_n$ and $\mu_p C_{ox}(W/L)_p$ are matched, the bottom-plate CMOS on-resistance is independent of the input within the range covering the supply minus both NMOS and PMOS threshold voltages as shown in Figure 4.23.

The top-plate switch is in linear mode because it is grounded, and the voltage drop across the top-plate switch on-resistance R_{top} is linear. Therefore, the top plate switch size can be minimized. However, the linear input range shrinks as the supply voltage is lowered in scaled CMOS. The nonlinearity of R_{on} remains as the most critical issue in bottom-plate sampling. Therefore, for linear sampling, it is necessary that the voltage drop across R_{on} should be minimized. That is, the bottom switch should be sized large so that the voltage drop across it

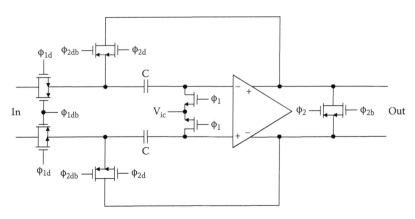

FIGURE 4.24
Fully differential bottom-plate S/H.

can be small. The on-resistances of both NMOS and PMOS switches should also be matched as closely as possible to keep R_{on} constant. However, their mobilities and threshold voltages are not well matched. Because the N-channel mobility is about 3 ~ 3.5 times higher than the P-channel mobility, PMOS switches are sized proportionately larger, although too large switches add more nonlinear junction capacitances at the sampling node. In low-frequency voice and audio systems, degenerating the CMOS switch even with a series resistance greatly improves the sampling linearity although the input sampling bandwidth is narrowed.

The standard version of the bottom-plate S/H using CMOS bottom-plate switches is shown in Figure 4.24 [1]. For this, two nonoverlapping clocks ϕ_1 and ϕ_2 are used, and the subscripts d and b denote delay and bar for delayed and inverted clocks, respectively. Differential sampling topology using two in parallel reduces the even harmonics. It samples the signal on the bottom plate and the common-mode voltage on the top plate. The top-plate switch is turned off early to reduce the V_{th}-dependent nonlinear charge injection and feed-through error. The sampling aperture effect also disappears because the top plate always samples the constant voltage. The bottom-plate switch is turned off later with a delayed clock. The outputs are shorted together to prevent the opamp output from wandering while sampling.

4.2.4 Clock Boosting

The key design aspect of the bottom-plate sampling is to make the bottom-plate switch on-resistance R_{on} either small or constant for high linearity so that the voltage drop across R_{on} can be either minimized or linearized. However, the top-plate on-resistance can be set high to minimize the nonlinear charge injection and feed-through error. Therefore, the input bandwidth of the bottom-plate sampling circuit is limited by the time constant given by $R_{top}C$, which should be shorter than a fraction of half the sampling period. Because R_{on} should be made much smaller than R_{top}, it is easy to make $R_{on} \sim R_{top}/10$, for example, at low sampling rates with high supply voltages. However, it is impossible to get the same high ratio at high sampling rates with low supply voltages. The bottom-plate switch size becomes prohibitively large to keep the R_{top}/R_{on} ratio close to 10 in wideband sampling as in the example shown in Figure 4.25.

Assume the sampling capacitors are the same 5pF, and both bottom- and top-plate switches are made of NMOS switches with an overdrive voltage ($V_g - V_i - V_{th}$) of 1 V and

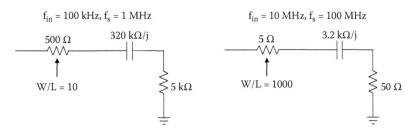

FIGURE 4.25
Two cases for sizing the bottom-plate switches.

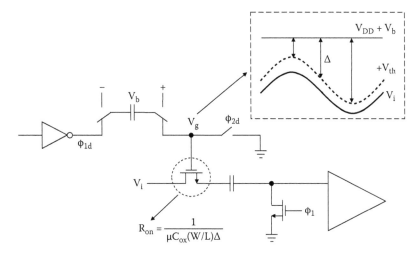

FIGURE 4.26
Clock boosting to lower R_{on}.

$\mu C_{ox} = 200\ \mu A/V^2$. Also, allow the sampling network to have a short time constant $R_{top}C$ of $T/40$ or $1/40 f_s$. In the low-frequency sampling of 100 kHz input at 1 MHz, $R_{top} < 5\ k\Omega$. Then the top-plate switch should be sized to have a W/L ratio of 1. That is, it is easy to make even $W/L = 10$ for the bottom-plate switch. However, as shown on the right, to sample 10 MHz at 100 MHz, the top-plate switch on-resistance is only 50 Ω. The top-plate switch needs the W/L ratio of 100, and demands the bottom-plate switch be made into a ratio with $W/L = 1000$ to make the bottom-plate switch on-resistance 5 Ω. Also note that the impedance of the sampling capacitor changes from 320 kΩ at 100 kHz to 3.2 kΩ for 10 MHz, and keeps the same attenuation factor.

Although using low-threshold devices helps to reduce R_{on}, CMOS switches are not linear enough at low supply voltages. An effective way to further reduce R_{on} is to increase the overdrive voltage Δ of the switch as shown in Figure 4.26 [2]. It is the capacitive clock boosting technique commonly used in digital circuits such as in dynamic random access memory (DRAM). Assume a capacitor is precharged to a boosting voltage V_b. If the capacitor voltage is added to the standard logic output level of V_{DD}, the switch gate drive voltage V_g is raised to $(V_{DD} + V_b)$ when the logic is high, which implies that the switch on-resistance R_{on} is reduced due to the increased overdrive voltage Δ, which is $(V_g - V_i - V_{th})$. R_{on} can be made smaller this way, but it still varies with signal. Also, the charge injection and feed-through error is nonlinear due to the signal-dependent threshold voltage V_{th}. As

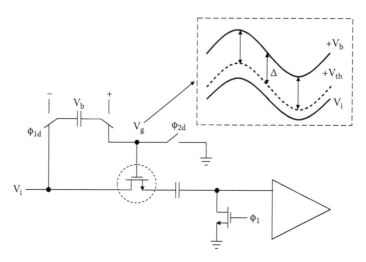

FIGURE 4.27
Clock boosting for constant overdrive Δ.

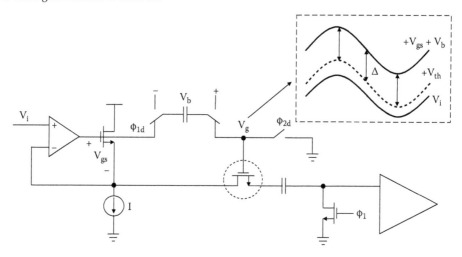

FIGURE 4.28
Clock boosting with tracking V_{th}.

a result, this constant-voltage clock boosting is simple but somewhat nonlinear mainly because of the signal-dependent overdrive voltage Δ.

There are two variations of the clock-boosting scheme to make it less sensitive to the signal. One in Figure 4.27 is to make the overdrive voltage constant by adding the boosting voltage to the input signal rather than to the clock [3]. Although this boosting can keep the boosting voltage Δ constant, the threshold voltage V_{th} of the switch still depends on the signal, which results in the nonlinear charge injection and feed-through error. In another boosting shown in Figure 4.28, the threshold-voltage nonlinearity can be further canceled out by including the replicated threshold voltage [4]. The boosting voltage consists of a constant boosting voltage plus the gate-source voltage V_{gs} of a transistor with its source voltage tracking the signal, thereby canceling the nonlinearity of V_{th}. However, it is difficult to make the V_{gs} adder due to the delay and the limited bandwidth of the opamp connected in the unity-gain feedback at the input.

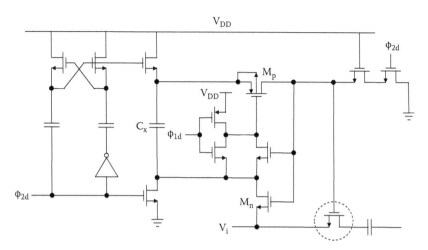

FIGURE 4.29
Clock boosting circuit example.

If the clock boosting is not properly implemented, switches are forced to operate with higher voltages than the maximum allowed voltages between their terminals. Extreme care must be taken during the design phase to ensure that all voltages between terminals are within the limits set by the process technology. Operating devices over limits cause many reliability issues such as oxide breakdown, gate-induced drain leakage, hot-electron effects, and punch-through. For reliable operations, all interterminal voltages V_{gs}, V_{gd}, and V_{ds} should be limited to be lower than the supply V_{DD} in any switching conditions as shown in Figure 4.29 [5].

Two capacitors on the left are constantly charged up to V_{DD}, and update C_x with V_{DD} during ϕ_2. During the sampling phase ϕ_1, C_x is isolated from the charging circuit on the left, and connected between the source and the gate of the bottom-plate sampling switch on the right through M_n and M_p. Note that the body of the PMOS switch M_p goes above the supply rail V_{DD}, and its body terminal should be connected to its source. M_n and M_p are switched on and off only by the gate voltage of M_p. The gate of the sampling switch can be raised above V_{DD}. After sampling, the gate of the sampling switch is reset. Switches are operated with all terminal voltages of V_{DD} or less except for the NMOS source or drain junction, which can sustain higher voltages than V_{DD} relative to the substrate without breakdown. By just boosting the switch on-resistance and keeping it constant, the bottom-plate switched-capacitor approach in CMOS enables the most accurate high-linearity sampling in broad bandwidth.

Last, as switches get larger in high-frequency sampling, the nonlinear source-drain junction capacitances can distort the charge to be stored in the capacitor. No clear solution to cancel high-order errors such as the nonlinear junction capacitance is present for now, but using dummy diodes with opposite polarities or boosting both NMOS and PMOS switches may be effective to alleviate some of the sampling charge nonlinearity error contributed by the junction capacitance.

4.2.5 Clock Feed-Through Effect on Offset

After sampling, the charge injection and feed-through from the top-plate switches lower the input common mode. When the top-plate switch is turned off, the opamp input common mode shifts down by a few 10 ~ 100s of mV depending on the size of the switch and the sampling capacitor. This common-mode shift does not affect the S/H performance as

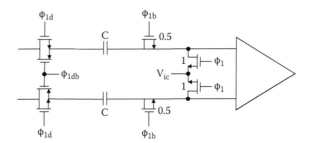

FIGURE 4.30
Charge injection and feed-through cancellation.

long as the opamp input differential pair stays in saturation. If the common-mode rejection ratio (CMRR) of the S/H is limited, the common-mode shift shows up at the differential S/H output after being divided by the CMRR.

If the input common-mode shift becomes an issue, dummy transistors of half the size can be connected and driven using the inverted sampling clock to absorb the charge from the top-plate switch as shown in Figure 4.30. This is based on an assumption that the channel charge of the top-plate switch is equally split. It is not true if both sides of the impedances seen by the switch are different, but the common-mode shift can be restored, though not exactly. However, the difference of the common-mode shifts on the positive and negative inputs also contributes to the offset of the S/H, and the offset gets larger because now mismatches of four transistors contribute to the offset. Leaving these dummy switches on has another implication. The capacitive coupling directly into the summing nodes through the dummy switches can aggravate the power-supply rejection.

4.2.6 *kT/C* Noise and Clock Jitter

Other than the nonlinear errors discussed, there are many other error sources to consider in sampling. The S/H output can also be corrupted by noises directly coupled from bouncing power supplies, bias lines, and poor ground. To reduce such noises, it is necessary to carefully separate analog and digital supplies and to eliminate sets of ground loops using a star-ground configuration. The use of the fully differential architecture also helps to reduce such coupling noises. However, the thermal and 1/f flicker noises are fundamental in all analog circuits. The S/H performance is affected by two distinct random noises: *kT/C* sampled noise and sampling clock jitter.

Neglecting the noise from the source, the dominant noise in the S/H is from the finite switch on-resistance when it is turned on. No noise is generated when off. When the sampling switch is turned off, the thermal noise across the sampling capacitance will be sampled. Because the sampling network is band limited, the sampled noise power will be

$$v^2 = 4kTR_{on} \times \frac{1}{2\pi R_{on}C} \times \frac{\pi}{2} = \frac{kT}{C}. \tag{4.17}$$

The factor $\pi/2$ is the portion of the out-of-band noise power shaped by a single-pole roll-off low-pass filter. Note that the sampled noise power is independent of the switch on-resistance. The root-mean-square (RMS) noise sampled on a capacitor C is therefore $(kT/C)^{1/2}$, which is independent of the switch on-resistance. Regardless of the bandwidth of the sampling network, about 64 μV_{rms} is sampled on a 1pF sampling capacitor at room temperature. However, the S/H bandwidth will affect the amount of the input noise coming from the external source to be sampled.

Jitter is defined as a timing error in sampling. The uncertainty in timing is directly translated into the uncertainty in the sampled voltage. The jitter effect is prominent in all sampled-data processing, such as analog-to-digital and digital-to-analog conversions. The clock jitter greatly affects the noise performance of the sampled signal and is the most critical factor to consider when sampling and reconstructing high-frequency components. For example, the right signal sampled at the wrong time is the same as the wrong signal sampled at the right time. Similarly, precise timing is critical to correctly reconstruct analog waveforms. Unless analog waveform is reconstructed with the identical timing when it was sampled, the waveform will be poorly reconstructed because the DAC output changes at the wrong time. This, in turn, introduces either spurious components related to the jitter frequency or a higher noise floor.

If the jitter has a Gaussian distribution with an RMS jitter of Δt_{rms}, the worse SNR resulting from this random clock jitter can be easily derived as explained in Figure 4.31.

$$SNR = 10 \times \log \frac{M}{\left(2\pi f_{BW}\Delta t_{rms}\right)^2},$$ (4.18)

where f_{BW} is the signal bandwidth, and the oversampling ratio M of the sampling frequency f_s to the Nyquist sampling frequency is defined as

$$M = \frac{f_s}{2f_{BW}}.$$ (4.19)

The jitter requirement gets more stringent as more bits are resolved at higher sampling rates. For the jitter power to be lower than the quantization noise power, the upper limit for the tolerable worse-case clock jitter for an N-bit ADC/DAC can be set as follows:

$$\Delta t_{rms} \leq \frac{1}{2\pi f_{BW}2^N}\sqrt{\frac{2M}{3}}.$$ (4.20)

Figure 4.31 shows the worst-case jitter for the Nyquist rate ($M = 1$). The worst-case jitter is < 80 fsec to sample with 15b resolution at 100 MS/s rate, and it is < 90 psec to sample with 16b resolution at 44kS/s.

Virtually everything is to blame for poor clock jitter. Voltage-controlled oscillator (VCO), synthesizer, power and ground noise, slow technology, limited bandwidth, nonlinearity, and even group delay can affect the jitter performance. Note that nonlinearity and limited bandwidth significantly contribute to harmonics, and narrow bandwidth delays low-frequency clock components differently from high-frequency ones as explained in Figure 4.32. Jitter also accumulates even when the clean jitter-free clock goes through inverter stages. Therefore, the steep slope of the clock transition and the minimum delay in the clock generator help to reduce jitter accumulation.

It is challenging to supply high-frequency square-wave clocks externally due to the difficulty in broadband impedance matching at a board level. The impedance mismatch causes excessive ringing and degrades jitter performance. Therefore, unless the clock is generated on chip, a low-jitter differential sinusoidal clock filtered by a crystal filter is supplied

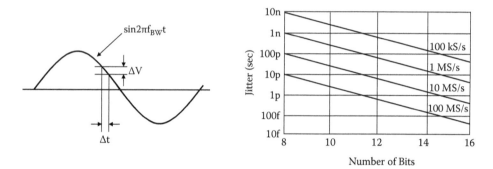

FIGURE 4.31
RMS jitter requirement for Nyquist sampling.

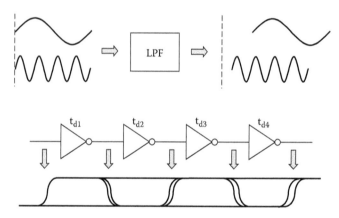

FIGURE 4.32
Delay is the source of clock jitter.

externally, and the square-wave clock should be generated on chip for high-frequency sampling. Figure 4.33 shows a two-phase nonoverlapping clock generator for the bottom-plate S/H with a minimized sampling clock delay.

To sum up, the bottom-plate S/H is free from the charge injection and clock feed-through error, but it suffers from the secondary effects such as the switch nonlinearity and common-mode shift in addition to the fundamental kT/C noise and the clock jitter. Most of these errors are not deterministic but random and cannot be added together. Only for the explanation purpose, the following relations are formulated for $V_{sampled}$ that is to be sampled on the sampling capacitor and V_{held} to be held in the opamp feedback.

$$V_{sampled} \approx V_i - \alpha V_i + \frac{\Delta Q}{C} + \sqrt{\frac{2kT}{C}} + \Delta V_{jitter}.$$

$$V_{held} \approx \frac{V_{sampled}}{1 + \frac{1}{a_o}\frac{\left(C + C_p + C_i\right)}{C}} + \frac{1}{CMRR} \times \Delta V_c. \tag{4.21}$$

The voltage to be sampled and held is far off from the original input voltage to sample even without considering the group delay effect. The sampled voltage is most affected by

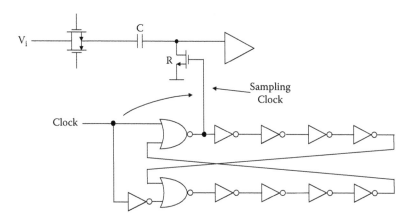

FIGURE 4.33
Nonoverlapping two-phase low-jitter sampling clock.

the switch nonlinearity and the attenuation at high frequencies as marked by αV_i. The differential charge error due to the charge injection and clock feed-through can be considered as a constant offset. The last two terms of the kT/C noise and the clock jitter are random and degrade the SNR of the S/H. The held voltage V_{held} is modified further by the finite gain a_o of the opamp. If the gain is linear, only the linear gain error results. Otherwise, it also affects the linearity of the S/H. The parasitic capacitance C_p at the opamp summing node and the opamp input capacitance C_i also reduce the sampled voltage because they decrease the feedback factor. The common-mode shift occurring when the top-plate sampling switches are turned off is not counted but contributes to the offset after being rejected by the CMRR. If the common-mode shift is compensated with dummy switches, offset will increase, and S/H will be more sensitive to the power-supply noise. All linear and constant errors are acceptable, but the nonlinearity and random noises should be minimized as they limit the performance of the S/H.

4.3 Flash Analog-to-Digital Converter

The simplest 1b ADC is an analog comparator, which detects the polarity of the input. The function of the comparator is to take an analog input, compare it with a reference (or threshold), and output a logic value 1 when the input is higher than the reference value or 0 otherwise. Multibit ADCs can be considered as a comparator bank connected to a set of divided references. Therefore, ADC architectures vary depending on how multiple-level decisions are made.

The comparator performance can be measured by the ADC resolution. For example, a 10b ADC can resolve 1024 levels. The accuracy of the comparator depends on the comparator offset and the accuracy of the reference divider usually made of components such as resistors, capacitors, or currents. So it is difficult to achieve higher resolution than 8 ~ 10b with bare matching accuracies of active and passive components. The ADC conversion speed is limited by the time needed to complete all comparator decisions. Slope-type ADCs are slow because they need a maximum of 2^N comparator decisions while flash ADCs make all

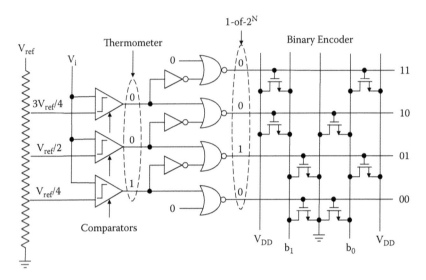

FIGURE 4.34
2b flash ADC example.

comparator decisions in one shot. In the middle range, successive approximation register (SAR) ADCs make one bit decision at a time. Although it is fast, the complexity of the flash ADC grows exponentially. On the other hand, the SAR ADC is simple but slow. Between these two extremes, there exist many methods resolving a finite number of bits at a time, such as pipelined and multistep ADCs.

The most straightforward way of making an ADC is to compare the input with the divided levels of the reference simultaneously. The conversion occurs at once like a flash. The flash architecture is the fastest among all ADCs, and its concept is explained with a 2b example in Figure 4.34. Three reference levels of $3V_{ref}/4$, $V_{ref}/2$, and $V_{ref}/4$ are generated with a resistor-string divider. Assuming that three comparators compare the sampled input with them, the comparator outputs from the top will be 0, 0, and 1, respectively, if $V_{ref}/4 < V_{in} < V_{ref}/2$. This thermometer-coded digital output is converted into the one-out-of-2^N code first, and then encoded into the binary output. Only one comparison cycle is needed. However, in practice, the limit is the exponential growth in the number of comparators and resistors. For example, an N-bit flash needs $(2^N - 1)$ comparators and 2^N resistors.

To achieve high resolution using flash architectures at Nyquist-rate sampling, the input still needs an S/H amplifier that can freeze the input for all comparators. As the number of bits grows, the comparator bank presents a significant loading to the input S/H. This heavy load on the input S/H is difficult to drive, and requiring fast S/H at the input contradicts with the speed advantage of the flash architecture. Also, the control of the reference divider accuracy and the comparator resolution is getting elusively difficult to achieve, and the power consumption becomes prohibitively high. As a result, flash converters with more than 10b resolution are rare as stand-alone ADCs. However, flash converters find most applications as low-bit sub-ADCs in other ADCs such as pipelined or oversampling ADCs.

4.3.1 Kickback and Sparkle Noise

Because the flash ADC operates many comparators in parallel with the common input, it suffers from the kickback effect as shown in Figure 4.35. If the source resistance R_s is

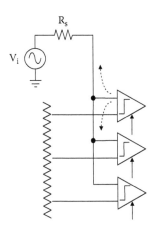

FIGURE 4.35
Kickback in flash ADCs.

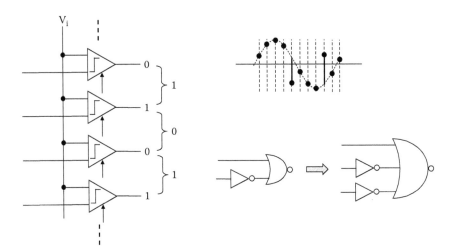

FIGURE 4.36
Sparkle noise in the flash ADC.

finite, the comparator/latch transient can be capacitively coupled to the input side and affect either the source or neighboring comparators. This effect can be alleviated using a low-impedance input buffer, or by keeping the preamplifier input capacitance constant. In switched-capacitor versions of comparators, the kickback effect can be localized.

Another unique problem in the flash ADC is the sparkle noise as shown in Figure 4.36. Large impulse errors in the digital output codes occur in time domain. It is because time delays from the source to all comparator outputs are not uniform, and also because the meta-stability of comparators can cause more than one 1-0 transition in the thermometer-coded output. If the thermometer-coded output has multiple 1-to-0 transitions, the 1-out-of-2^N encoder using a two-input NOR gate can generate multiple 1's. This localized error of a 1 LSB order can result in the change of the MSB bits in the binary encoding like an impulse error in the output digital codes.

The sparkle noise can be reduced using either multiple-input NOR gates or Gray number coding. Figure 4.37 shows that the Gray number coding can alleviate the impact of the

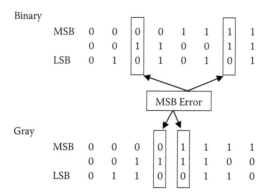

FIGURE 4.37
Gray coding to reduce the sparkle noise.

sparkle noise because even MSB change only leads to the neighboring digital output code. Gray coding is relatively immune to the sparkle noise and gradually degrades the flash ADC performance.

4.4 Comparator

The function of the comparator is to make a high or low digital output decision on the small analog seed input. It is made of two functional blocks as shown in Figure 4.38. One is a preamplifier, and the other is a regenerative positive feedback latch. Typical positive feedback digital latches exhibit a high level of hysteresis or uncertainty in offsets on the order of 10 ~ 20 mV. Therefore, the preamplifier should amplify the analog input seed signal Δ smaller than 1 LSB with a sufficient gain A so that the digital latch can make correct decisions on the amplified seed of a few 10s of mV level.

4.4.1 Preamplifier

A preamplifier is an open-loop wideband amplifier. For high resolution, it should be designed in multiple stages for high gain because it is difficult to achieve high gain and wide bandwidth simultaneously with a single stage due to the inevitable gain-bandwidth trade-off. The input of the preamplifier is the difference of the signal input and the reference level. If the signal is single ended, just a simple differential pair can be used as a preamplifier. However, in most integrated data converters the signal is fully differential, and it is not straightforward to amplify the difference of two differential signals.

Figure 4.39 shows two differential pairs as differential preamplifiers. At the top, two differential pairs convert the differential input V_i and the reference V_{ref} into currents first. Then, the output current difference becomes the signal difference.

$$I_o = G_{m1}V_i - G_{m2}V_{ref} = G_m(V_i - V_{ref}),\tag{4.22}$$

where $G_m = G_{m1} = G_{m2}$ is assumed. This method requires two differential pairs with very wide linear ranges covering the full magnitude input and reference, and their offsets

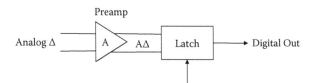

FIGURE 4.38
Comparator.

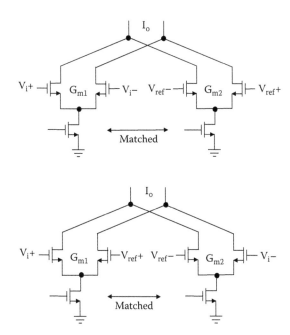

FIGURE 4.39
Two versions of differential preamplifiers.

and transconductances G_{m1} and G_{m2} should be well matched. That is, the input device should have a very large well-matched overdrive voltage ($V_{GS} - V_{th}$) that can cover the signal range. This requirement reduces the transconductance value and raises the output load resistance for high gain, which, in turn, narrow-bands the comparator. One way to avoid these wide linear range and matching requirements is to detect the positive and negative sides of the signal separately as shown at the bottom of Figure 4.39. Then the output current is

$$I_o = G_{m1}\left(V_i^+ - V_{ref}^+\right) - G_{m2}\left(V_i^- - V_{ref}^-\right) = G_m(V_i - V_{ref}),$$ (4.23)

where the same assumption of $G_m = G_{m1} = G_{m2}$ is made. The difference is that differential pairs do not need to be linear over the full signal range. The problem of this preamplifier is that both input and reference should have the same common-mode voltage, and both differential pairs should have the common-mode swing equivalent to the full input swing.

As shown, the difference circuit to detect two differential signals is difficult to implement in continuous time. Furthermore, comparators need to sample inputs with high accuracy for latching; therefore, it is logical for most comparators in CMOS to use switched-capacitor sampled-data preamplifiers as shown in Figure 4.40.

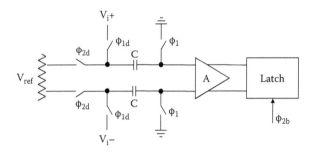

FIGURE 4.40
Comparator using switched-capacitor preamplifier.

The bottom-plate sampling, discussed earlier, is basic to the linear sampling, which turns off the top plate first to sample the signal at the bottom plate. If the total parasitic capacitance at the preamplifier input node is C_x, the gain of this preamplifier is slightly reduced. After amplifier settles, the amplified seed at the preamplifier output V_o becomes

$$V_o = A\frac{C}{C+C_x}\left(V_i - V_{ref}\right). \tag{4.24}$$

The performance of this comparator is limited by the offset of the preamplifier, which results from two sources. One is the clock feed-through offset V_{ft} that occurs due to the switch mismatch when the switches at the preamplifier inputs are turned off first at the falling edge of ϕ_1. The other is the offset of the preamplifier differential pair V_{pre}, which results from the random mismatch of the differential pair. Typical random offsets with low $(V_{GS} - V_{th})$ are about a mV or two. The comparator offset V_{comp} is approximated as

$$V_{comp} = \frac{C+C_x}{C}\left(0.1 \times V_{ft} + V_{pre}\right). \tag{4.25}$$

The switch mismatch is assumed to be about 10%. The feed-through voltage can be estimated as follows. Consider typical process parameters with $V_{DD} = 2.5$ V, $V_{th} = 0.5$ V, $C_{ox} = 4$ fF/m², and $C_{gdo} = 600$ pF/m. If the top-plate switches have $W/L = 5$ μ/0.25 μ, and total capacitance plus parasitic is $(C + C_x) = 0.1$ pF, the top-plate feed-through voltage initialized to the preamplifier input common-mode voltage $V_{ic} = 1.25$ V will be

$$V_{ft} \approx \frac{C_{ox}WL\left(V_{DD} - V_{ic} - V_{th}\right)}{2\left(C+C_x\right)} + \frac{C_{gdo}W\left(V_{ic} + V_{th}\right)}{\left(C+C_x\right)}$$

$$\approx \frac{4\,fF/m^2 \times 5\mu \times 0.25\mu \times 0.75V}{2 \times 0.1pF} + \frac{600pF/m \times 5\mu \times 1.75V}{0.1pF} \approx 70\,mV. \tag{4.26}$$

Assume that the body effect on the threshold voltage is negligible, and the channel charge is split evenly into the source and the drain sides of the switch. That is, the common-mode input of the preamplifier drops by about 70 mV after initialization, and the residual differential offset will be about 7 mV. As a result, the feed-through mismatch can be far larger than the random mismatch.

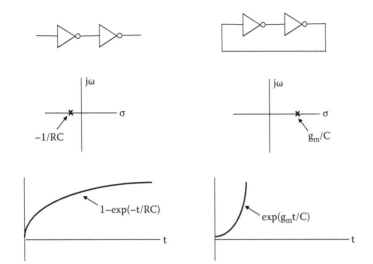

FIGURE 4.41
Two-stage amplifier versus positive feedback latch.

Operating this comparator at high sampling rates requires capacitors and switches to be sized to be smaller and larger, accordingly. Furthermore, the preamplifier input devices should also be made larger, and the parasitic at the comparator input node would increase significantly. This would, in effect, reduce the comparator resolution as feed-through voltages get larger, and the gain gets lower as the parasitic increases.

4.4.2 Regenerative Latch

The latch is a positive feedback system with two amplifier stages connected back to back to make a decision giving a high or low (1 or 0) digital output as shown in Figure 4.41. Once the minimum seed signal is amplified to a level higher than the latch offset, a regenerative latch can produce the digital output.

For example, the two-stage linear amplifier on the left has a pole on the negative real axis usually at $-1/RC$, where R and C are the total output resistance and capacitance of each amplifier, respectively. With this negative real pole, preamplifier amplifies the seed signal only with an exponential time constant of RC. On the other hand, the positive feedback latch has a pole on the right-half plane and is unstable. If the amplifier gain is assumed to be $g_m R$, the pole is placed at g_m/C as shown. The output of the unstable system grows exponentially responding to the seed signal. Note the time constant C/g_m is now much shorter than that of the amplifier by the gain factor $g_m R$. Therefore, for latching purposes, amplifiers with just high gains are not good enough for comparators.

The single-bit comparator is most widely used in memory sensing, such as in DRAM (dynamic random access memory), to detect whether the stored charge in a memory cell is digital 1 or 0 as shown in Figure 4.42. It is two digital inverters connected back to back. Because the charge on the memory capacitor leaks, the DRAM cell should be refreshed every once in a while before losing its data. This charge can be refreshed to be full simply by reading the memory cell periodically. The latch outputs are capacitive nodes. When the memory is accessed using a dummy cell, the charge difference on the latch outputs becomes the seed signal, and the positive feedback regenerates the memory cell.

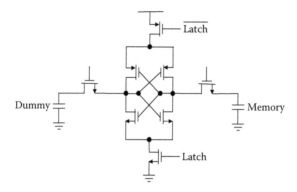

FIGURE 4.42
Regenerative latch for dynamic random access memory (DRAM).

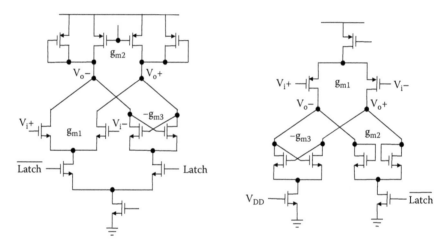

FIGURE 4.43
Comparator with preamplifier and latch combined.

This positive feedback system is bistable with either high or low output. Because comparators should make definite decisions either high or low, even one latch is not fast enough, and the double latching by cascaded latches is better when the time for comparator decisions is limited as in high-speed ADCs. The logic behind the double latching is that two latches cannot have identical offsets. So even with a small seed signal close to the first latch offset, the second latch can be fully regenerated. Another important reason behind the double latching is that wrong decisions are better than no decisions. The former can be corrected digitally in ADCs, but the latter meta-stability or indecision tends to make ADC very noisy.

Latches for low-resolution flash ADC consist of a simple preamplifier followed by a digital latch. The preamplifier and latch can be combined as shown in Figure 4.43. The one on the left side is a CMOS version of the common bipolar junction transistor (BJT) comparator. When the Latch clock is low, it works as a linear amplifier with a small signal gain of g_{m1}/g_{m2} to amplify the small input seed signal to a level that can flip the latch reliably. If the Latch clock goes high, the negative conductance $-g_{m3}$ given by the positive feedback is added in parallel to the load. That is, if $g_{m3} > g_{m2}$, the load resistance becomes negative, and the latch becomes regenerative. Therefore, it flips to one of the bistable states with a positive exponential time constant of $C/(g_{m3} - g_{m2})$, where C includes all parasitic capacitances and loading at the output node. The one on the right is modified to switch the load. When the Latch clock is low, the positive conductance diode load g_{m2}, which is larger than the negative conductance g_{m3}, is already on, and the positive high gain can be achieved as follows [6]:

$$\frac{V_o}{V_i} \approx \frac{g_{m1}}{g_{m2} - g_{m3}}. \tag{4.27}$$

If the Latch clock goes high, the load becomes negative due to the negative conductance $-g_{m3}$, and the latch starts to flip to one of the bistable states with a positive exponential time constant of C/g_{m3}. The disadvantages of these comparators are the kickback effect and their limited signal swing that requires another differential amplifier or inverters to make the CMOS-level digital output.

4.4.3 Comparator Design

Commonly used comparators in CMOS are dynamic comparators. Comparators should make decisions within a given time interval without suffering from any kickback and common-mode problems. The meta-stability of a single latch can be avoided by double latching. A comparator with a switched-capacitor preamplifier and a purely dynamic latch is shown in Figure 4.44, where the symmetric negative portions of the input and output are omitted for simplicity.

It operates with two standard nonoverlapping clocks. At the end of ϕ_1, the input is sampled, and the input is switched to the reference during ϕ_2 so that the preamplifier can amplify the seed signal ($V_i - V_{ref}$). This amplified seed signal is sampled on the latch input at the end of ϕ_2. During ϕ_2, the latch is turned off, and both its outputs are parked high at the digital supply. As the latch samples the amplified seed signal from the preamplifier, the latch is also turned on. Initially, both latch outputs fall together toward the middle of the supply, but depending on the seed signal, one side of the latch is more strongly pulled down than the other output.

This output difference grows exponentially with a time constant of C/g_m, which is mainly set by process technology. Most digital latches exhibit a hysteresis of tens of mV and flip in less than a fraction of nsec. The last stage is the additional digital dynamic

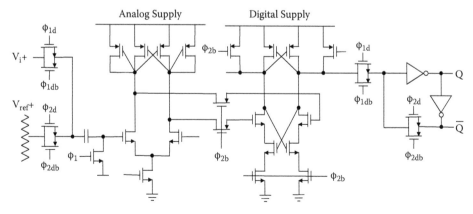

FIGURE 4.44
Switched-capacitor comparator with a dynamic latch.

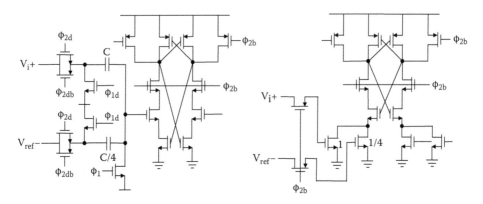

FIGURE 4.45
Two dynamic latches with reference dividers.

latch to get the digital output before the latch is reset. While the latch is operating, the preamplifier is reset to sample the new input. Therefore, the latch does not kick back to the input side. The size of the input sampling capacitor should be chosen carefully. If too small, the DC gain of the preamplifier gets significantly lowered due to the input capacitance of the preamplifier.

When the flash ADC is used as a sub-ADC in multistep or pipelined architectures, only a low number of bits are resolved per stage. The simplest one is the 1.5b/stage pipelined ADC case. For 2b decision, a simple latch can offer sufficient accuracy because no high gain preamplifier is required. In such a case, the subtraction of the reference from the input can be achieved without using a resistor divider. Figure 4.45 explains two examples. The example on the left side uses the capacitor ratio [7], and that on the right side uses current ratio [2] to make the same decision of $(V_i - V_{ref}/4)$.

Note that the three latches shown in Figures 4.44 and 4.45 are slightly different. First look at the latch shown on the right in Figure 4.45, which injects the seed signal using a current source. The current source is in triode mode when the seed signal is sampled on its gate. One drawback of this current source input is that it is not completely dynamic. One branch of the latch with the low output still conducts current after latched, while the latch shown on the left is purely dynamic and consumes no static current after latched because all currents are turned off. Without this difference, they are the same dynamic latch. If the seed input is raised farther above the latch clock as shown in Figure 4.44, the seed input is biased on the boundary between saturation and cutoff modes. This is obtained by swapping the seed input and switch locations from the current source input case. Operating the seed input in saturation mode achieves lower latch offset than operating the transistor in triode mode.

4.5 ADC Testing

Although it is straightforward to test the noise and distortion of ADCs using Fourier transform in the digital domain, it is challenging to test the static performance of ADCs such as DNL and INL. Due to the difficulty of generating a linear ramp or triangle waveform,

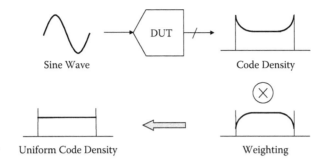

FIGURE 4.46
Code density test of ADC.

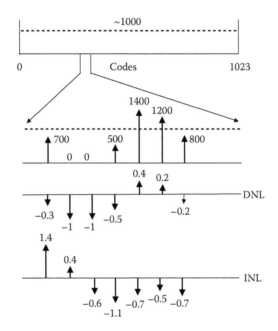

FIGURE 4.47
Differential nonlinearity (DNL) and integral nonlinearity (INL) definitions using the code density.

it is easier to apply a purely sinusoidal input for testing, and to find the code density of the digital output.

The code density test concept is explained in Figure 4.46 [8]. The only problem is that the code densities at both extremes are higher when the sinusoidal input is used. The code density obtained this way is weighted using the inverse sine function to obtain the uniform code density.

DNL and INL are defined as any deviations from the ideal count of the code outputs as shown in Figure 4.47. The example is when about 1 M samples collected from a 10b ADC are distributed over 1024 code bins. Then each code bin has an ideal code count of about 1000. DNL is defined as the normalized deviation of the bin counts from the ideal one, and INL is obtained by integrating DNL. In practice, noise and dither improve the DNL measurement, and missing codes are difficult to detect in the code density testing as

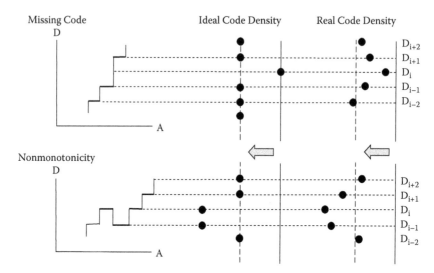

FIGURE 4.48
Noise effect on the code density.

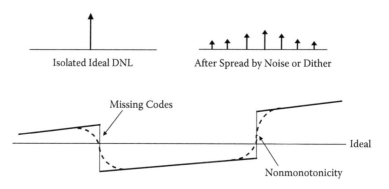

FIGURE 4.49
Spread DNL and smoothed-out INL in code density testing.

explained in Figure 4.48. When noisy, code bins are overflowing into neighboring bins, and even the missing code bin at D_i can be filled up. Similarly, the nonmonotonicity at codes D_i and D_{i-1} cannot be detected. It appears as a positive DNL with larger steps. Far more samples are needed for INL than for DNL because DNL errors accumulate when integrated.

Both noise and dither have the same effect on the DNL measurement. Because the DNL is spread and INL is smoothed out as shown in Figure 4.49, measured DNL results are always better in the code density test and are grossly misleading.

The noisier the ADC, the better is the DNL performance. For most high-resolution ADCs, dithering is mandatory to reduce the low-level distortion caused by DNL. Missing codes and nonmonotonicity can be estimated from the sudden irregular step in the INL measurement. If they exist, the INL measurement exhibits sudden up and down steps larger than 1 LSB. It is likely that the former is due to the nonmonotonicity, and the latter results from missing codes.

4.5.1 ADC Figure of Merit

The ADC performance is often measured by a common figure of merit (FOM). It is the power consumed for a one-level decision per unit time, and the unit is J (Joules). Joules is a unit of work required to continuously produce 1 watt of power for 1 second.

$$FOM = \frac{P}{2^N \times f_s},$$

(4.28)

where P is the power consumption, N is the number of bits, and f_s is the sampling rate. The lower the number is, the less power is consumed for a level decision per unit time. For example, if a 10b converter sampling at 50 MS/s consumes 50 mW, the number is about 1 pJ.

For low FOM, power should be minimized using low supply voltages. When operating ADCs well within the technology limits of resolution and speed, architectures based on comparators exhibit an advantage over others using opamps, because comparators can be operated with lower power supplies. However, when limited by resolution and speed requirements, FOM can be improved only by either faster processes or new architectures. Therefore, for example, a SAR ADC can achieve very low FOM because it uses one comparator though its speed and resolution are somewhat sacrificed.

4.6 Averaging and Interpolation Techniques

To achieve high resolution with flash ADCs, many high-speed, high-resolution comparators should be operated in parallel. As the complexity grows exponentially, two major issues arise as to the comparator offset and the number of comparators. Two techniques help to alleviate some of the problems. They are averaging and interpolation and are often used together to make flash and folding ADCs. The former spatially averages the offsets of neighboring preamplifiers, while the latter produces more zero-crossings for comparators.

4.6.1 Offset Averaging

It is important to warrant the zero-crossing accuracy of the comparator in the flash or folding ADC design. Each input differential pair amplifies the difference between the input and the reference voltage generated by a resistor string as shown in Figure 4.50. The zero-crossings of the preamplifiers are shifted due to preamplifier offsets and inaccurate reference voltages.

Preamplifiers can be offset-canceled individually, but a simpler method is to use a spatial averaging concept assuming all offsets are not systematic but random. The outputs of the preamplifiers separated in space can be coupled using resistors. This is to average out the offsets of the neighboring preamplifiers as well as to enhance the accuracy of the reference divider. The idea is to couple the outputs of the preamplifier transconductance (g_m) stages so that the offset errors can be spread over the adjacent preamplifier outputs as shown in Figure 4.51 [9].

Assume that $R_L = 2R$. If the preamplifier array repeats, the preamplifier gain can be approximated using the R-2R DAC concept. Because each preamplifier sees three parallel resistive loads of 2R, each stage gain is approximately

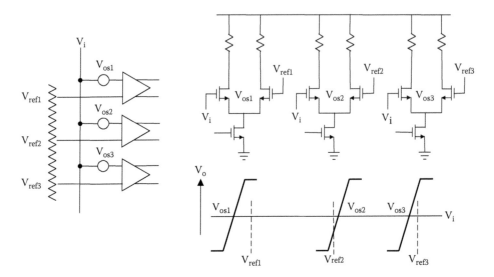

FIGURE 4.50
Shift of zero-crossings due to offsets.

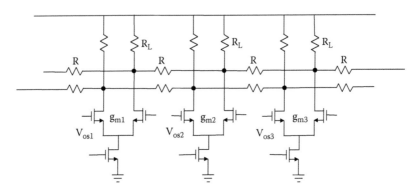

FIGURE 4.51
Spatial averaging of preamplifier offsets.

$$Gain = g_m \left(2R\,||\,2R\,||\,2R\right) = \frac{2}{3}g_m R. \tag{4.29}$$

Considering the attenuation in the R-2R DAC, the weighted sum of the offsets is derived adding offsets of the adjacent (±1) and alternate (±2) preamplifiers, and so on.

$$Weighted\ V_{os} = V_{os} + \frac{\left(V_{os+1} + V_{os-1}\right)}{2} + \frac{\left(V_{os+2} + V_{os-2}\right)}{4} + \cdots, \tag{4.30}$$

Two examples of DNL and INL before and after averaging are compared in Figure 4.52 to show the averaging effect.

For example, if the coupling resistance is infinite or far larger than the load resistance ($R \gg R_L$), all preamplifiers are independent, and their offsets are purely random. However,

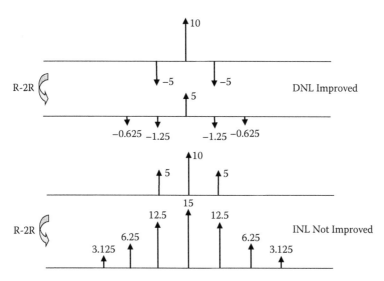

FIGURE 4.52
Examples of DNL and INL before and after averaging.

if R is finite and comparable to the load resistance ($R \sim R_L$), one preamplifier output becomes the weighted sum of the outputs of its neighboring preamplifiers. However, for the case in which errors to average have the same polarity, averaging is not that effective, and the INL improvement is minimal. The overall DNL improvement can be significant, and the smaller the R is, the more offsets are averaged. Note that the preamplifiers at the top and the bottom ranges should be terminated properly.

4.6.2 Interpolation

Flash ADCs are implemented with multistage preamplifiers. Interpolation reduces the number of preamplifiers. Intermediate preamplifier outputs can be interpolated using the adjacent preamplifier outputs as shown in Figure 4.53. Two preamplifier outputs V_a and V_b are used to generate three more outputs V_1, V_2, and V_3 using a resistor divider. The preamplifiers marked in dotted lines can be left out because the comparators can still make decisions on the interpolated outputs. Therefore, three out of four preamplifiers can be eliminated by interpolation. Interpolation improves DNL within the interpolated range due to the monotonic nature of the divider, but the overall DNL and INL contributed by preamplifiers are not improved.

Interpolation is usually done using resistor dividers, but it is also possible using capacitor dividers or in the current domain. Capacitor dividers can be used only with switched-capacitor circuits because the capacitive nodes should be reset periodically. Conceptually, current interpolation is possible if current sources are duplicated, but it does not improve DNL because they are not monotonic. An example of the capacitor interpolation is shown in Figure 4.54.

Interpolating any arbitrary number of intermediate levels is possible by tapping more divider outputs as long as preamplifiers operate linearly within the interpolation range. However, the linearity will be affected if interpolating wide ranges, because it is inevitable that the nonlinear range of the preamplifier will be used. Accuracy of zero-crossings is degraded by four main error sources: reference error, preamplifier offset, interpolation error, and preamplifier gain mismatch.

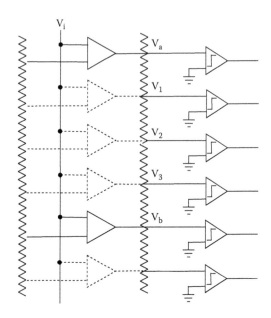

FIGURE 4.53
Interpolating preamplifier outputs.

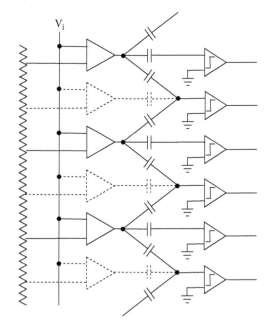

FIGURE 4.54
Interpolation using capacitor dividers.

Four interpolation errors are explained in Figure 4.55. The reference and offset errors only shift the preamplifier zero-crossings, and the interpolation error makes the zero-crossings irregular. The preamplifier gain mismatch error results in nonuniform gains in the interpolated outputs. Assume that two preamplifier outputs with zero-crossings separated by T have mismatched normalized gains of 1 and $(1 + \alpha)$, respectively. A total N

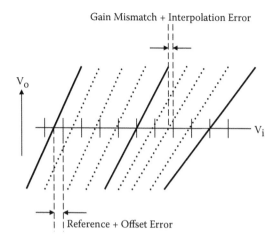

FIGURE 4.55
Interpolation errors.

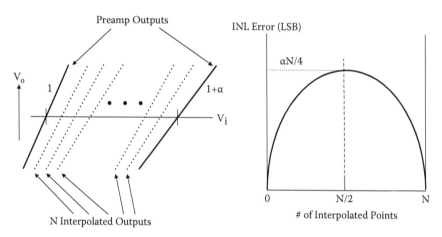

FIGURE 4.56
Preamplifier gain matching requirement.

intermediate outputs will be interpolated in between. The gain matching requirement is illustrated with preamplifiers zero-crossings in Figure 4.56.

The zero-crossing of the nth interpolated output can be estimated as

$$t_n = \frac{nT}{N(1+\alpha) - n\alpha}. \tag{4.31}$$

The maximum INL error is $\alpha(N/4)$ occurring at $n = N/2$ as shown on the right side. The gain matching requirement to achieve INL smaller than half LSB is, therefore,

$$\alpha < \frac{1}{2} \times \frac{4}{N} = \frac{2}{N}. \tag{4.32}$$

For example, to resolve 50 more levels ($N = 50$) within the interpolated range, the gains of two preamplifiers should be matched better than 4% ($\alpha < 0.04$).

4.7 Low-Voltage Circuit Techniques

Digital circuits in nanometer CMOS work even with sub-1V supplies, but analog circuits are not even operable at supplies below 1V. Amplifiers run out of the signal swing, and switches cannot be turned on due to the lack of the overdrive voltage. To operate circuits with low supplies, scaled CMOS offers the low-threshold device option. Also, many efforts have been made to overcome the problems at the circuit level, such as clock boosting for switches to operate nominal threshold devices at low supplies. Although low-threshold devices and clock boosting solve the sampling switch problem, the limited signal swing with poor linearity significantly degrades the analog system performance unless system architectures are configured without relying on a large linear range of the amplifier. Although the performance is sacrificed, the main incentive to design low-voltage circuits is to achieve low power for battery operation with medium to low resolution.

4.7.1 Low Bound of Analog Supply

A good example is a switched-capacitor circuit with opamps and switches as shown in Figure 4.57. Assume that Δ is the overdrive voltage V_{DSsat}, which is defined as $V_{DSsat} = (V_{GS} - V_{th})$, and all devices are assumed to have the same thresholds and are biased with the same Δ, for simplicity. As shown with the simplest opamp with the PMOS differential pair, the minimum required power supply voltage V_{DD} is $(V_{GS} + 2\Delta)$. Then the maximum opamp output swing $V_o|_{max}$ is $(V_{DD} - 2\Delta)$ or simply V_{GS}. The input common-mode voltage V_{ic} should be $(V_{DD} - \Delta - V_{GS})$ or just Δ above the most negative supply V_{SS}.

$$V_{DD} > V_{GS} + 2\Delta. \quad V_o|_{max} < V_{GS}. \quad V_{ic} < \Delta. \tag{4.33}$$

No opamp with a differential pair can operate with any lower supply than this unless the transistors are biased in the subthreshold region.

As shown in Figure 4.57, the top plate can be initialized to a voltage close to the most negative supply V_{SS}. However, the bottom-plate switches are not operable with the gate

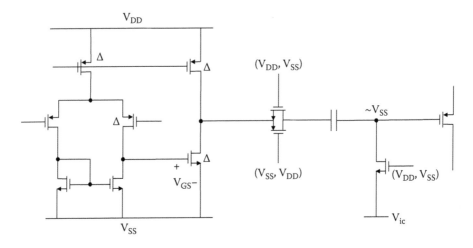

FIGURE 4.57
Low-voltage design issues.

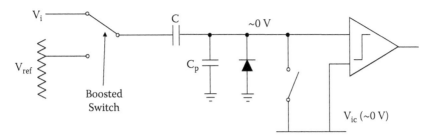

FIGURE 4.58
Risk in operating comparators at low voltage.

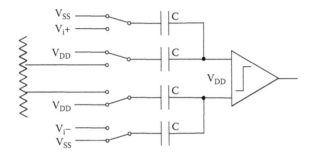

FIGURE 4.59
Low-voltage comparator.

voltages of V_{DD} and V_{SS} alone without using low-threshold device or boosted clock. Note that initializing the input common-mode voltage V_{ic} close to the negative supply is risky. In the switched-capacitor comparator example as shown in Figure 4.58, the input common mode is initialized to almost 0 V, which is the most negative supply V_{SS}.

Then the NMOS switch has a source-drain diode that is barely reverse biased. If the reference voltage is far lower than the sampled input, the comparator input node experiences a sudden voltage step down when the reference is selected. This step is attenuated in the capacitive divider made of C and the parasitic C_p, but if this sudden transient goes lower than V_{SS}, it will trigger the diode to be forward biased, and the initialization charge at the input node will be lost. This is the most common cause that fails low-voltage switched-capacitor circuits. A low-voltage comparator with the input common mode set to V_{DD} as shown in Figure 4.59 avoids this forward-firing diode problem.

From the opamp shown in Figure 4.57, the tail current can be removed to squeeze the supply voltage further. The first thing to lose from this arrangement is the differential pair, and the opamp input works as a difference amplifier, which is the difference of two single-ended amplifiers. The signal swing can be the device threshold V_{th}, but the common-mode input has no room and should be at the most negative supply. Equation (4.33) can be modified as follows:

$$V_{DD} > V_{gs} + \Delta. \quad V_o\big|_{max} < V_{th}. \quad V_{ic} \approx 0. \tag{4.34}$$

In such a low-voltage environment, the lack of a differential pair limits the design flexibility, and only dynamic techniques and current-mode circuits may perform with compromised performance.

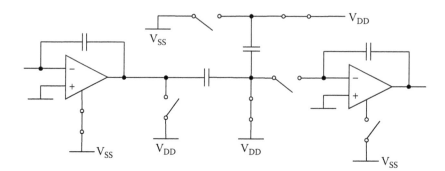

FIGURE 4.60
Switched-opamp technique.

4.7.2 Switched-Opamp Technique

If the supply approaches the lower limit of $(V_{GS} + \Delta)$ or $(V_{GS} + 2\Delta)$, switching only between V_{DD} and V_{SS} is possible [10], which is, in fact, equivalent to switching opamps on and off. This technique is effective in the low-supply environment, but it can suffer from parasitic effects of many switches and on/off devices at the opamp input.

Figure 4.60 shows that the signal moves through the capacitor divider network, which is charged to V_{DD} when sampling the signal. After sampling on the capacitors, the next stage power is turned on. Then its input node remains at V_{DD}, but the charge transfer is made. It is possible to implement most functional blocks such as $\Delta\Sigma$ modulator, multiplying digital-to-analog converter (MDAC), and integrators for filters using switched-opamp techniques. Band-pass circuits are easier to implement at low voltages due to the high-pass input coupling, and in some cases, continuous-time input or voltage-to-current converter can be used. However, it is not simple to implement a switched-opamp S/H.

One switched-opamp S/H example is shown in Figure 4.61, where initialization voltages are marked as either V_{DD} or V_{SS}. Two series input capacitors are initialized as shown, and the input is sampled on them. During the hold phase, the opamp is turned on, and the charge is transferred to the output. Note that the input stays at V_{DD} all the time.

4.7.3 Current-Mode Circuits

Current-mode circuits have been studied extensively to find alternative low-voltage solutions. However, they suffer from their own unique problems that do not exist in normal voltage-mode circuits. They also require the voltage-to-current conversion. In the current-mode circuit, the signal current is defined as a small fluctuation of the bias current and should be made much smaller than the standby current for linearity purposes. Because the bias current noise is far higher than the signal current noise, the current-mode circuit is noisier and more nonlinear than the voltage-mode circuit. The current-mode circuit can operate with supply voltages as low as $(V_{GS} + 2\Delta)$ as in the current-mode adder and interpolator example shown in Figure 4.62.

The resistive voltage-to-current converter injects the input current I_i, which can be mirrored into other current sources.

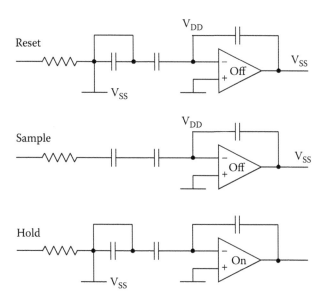

FIGURE 4.61
Switched-opamp sample and hold (S/H).

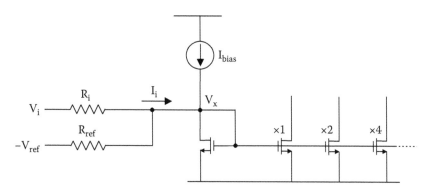

FIGURE 4.62
Current-mode adder and interpolator.

$$I_i = \frac{V_i}{R_i} - \frac{V_{ref}}{R_{ref}} - V_x\left(\frac{1}{R_i} + \frac{1}{R_{ref}}\right). \tag{4.35}$$

The input impedance of this circuit is low. If a high-impedance input buffer is added, the complexity grows. The V_x error can be reduced by feedback such as the G_m-boosting technique, but again it adds to the overall complexity. The current interpolation is also nonmonotonic due to the random current mismatch. For better matching, V_{GS} should be made higher using longer channel devices. Last, the current sampling is somewhat nonlinear and inaccurate compared to the voltage-mode bottom-plate sampling. If all errors are taken care of, the current-mode circuit exhibits the complexity of the voltage-mode circuit, and will likely lose its low-voltage advantage.

4.8 Digital-to-Analog Converter Basics

Digital-to-analog converters (DACs), called decoders in communications terms, are devices through which digital processors communicate with the outside analog world. Although they are key elements that make ADCs, they also find numerous applications in stand-alone devices such as panel displays, voice/music decoders, test systems, waveform generators, and digital transmitters in modern digital communications, and so forth.

The basic function of the DAC is to convert digital numbers into analog values. An N-bit DAC gives a discrete analog output level, either in voltage or in current, for every level of 2^N digital codes applied to the input. Therefore, an ideal voltage DAC generates 2^N discrete analog output voltages for digital inputs varying from 000...00 to 111...11, and the output has one-to-one correspondence with the digital input D_i for $i = 0, 1, 2, ..., 2^N - 1$.

$$V_o(D_i) = V_{ref}\left(\frac{b_N}{2} + \frac{b_{N-1}}{2^2} + \frac{b_{N-2}}{2^3} + \cdots + \frac{b_2}{2^{N-1}} + \frac{b_1}{2^N}\right), \tag{4.36}$$

where V_{ref} is a reference voltage setting the output range of the DAC, and $b_N b_{N-1} b_{N-2}...b_2 b_1$ is the binary representation of the input digital word D_i. Figure 4.63 shows an ideal DAC transfer function.

In the unipolar DAC, the reference point is 0 when the digital number is 000...00 as shown, but in the bipolar or differential DAC, the reference point is the midpoint of the full scale when the digital input is 100...00, and the output range is defined from $-V_{ref}/2$ to $V_{ref}/2$. The static DAC output should be spaced uniformly with a constant step Δ, which corresponds to the ADC quantization step. It is the inverse process of the quantization. Therefore, all definitions such as INL, DNL, total harmonic distortion (THD), SNR, clock jitter, and monotonicity for ADC are equally applied to DAC. However, unlike ADC that works on sampled-and-held DC analog values, DAC is more prone to analog dynamic

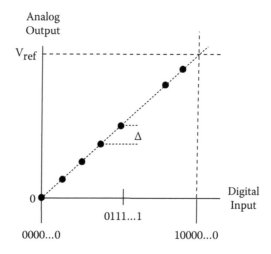

FIGURE 4.63
Ideal digital-to-analog converter (DAC) transfer function.

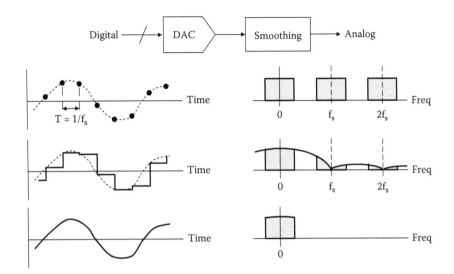

FIGURE 4.64
Basic DAC function.

errors since its output is an analog waveform. The digital-to-analog conversion process that creates analog waveform in time is illustrated in Figure 4.64.

DAC can be classified either as Nyquist-rate or oversampling DAC depending on its sampling ratios. In the digital domain, the spectrum repeats at multiples of the sampling frequency as shown, because it is equivalent to the sampling by the impulse sequence. However, the impulse waveform is not feasible in the analog domain, and the closet is the ideal sampled-and-held rectangular gating function as shown. Due to the finite gating period, the spectrum is then modified by the sinx/x (SINC) function, which is nulled at every sampling frequency. Because the signal is only at the baseband, the DAC output spectrum should be band-limited to the Nyquist bandwidth using a smoothing filter. Note that the broadband baseband DAC output experiences a gain droop at the pass-band edge, which can be compensated for by using either a digital or an analog filter.

4.8.1 DAC Accuracy Considerations

A distortion-free DAC instantaneously produces an output voltage that is proportional to the input digital number. In reality, if the input digital number changes from one value to another, the DAC output voltage reaches a new value later. Therefore, DAC suffers from both static and dynamic errors as shown in Figure 4.65.

The former is the error in the final settled voltage given by the DAC non-linearity error defined by DNL and INL, but the latter results from non-linear settling of the output.

4.8.2 Limited Slew Rate

Figure 4.66 illustrates step responses of a DAC when it settles exponentially with a time constant and when it slews.

An ideal transient response of a DAC to a step input is an exponential function, which only generates an error growing linearly with the signal. The transient errors are shaded.

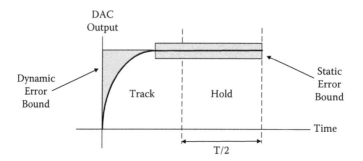

FIGURE 4.65
Static and dynamic errors of no-return-to-zero (NRZ) DAC.

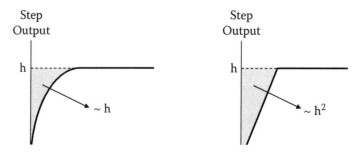

FIGURE 4.66
Linear and nonlinear DAC settlings.

If the time constant is τ and the slew rate is S, the step responses are given as follows, respectively:

$$V_{step}(t) = h\left(1 - e^{-\frac{t}{\tau}}\right). \quad V_{step}(t) = St. \tag{4.37}$$

Then the errors represented by the shaded areas in both cases are obtained by integration as follows:

$$Error = h\tau. \quad Error = \frac{h^2}{2S}. \tag{4.38}$$

This implies that an exponential settling of the former case with a single time constant generates a linear error in the output, which does not contribute to the DAC nonlinearity.

Any other transient responses give rise to dynamic errors that have no bearing on the input signal. The limited slew rate is a significant source of nonlinearity in most high-speed current DACs because the error is proportional to the square of the signal. The worst-case harmonic distortion (HD) when generating a signal with a magnitude V_o with a slew rate S is given from [11] as

$$HD_k = 8\frac{\sin^2\left(\dfrac{\omega T}{2}\right)}{\pi k(k^2 - 4)} \times \frac{V_o}{ST}, \quad k = 1, 3, 5, 7, \cdots. \tag{4.39}$$

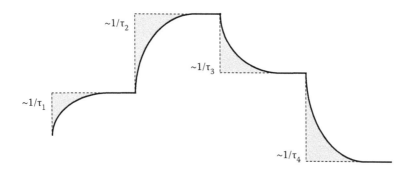

FIGURE 4.67
Code-dependent nonlinearity error.

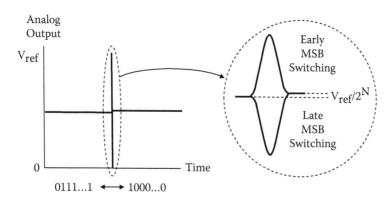

FIGURE 4.68
DAC output glitch.

The minimum slew rate of $\omega_o V_o$ is required to generate a peak signal V_o within a bandwidth ω_o. If $S \gg \omega_o V_o$, the DAC system will exhibit no slew-related distortion.

4.8.3 Code-Dependent Time Constant

Other than slewing, the shape of the transient response is governed by the code-dependent settling time constant as shown in Figure 4.67. The code-dependent settling time constant can introduce a nonlinearity error because the settling error is a function of the time constant τ given by Equation (4.37).

In the current-steering DAC, unless the DAC output is buffered, DAC current elements are directly switched to the output load back and forth, and the number of elements switched to the output varies depending on the digital input code. As a result, the parasitic at the output node varies. Similarly, the resistor-string DAC exhibits severe code-dependent settling because the output resistance of the DAC varies with the digital input code.

4.8.4 Glitches

DAC output glitches are caused by small on and off time differences when binary-weighted DAC elements are switched as shown in Figure 4.68.

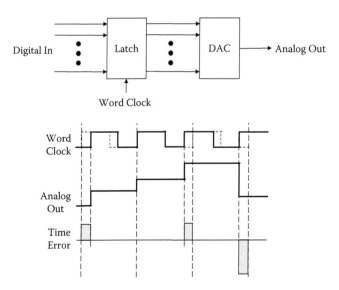

FIGURE 4.69
Word clock jitter effect.

For example, take the major code transition at the half scale from 011...11 to 100...00. The MSB current source turns on while all other current sources are turned off. The small difference in the switching times can result in a narrow half-scale glitch.

To alleviate both glitch and slew-rate problems related to transients, a deglitcher can be used. The deglitcher stays in the hold mode while the DAC changes its output value. After the switching transient subsides, the deglitcher gets into the hold mode. During the hold time, the output of the deglitcher can be made independent of the DAC output transients. However, note that the slew rate of the deglitcher is also on the same order as that of the DAC, and the transient distortion may now remain as an artifact of the deglitcher. The glitch impulse energy is measured using the unit of pV*sec.

4.8.5 Word Clock Jitter

Even though DAC is ideally linear, it needs precise timing to correctly reproduce an analog output. If DAC is not timed with the identical clock the waveform was sampled with, the timing jitter error will result as shown in Figure 4.69. The timing jitter in the word clock produces an amplitude variation in the DAC output, causing the waveform to change its shape. This, in turn, introduces spurious tones related to the jitter frequency or raises the DAC noise floor if jitter is random. The jitter timing error affects both ADC and DAC similarly. If jitter is Gaussian-distributed, the worst-case SNR and the upper limit of the tolerable word clock jitter are given by Equations (4.18) and (4.20), respectively.

References

1. P. W. Li, M. J. Chin, P. R. Gray, and R. Castello, "A ratio-independent algorithmic analog-to-digital conversion technique," *IEEE J. Solid-State Circuits*, vol. SC-19, pp. 828–836, December 1984.

2. T. B. Cho and P. R. Gray, "A 10-b, 20-M sample/s 30mW pipeline A/D converter," *IEEE J. Solid-State Circuits*, vol. SC-30, pp. 166–172, March 1995.

3. T. L. Brooks, D. H. Robertson, D. F. Kelly, A. Del Muro, and S. W. Harston, "A cascaded sigma-delta pipeline A/D converter with 1.25-MHz signal bandwidth and 89dB SNR," *IEEE J. Solid-State Circuits*, vol. SC-32, pp. 1896–1906, December 1997.

4. H. Pan, M. Segami, M. Choi, L. Cao, and A. A. Abidi, "A 3.3-V 12-b 50-MS/s A/D converter in 0.6-μm CMOS with over 80-dB SFDR," *IEEE J. Solid-State Circuits*, vol. SC-35, pp. 1769–1780, December 2000.

5. A. M. Abo and P. R. Gray, "A 1.5-V, 10-bit, 14.3-MS/s CMOS pipelined analog-to-digital converter," *IEEE J. Solid-State Circuits*, vol. SC-34, pp. 599–606, May 1999.

6. B. S. Song, S. H. Lee, and M. F. Tompsett, "A 10-b, 15-MHz CMOS recycling two-step A/D converter," *IEEE J. Solid-State Circuits*, vol. SC-25, pp. 1328–1338, December 1990.

7. W. C. Song, H. W. Choi, S. U. Kwak, and B. S. Song, "A 10-b, 20-Msample/s low-power CMOS ADC," *IEEE J. Solid-State Circuits*, vol. SC-30, pp. 514–521, May 1995.

8. J. Doernberg, H. S. Lee, and D. A Hodges, "Full-speed testing of A/D converters," *IEEE J. Solid-State Circuits*, vol. SC-19, pp. 820–827, December 1984.

9. K. Kattman and J. Barrow, "A technique for reducing differential non-linearity errors in flash A/D converters," *Dig. Tech. Papers, IEEE Int. Solid-State Circuits Conf.*, pp. 170–171, February 1991.

10. J. Crols and M. Steyaert, "Switched-opamp: An approach to realize full CMOS switched-capacitor filters at very low power supply," *IEEE J. Solid-State Circuits*, vol. SC-29, pp. 936–942, August 1994.

11. D. Freeman, "Slewing distortion in digital-to-analog conversion," *J. Audio Eng. Soc.*, vol. 25, pp. 178–183, April 1977.

5

Nyquist-Rate Data Converters

There exist many data converters for analog–digital interfaces with varied resolution and at different sampling rates from voice and audio systems for the kHz-range low-end of the spectrum to disk drive and optical systems for the GHz-range high-end of the spectrum. In the middle MHz range, there are video and image digitizers. The design strategies of analog-to-digital converters (ADCs) and digital-to-analog converters (DACs) depend heavily on the system requirements. Recently, as wireless devices became ubiquitous, two types of converters started to draw more attention. One is with high spurious-free dynamic range (SFDR) for the large blocker environment, and the other is with low-power consumption for portable devices.

5.1 Analog-to-Digital Converter Architectures

The conversion rate of the ADC is limited by the time to complete all quantization-level decisions. It varies widely from the slowest slope-type ADC that in effect resolves one quantization level at a time to the fastest flash-type ADC that resolves all levels at once. The successive-approximation register (SAR) ADC is simple and straightforward but somewhat slow because it resolves one bit at a time. Between the two extremes, there are architectures that resolve a finite number of bits at a time, such as multistep or pipeline. The oversampling $\Delta\Sigma$ ADC achieves higher resolution by applying feedback.

Figure 5.1 shows the ADC resolution spectrum. As expected, resolution is directly traded for speed. As both semiconductor process and design technologies advance, the performance envelope has been constantly pushed out, mostly driven by the demand for high resolution at increasingly higher sampling rates. Slope-type ADCs are slow because they use a voltage-to-time converter and implement the quantizer in the time domain. They have been used for slow high-resolution digital panel meters. At the low end of the spectrum, $\Delta\Sigma$ ADCs benefit greatly from the feedback topology, and find a broad range of uses in voice, audio, and even instrumentation.

As digital technology advances, $\Delta\Sigma$ ADCs find new applications as high-SFDR intermediate-frequency (IF) quantizers in wireless receivers. Its continuous-time version needs no anti-aliasing filter and greatly simplifies RF receiver architectures. SAR ADCs perform well as medium-resolution converters. Their figure-of-merit (FOM) number can be very low because they use only one comparator. They are suitable for low-power or portable applications. Pipelined ADCs are useful for most medium-resolution high-speed applications, such as video, imager, and wireless/wireline communications. At the top end of the spectrum, only flash ADCs can survive. They have low resolution but offer the speed advantage critical in such high-speed applications as in magnetic and optical storages and optical networking.

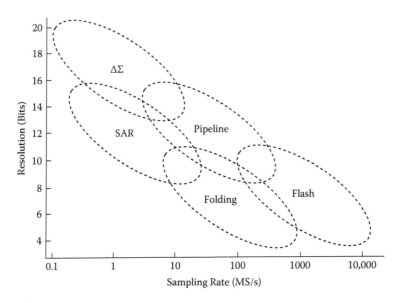

FIGURE 5.1
Resolution spectrum of ADCs.

Purpose	Method
To save number of preamps	Interpolation
To save number of comparators	Folding
To save number of both	Subranging
To reduce preamp offsets	Averaging
To speed up conversion	Interleaving
To speed up subranging	Pipelining
To improve linearity	Calibration

FIGURE 5.2
ADC design methods.

Although flash is the fastest, its complexity and power consumption exponentially grow as the number of bits increases. To mitigate the complexity and power problem, various design methodologies have evolved, as listed in Figure 5.2. Design concepts such as folding, subranging, and interpolation somewhat compromise the speed of the flash but can simplify the flash ADC and help to significantly reduce power and complexity. Others are more generic and applicable to any other ADC architectures.

Figure 5.3 shows the Nyquist-rate ADC family branched out from the flash ADC. Three basic design concepts can be combined to make subranging-interpolation, folding-interpolation, subranging-folding, and even subranging-folding-interpolation to achieve certain design goals. Note that the multistep or pilelined architecture is based on the subranging concept because later stages can work on the subranged residues. Due to the concurrent operation of the multiple stages, the pipelined architecture can achieve high throughput rate and save both power and chip area, though it requires a long initial latency period.

ADC architectures are chosen mainly to meet the resolution and speed requirements, but many other factors such as power, chip area, supply voltage, latency, complexity, operating environment, or technology also limit the range of choices. The current trend is

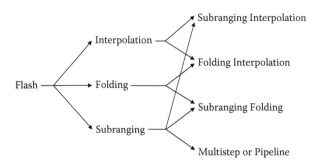

FIGURE 5.3
Nyquist-rate ADC family tree.

toward low-cost integration without using expensive discrete technologies such as thin film and laser trimming. A growing number of ADCs are being implemented using the mainstream scaled digital complementary metal-oxide semiconductor (CMOS) technologies. Some ADC architectures are often preferred to others for certain applications. Three architectures stand out among many. The oversampling $\Delta\Sigma$ ADC is exclusively used to achieve high resolution above the 14 ~ 16b level at low sampling rates. The difficulty in achieving higher resolution over 12b using conventional methods gives a fair advantage to the oversampling technique. For medium resolution at high sampling rates, the pipelined ADC is preferred due to its simplicity. However, the flash ADC offers extremely high premium sampling rates only with low resolution.

5.2 Slope-Type ADC

The simplest ADC is just using a digital counter based on the time resolution. It is to resolve one quantization level at a time. So it takes time to resolve the total 2^N levels. For the time-domain data conversion, it is necessary to convert the input voltage either into frequency or time so that a digital frequency counter can be used to find its digital equivalence. The former is called a voltage-to-frequency or voltage-controlled oscillator (VCO), and the latter is a slope or ramp generator. However, it is necessary that the VCO or the ramp generator be very linear to meet the ADC linearity requirement. The time-to-digital converter (TDC) gains momentum as scaled nanometer CMOS technologies can produce very fine time resolutions equivalent to the digital inverter delay. In such low-voltage technologies, it is difficult to maintain the high linearity of the operational amplifier (opamp) output, but the high-speed technology can be used advantageously to achieve precise digital time resolution.

Traditionally, due to the limited time resolution, the slope-type ADC has been used only for slow digital instruments mainly because of their simplicity and inherent high linearity. There are many variations, but dual- or triple-slope techniques are used because the single-slope method is sensitive to errors related to switching. The resolution of this type of ADC depends on the accurate control of charge on the capacitor and the accuracy of the threshold detecting comparator. The dual-slope approach generates two positive and negative slopes that are proportional to V_{ref} and V_i, respectively, and compares the charging-up and charging-down intervals to get their ratio digitally, which is the digital representation

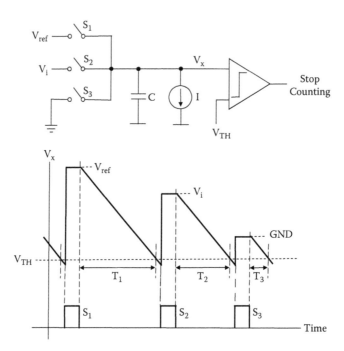

FIGURE 5.4
Triple-slope ADC concept.

of the analog input. To detect a single zero-crossing threshold, an opamp is used as a ramp generator to ramp up with V_{ref} and to ramp down with V_i.

On the other hand, the triple-slope concept as shown in Figure 5.4 uses a single comparator, and reduces the offset effect. Unlike the dual-slope method comparing two slopes, it measures three time intervals T_1, T_2, and T_3 by charging the capacitor with V_{ref}, V_i, and the ground GND, and discharging it with the same slope of I/C. Note that the three switches S_1, S_2, and S_3 initialize the capacitor and also reset the counter. The counter values of T_1, T_2, and T_3 are read when the comparator detects the zero-crossings. The comparator threshold V_{TH} can be set lower than GND. From three time-interval measurements, the ratio of V_i/V_{ref} can be computed as $(T_2 - T_3)/(T_1 - T_3)$.

The comparator is a wideband amplifier without the latch clock. The delay of the comparator does not matter much as long as it is constant. The switch offsets resulting from the charge injection and feed-through affect the converter linearity and should be minimized. The current source should be cascoded for high output resistance, and should remain constant over the wide output swing. The accuracy of the triple-slope ADC is mainly limited by the digital counter, which inevitably makes a counter error of one. Because one count corresponds to one quantization step, the total quantization noise when computing the ratio of $(T_2 - T_3)/(T_1 - T_3)$ will be quadrupled. Therefore, the equivalent number of bits is

$$N = \log_2\left(\frac{1}{2} \times \frac{f_{clk}}{3f_s}\right),$$ (5.1)

where the factor 3 is for three slope measurements per sampling. For example, counting a 10 MHz clock within a 10 msec time window gives an average count of 100 thousands, which is equivalent to using a quantization step of a 14b quantizer. The slope-type ADC is inherently slow because it requires the number of time slots to be 2^N. This handicap can be overcome in nanometer CMOS by operating the clock at GHz rates and also with high oversampling rates, though the resolution would be limited by the ramp generator linearity. The time-domain circuit concept is also applied to TDCs that implement digital phase detectors in digital phase-locked loops (PLLs).

5.3 Successive Approximation Register ADC

The successive-approximation concept is to perform the straightforward binary search algorithm. Each bit decision is made one at a time starting from the most significant bit (MSB). The sampled input is compared with an output of a DAC, and the comparator output is fed back to reconstruct the input with more accuracy as shown in Figure 5.5. The full range is progressively scaled down by half depending on which half range the input belongs to.

As shown in the DAC output V_x, the first MSB decision starts by comparing the S/H output with $V_{ref}/2$ by setting the MSB of the DAC to be 1. If the input is higher, the input belongs to the upper half of the range, and the MSB stays as 1. Otherwise it is reset to 0. In the second bit decision, the input is compared with the DAC output of $3V_{ref}/4$ in this

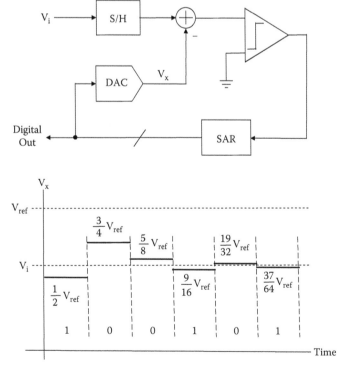

FIGURE 5.5
SAR architecture.

example by setting the second bit to be 1. If the first MSB bit is reset to 0, the input is then compared with $V_{ref}/4$. If the input is lower than the DAC output as in this example, the second bit is reset to 0, and the third bit is set to be 1 for the third bit decision with the DAC output of $5V_{ref}/8$. The bit decisions continue until all the bits are obtained. Therefore, the N-bit SAR ADC requires $(N + 1)$ clock cycles including the S/H phase to complete one sample conversion.

The performance of the SAR ADC is limited by the DAC resolution as long as the comparator offset stays constant during the conversion period. The commonly used DACs for SAR ADCs are resistor-string, capacitor-array, and current DACs. In general, thermometer-coded capacitor-array, and current DACs are monotonic like resistor-string DACs, but binary-weighted ones exhibit poor differential nonlinearity (DNL). Resistor-string DACs are monotonic and have poor integral nonlinearity (INL), but differential resistor-string DACs exhibit better INL than the single-ended ones. Among these, the binary-weighted capacitor array has been widely used for SAR ADCs. The advantage is that it consumes no power, and capacitors are far superior in matching to resistors or currents.

The standard charge-redistribution SAR DAC is shown in Figure 5.6 [1]. For N-bit conversion, it requires a total of 2^N unit capacitors, a comparator, and the SAR logic. All top plates are connected together, but the bottom plates are grouped in binary ratios. The largest one is made of half the total capacitors. The second largest one is then made of half the remaining ones, and so on. The top plate is connected to the ground through one switch, and all bottom plates are made switchable to the input, V_{ref}, or the ground.

During the initialization phase of the top plate, the input is sampled on the bottom plates of all 2^N capacitors as in the standard bottom-plate sampling as discussed in Chapter 4. For now, assume that the top-plate parasitic capacitance is negligible. After all bottom plates are switched to the ground, the top-plate voltage V_x becomes the sampled voltage of $-V_i$. The MSB decision starts by flipping the bottom of the MSB switch to V_{ref}. This is equivalent to setting the MSB bit of the DAC to be 1 ($b_N = 1$) with all other bottom plate switches still

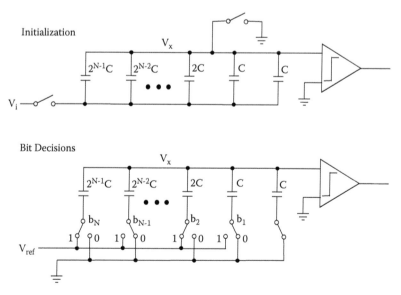

FIGURE 5.6
Charge-redistribution SAR ADC.

connected to the ground ($b_{N-1} = b_{N-2} \dots = b_1 = 0$). Then due to the charge redistribution, the top-plate voltage V_x becomes

$$V_x = -V_i + \frac{2^{N-1}C}{2^N C}V_{ref} = -V_i + \frac{V_{ref}}{2}. \tag{5.2}$$

The comparator makes the MSB decision on this, and the SAR operation continues until all bits b_N, b_{N-1}, b_{N-2}, $\dots$, and b_1 are obtained, and the residual top-plate voltage can be smaller than one least significant bit (LSB).

$$V_x = -V_i + \sum_{i=1}^{N} b_i \frac{2^{i-1}}{2^N}V_{ref}. \tag{5.3}$$

That is, because the top-plate voltage converges back to the initial precharged voltage, the total charge in and out of the parasitic capacitance at the top plate is close to zero. Therefore, it does not affect the performance of the SAR.

5.3.1 Accuracy Considerations

Capacitors made of the double-poly or metal-metal structure in CMOS are considered to be one of the most accurate passive components comparable to the film resistors in BJT both in matching accuracy and also voltage and temperature coefficients. The disadvantage is to rely on the charge redistribution principle, which is not useful to make stand-alone DACs without output S/H or deglitcher. The switched-capacitor counterpart of the resistor-string DAC is a capacitor array made of 2^N unit capacitor array. The thermometer-coded capacitor-array DAC has a distinct advantage of monotonicity, but the thermometer-coded capacitor array is prohibitively complicated to handle.

The common grouping of 2^N unit capacitors in binary ratio values results in the largest error when the MSB is selected. As in Equation (5.2), the error at the midpoint $V_{ref}/2$ is limited by the ratio mismatch between the half sum of $2^{N-1}C$ grouped for MSB and the total sum of the array as follows:

$$MSB\ Error = \frac{\Delta(2^{N-1}C)}{2^N C}V_{ref}. \tag{5.4}$$

If we assume that all unit capacitors are matched with an accuracy of $\Delta C/C$, the root-mean-square (RMS) mismatch error of a large array gets smaller because the capacitor size increases linearly, but the mismatch error increases as a square root function. Similarly, other errors at $V_{ref}/4$, $V_{ref}/8$, and $V_{ref}/16$, ... are smaller than $\Delta C/C$. The unit capacitor in current scaled CMOS technology is usually made of about 20 ~ 30 μ^2, and the ADC performance of 12 to 14b level can be achieved. The matching accuracy of the capacitor depends on the geometry sizes of the capacitor width and length and the dielectric thickness.

5.3.2 SAR ADC with R + C, C + R, or C + C Combination
Digital-to-Analog Converter

Both resistor-string and capacitor-array DACs need 2^N unit elements for N bits, and the number grows exponentially. A need arises to reduce the DAC array complexity for high resolution and also for small chip area. A resistor-string DAC can be combined easily with the capacitor-array DAC to make a (R + C) combination DAC as shown in Figure 5.7 [2].

The resistor-string DAC covers the *M*-bit MSB range and supplies the reference voltages to the capacitor-array DAC for the additional *N*-bit subranging. After the top plate is initialized, all capacitor bottom plates are switched together for *M*-bit decisions using the resistor DAC. The binary search can be used to find the correct subrange segment. Once the correct subrange is found, the bottom plates are selectively switched to either upper or lower bounds of the segment depending on the LSB bit decisions. This segmented DAC approach gives an inherent monotonicity due to the resistor-string MSB DAC if the LSB DAC is monotonic within its resolution. However, the thermometer-coded resistor-string DAC suffers from poor INL. The fully differential implementation of this architecture benefits from the cancellation of the even-order nonlinearity, and thereby improves INL.

On the other hand, in the (C + R) combination DAC shown in Figure 5.8 [3], the operation of the capacitor-array DAC is the same. The resistor-string DAC adds the subrange through the smallest unit capacitor *C*, which samples the input but is tied to the ground during the MSB decision cycles. The MSB side reference voltage is fixed, but the subdivided references are supplied to the smallest unit capacitor *C* using the resistor string DAC for LSB decisions. The capacitor array MSB DAC improves INL in the (C + R) combination, but the DNL would be poor due to the binary-weighted capacitor array. Resistor-string and capacitor-array DACs are integrated together flawlessly without any buffer amplifiers between them, but the resistor-string DACs consume power.

A more logical choice is to split the capacitor array to make the (C + C) combination. Splitting capacitor arrays into two, one for MSBs and the other for LSBs, requires a buffer amplifier to interface between two arrays. However, one floating capacitor C_a can approximate the buffer amplifier function as shown in Figure 5.9 [4].

Two capacitor arrays of *M* and *N* bits each can be combined using the smallest capacitor. For two arrays to work together, the value of C_a can be set to be slightly larger than the

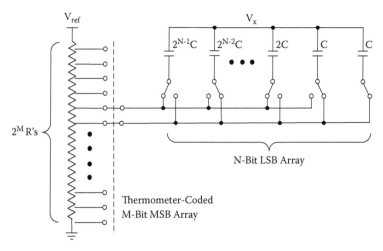

FIGURE 5.7
(R + C) combination DAC.

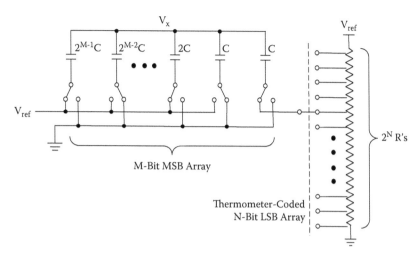

FIGURE 5.8
(C + R) combination DAC.

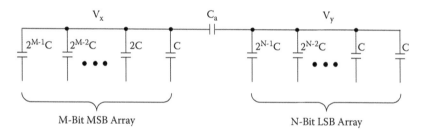

FIGURE 5.9
(C + C) combination DAC.

unit C so that the capacitance looking into C_a from the top plate of the MSB array should be approximately the unit capacitor C.

$$C_a = \frac{2^N}{2^N - 1} C. \tag{5.5}$$

After successive approximation, V_x approaches the initial precharged voltage, but the top plate V_y of the LSB array is floating. As a result, the parasitic at V_x matters little, but the error ΔC of C_a affects the MSB accuracy. The capacitance error when looking into the MSB array through C_a should meet the following:

$$\frac{1}{\dfrac{1}{C_a + \Delta C} + \dfrac{1}{2^N C}} - C \approx \frac{1}{\dfrac{1}{C_a} + \dfrac{1}{2^N C} - \dfrac{\Delta C}{C_a^2}} - C \approx \Delta C < \frac{C}{2^N}. \tag{5.6}$$

Similarly, the LSB array will be affected by the top-plate parasitic C_p. If it is smaller than the unit capacitor C, the error would be smaller than one LSB.

5.4 Subranging and Multistep ADC

The subranging concept is basically to save the numbers of both preamplifiers and comparators in the flash ADC. That is, rather than using $(2^N - 1)$ comparators, only a few number of comparators close to the input threshold can be activated like a sliding comparator array. However, an extra coarse decision cycle is required to approximate the input range, which slows the conversion rate.

Figure 5.10 shows the difference in the number of zero-crossings in the flash and subranging ADCs. In the former, $(2^N - 1)$ comparator thresholds should be closely spaced to cover the full input range, while in the latter only a few comparators can cover the input if the input can be estimated.

A straightforward subranging ADC is shown in Figure 5.11 [5]. One resistor-string DAC generates all 2^N reference levels for the flash ADC. The coarse comparator first selects the subrange with coarsely spaced reference taps. All the reference levels within the chosen subrange are then switched to the fine comparators. These dual ladder resistor strings are often connected in parallel to shorten the resistance-capacitance (RC) settling times at the comparator inputs. The subranging concept can be generalized to derive other more popular multistep and pipeline architectures.

Rather than making all bit decisions at once, resolving a few bits at a time makes the ADC system simpler and more manageable. It also enables us to correct digital errors made during earlier decisions.

The subranging system resolving total N bits is conceptually shown in Figure 5.12. The coarse M-bit ADC estimates the input and reconstructs the lower bound of the subrange to generate a residue for the fine $(N - M)$-bit ADC to quantize further.

5.4.1 Residue

The residue is defined as the unquantized portion of the signal. In the subranging family of ADCs such as multistep or pipeline, residues are generated to quantize further with

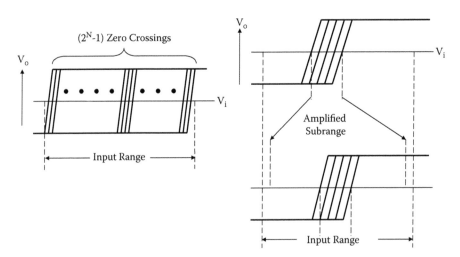

FIGURE 5.10
Flash versus subranging concept.

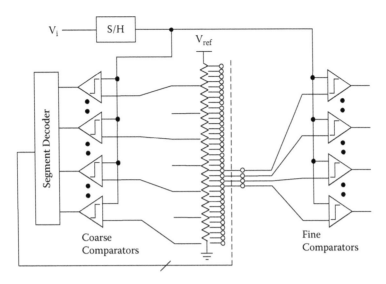

FIGURE 5.11
Subranging ADC using a resistor-string DAC.

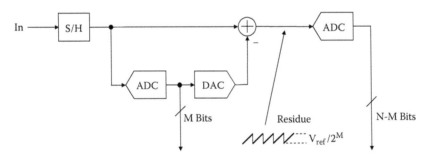

FIGURE 5.12
Subranging ADC concept.

finer ADCs in the later stages. The resolution of the subranging family ADC depends entirely on how accurate the residue is, and generating accurate residues is the most challenging part of the design. In the case of Figure 5.12, the residue varies like the sawtooth waveform as sketched when the input is ramped up linearly. The magnitude is the same as the coarse ADC step $V_{ref}/2^M$, which is the same as the quantization error from the coarse ADC. Therefore, both coarse and fine ADCs need the same N-bit resolution.

Although the complexity is reduced due to subranging, the stringent requirements for both coarse and fine ADCs stay the same, and the residue should be generated with the same resolution. Now assume that the residue can be made larger by adding a residue amplifier so that the resolution requirement of the fine comparators can be relieved. Then, the same subranging architecture can be modified as shown in Figure 5.13, where the fine ADC sees the larger amplified residue.

This modification opens up a new possibility of cascading multiple identical ADCs, and drastically changes the system requirements. Because the residue is larger, the fine ADC can resolve the residue with far lower resolution than before. If the gain of the amplifier is exactly 2^M, the full-scale residue covering the same range as the input is restored, and the

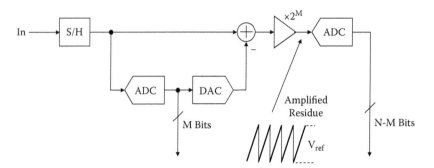

FIGURE 5.13
Subranging ADC concept modified with a residue amplifier.

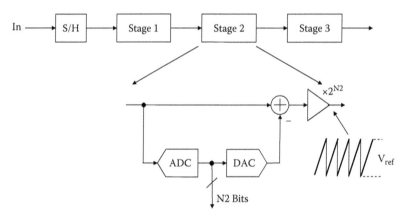

FIGURE 5.14
Multistep or ripple architecture.

fine ADC needs only $(N - M)$-bit resolution. However, the residue amplifier delay slows the conversion rate because the residue amplifier should settle with $(N - M)$-bit resolution. Subranging ADCs suffer from the same sets of error sources such as reference errors and comparator offsets, and the massive switching array adds to their complexity. They can be used for low resolution but relatively for high conversion rates.

5.4.2 Evolution of Multistep and Pipeline Architectures

The subranging tactic of making a few bit decisions at a time can be generalized with a residue amplifier. The obvious advantage is that fine comparators do not need to be accurate as the residue is amplified, but the disadvantage is that the high-gain residue amplifier should settle accurately. The subranging block consists of a coarse ADC, a DAC, a subtractor, and a residue amplifier. This block can be replicated in multiple stages as shown in Figure 5.14.

How many times the subranging stage is replicated represents the number of steps. In general, an n-step ADC has $(n - 1)$ subranging blocks. Also, ADCs and DACs in later stages require progressively less resolution. To complete one conversion in one cycle, subdivided polyphase clocks are needed. Due to the difficulty in clocking, the number of steps in the multistep architecture is usually limited to two, which does not incur a speed penalty and needs the standard two-phase clocking. There are many variations in the multistep architecture. If no polyphase clocking is used, it is a ripple ADC, because all subranging blocks make decisions one after another like the ripple propagates. The concept is similar, but if one subranging block is used repeatedly instead of using multiple blocks, it becomes a recycling ADC. The complexity of the two-step ADC, though manageable and simpler than the flash ADC, still grows exponentially as the number of bits to resolve increases. Specifically for high resolution above 12b, the complexity reaches the limit, and a need to further pipeline the subranging blocks arises.

5.5 Pipelined ADC

In the pipelined ADC, each stage resolves one or a few bits quickly and transfers the residue to the subsequent stage so that it can be quantized further in later stages.

The pipelined ADC architecture shown in Figure 5.15 is the same as the subranging or multistep architecture except for the interstage S/H driven by two-phase nonoverlapping clocks. Because the S/Hs are clocked by alternating clock phases, each stage can perform the decision and residue amplification in its own clock phase. As a result, the residue amplifier has longer time to settle than in the multistep or ripple ADC. Pipelining the residue also greatly simplifies the ADC architecture, and the complexity grows only linearly with the number of bits to resolve.

However, each pipelined stage needs three clock phases for sampling, comparison, and residue amplification. Because the next stage should sample during the amplification phase, both sampling and amplification phases should occupy almost a full clock phase. On the other hand, the comparison time is mainly for digital latching; therefore, the sub-ADC time can be squeezed in between the clock phases as shown in Figure 5.16 rather than assigning a separate clock phase.

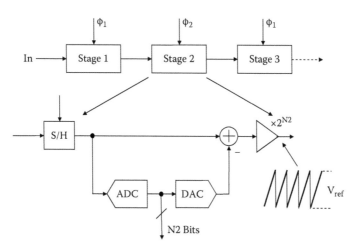

FIGURE 5.15
Pipelined ADC architecture.

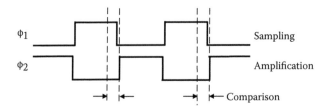

FIGURE 5.16
Nonoverlapping two-phase clocks.

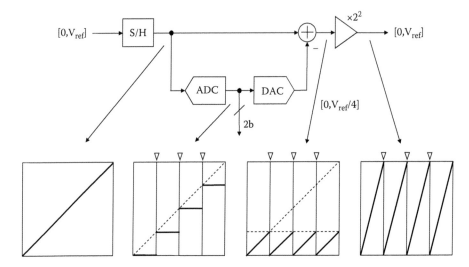

FIGURE 5.17
Residue plot of a 2b pipeline stage.

5.5.1 Residue Plot

The residue plot shows the output versus input transfer function of the residue amplifier when the input is ramped up, and it is very important in understanding the pipelined ADC. The residue from a 2b pipeline stage is sketched along with other plots of the internal nodes in Figure 5.17.

In the ideal case, as the input is swept from 0 to the full input range V_{ref}, the residue output also changes from 0 to V_{ref} each time V_{ref} is subtracted at the ideal locations of the 2b ADC thresholds, which are $V_{ref}/4$ apart and marked as triangles on the x-axis. In the 2b case, the residue output repeatedly covers the same range four times. In this stage, two bits are obtained, and the residue plot represents the four subrange inputs for the later stages to quantize further. Note that the residue in each subrange is fully restored to cover the same full range as the input in the digital domain. If four of the same stages are used to make an 8b pipelined ADC, how bit decisions are made and how the residue moves to the subsequent stages are explained in Figure 5.18.

Each stage resolves 2b, and the subsequent stages resolve the unquantized residue. The input and reference of all stages are drawn in vertical scales, and the triangles denote the ideal ADC comparator levels. One of the four subranges is amplified to fit into the full range of the later stage, and the amplified residue becomes the input. Depending on which segment the input belongs to, the 2b decisions are made, and the digital output of 01011011 is obtained from these four 2b decisions.

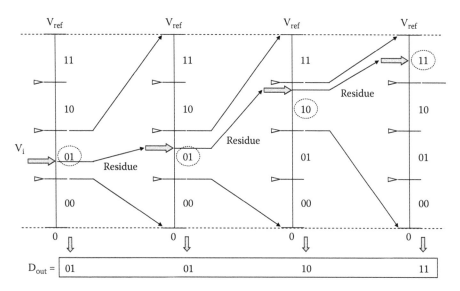

FIGURE 5.18
An 8b pipelined ADC example resolving 2b per stage.

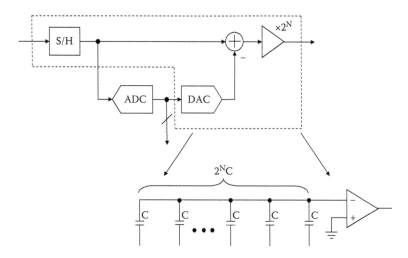

FIGURE 5.19
Capacitor-array multiplying digital-to-analog converter (MDAC).

5.5.2 Capacitor-Array Multiplying DAC

All pipeline stages are made of four functional blocks: a sample and hold (S/H), an ADC, a DAC, and a residue amplifier as shown in Figure 5.19.

The core three functional blocks included inside the dashed line except for the ADC can be replaced by a simple capacitor-array multiplying DAC (MDAC) [6], which can perform the three basic functions of input sampling, generating residue by subtracting DAC output, and residue amplification. The capacitor-array DAC is the same as the one used in the SAR ADC shown in Figure 5.6, but in the pipelined ADC, an opamp is used in place of the comparator.

The MDAC can be made with either the binary-weighted or thermometer-coded capacitor array. The former is simple to use and exhibits good INL but needs binary encoding. On the other hand, the latter is difficult to control due to its complexity but exhibits good DNL. In the SAR ADC, the binary-weighted array is preferred, but in the pipelined ADC, the thermometer-coded MDAC has an advantage because the ADC output is already thermometer coded. That is, there is no time to waste for binary encoding, and the handling complexity of the thermometer code is less severe because the number of bits used per sub-ADC stage is only a few bits.

The MDAC operation is explained in Figure 5.20 using the binary-weighted MDAC for simplicity. As usual, it starts with the initialization of the top plate, and the bottom-plate sampling is done at the same time. For example, during the clock phase ϕ_1, the top-plate switch is turned off earlier than the bottom-plate switches. During the amplification phase ϕ_2, the bottom plates are switched to either V_{ref} or the ground depending on the DAC digital bits, $b_N, b_{N-1}, b_{N-2} \ldots b_2, b_1$. At this time, the smallest unit capacitance is still connected in the opamp feedback path. If the opamp is ideal, charges on all capacitors are transferred to the feedback capacitor C, and the residue is amplified due to the charge conservation rule.

$$V_o = 2^N \times \left(V_i - \sum_{i=1}^{N} b_i \frac{2^{i-1}}{2^N} V_{ref} \right). \tag{5.7}$$

That is, the residue is generated by subtracting the DAC output from the input, and is amplified by 2^N. The next stage in the pipeline takes this full-scale residue to resolve more bits.

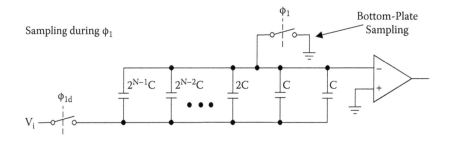

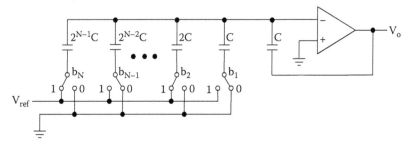

FIGURE 5.20
Capacitor-array MDAC operation.

5.5.3 Accuracy Considerations

The concept of amplifying the residue is to make the input larger so that less accurate comparators can be used. Therefore, if the input is amplified, the finite range of the next ADC cannot cover the amplified residue.

For example, as shown in Figure 5.21, the dotted line on the left side is the input amplified by 4 covering 0 to 4 V_{ref} range. Because the input range of the next ADC is still V_{ref}, the residue output should fit into the V_{ref} range. There are two ways to contain the amplified residue within the next ADC input range of 0 to V_{ref}. One is shifting, and the other is folding. The pipelined ADC is based on the former and shifts down the subranges. For that, three comparators are placed at $V_{ref}/4$, $2V_{ref}/4$, and $3V_{ref}/4$ as marked by the triangles, and the out-of-range segments above V_{ref} are shifted down to match the input range of the next ADC if V_{ref}, $2V_{ref}$, and $3V_{ref}$ are subtracted as shown, respectively.

To reconstruct the full-range digital output with the subranged digital outputs, the subtracted analog V_{ref} should be restored digitally as shown on the right side by adding MSB digital numbers 01, 10, and 11, respectively. This implies that if both analog V_{ref} and digital V_{ref} are completely matched, there are no errors occurring at the comparator thresholds when the full-range digital output is reconstructed. Otherwise, the output digital code cannot make smooth transitions at the comparator thresholds. This digital discontinuity resulting from the range mismatch of the analog and digital V_{ref} is the major nonlinearity source of the pipelined ADC.

Figure 5.22 explains that unless analog and digital V_{ref} ranges are perfectly matched, three residue errors can occur in the digital V_{ref} restoration process. The first DAC error results if the subtracted analog V_{ref} is not accurate as marked using dotted circles. That is, if it is larger or smaller than the input range of the next stage, the residue goes out of the range or falls short of the range as shown. This DAC error is random and varies from segment to segment. A similar effect can be observed if the residue gain of 2^N is not exact. The last ADC error is due to the nonideal comparator thresholds, which cause the analog V_{ref}

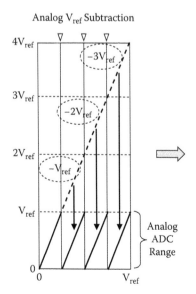
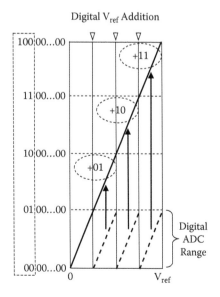

FIGURE 5.21
Digital restoration of analog V_{ref}.

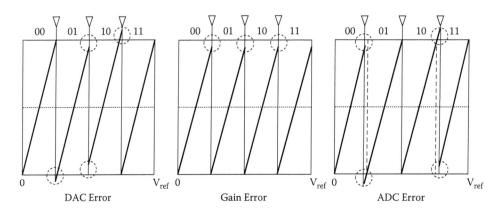

FIGURE 5.22
Three residue plots with errors.

shift to happen at thresholds away from the ideal ones. Even without DAC and gain errors, it results in irregular subranges.

Unless these errors are smaller than the LSB of the next-stage ADC, the digital output cannot make smooth transitions at the comparator thresholds, which directly translated into the DNL errors. The DAC and gain errors are fundamental because the subtracted analog V_{ref} vary due to the random DAC mismatch and the residue gain error. However, the ADC error can be corrected if the next ADC can be overranged.

5.5.4 Digital Correction

If the analog range V_{ref} matches the digital range V_{ref}, the ADC error does not matter because the digital restoration of V_{ref} will make the digital transitions across the comparator thresholds smooth without any discontinuities. Any subranging, multistep, or pipelined ADCs can be made insensitive to the ADC error if it is digitally corrected [7,8]. The problem occurs when the out-of-range residue is not quantized and lost. The residue that normally goes out of the full range can still be digitized by the next stage if the residue amplifier gain is reduced. If the residue gain is set to 2^{N-1} instead of 2^N, the residue can be contained within the full range. By overranging, the out-of-range residue can be quantized, and digital codes due to the ADC error can be recovered. The overranged residue plot for a 2b/stage pipeline stage example is shown in Figure 5.23.

Note that in the residue plot, the x-axis is the input from the previous stage, and the y-axis is the output of the current stage, which also becomes the input to the next stage. If the residue is bounded within the full range of 0 to V_{ref}, the inner range from $V_{ref}/4$ to $3V_{ref}/4$ becomes the normal range, and two redundant outer ranges cover the residue error resulting from the inaccurate coarse decision. That is, two subranges for digital codes 00 and 11 are redundant because the next ADC uses only the normal range of digital codes of 01 and 10. The errors at A and B occur due to the comparator threshold being lower than the ideal. Similarly, the errors at C and D are due to higher thresholds.

In every subrange, all digital codes repeat from all zeros to all ones. Therefore, the digital codes missing at A and D can be recovered from B and C, respectively. That is, the ADC errors are corrected digitally using the next ADC digital output as shown in the example. Decisions at A and C are obtained as shown when the thresholds are at ideal positions. Otherwise, decisions at B and D result. Therefore, if two results are added with one bit

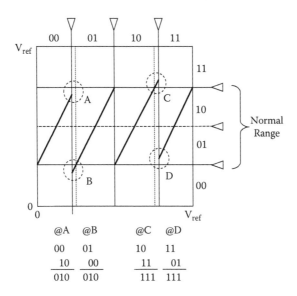

FIGURE 5.23
Overranged residue for digital correction.

overlapped, the resulting identical 3b digital outputs are obtained. Because the redundancy requires extra ADC resolution to cover the overrange, one extra redundant bit in the example covers only the out-of-range ADC error and does not contribute to the ADC resolution. That is, two half ranges are used for redundancy in this case. However, the amount of redundancy depends on the magnitude of the ADC error to correct in the multibit cases. In general, it is the trade-off between the comparator accuracy and redundancy range. Any ADC errors larger than the redundant range cannot be corrected digitally. Unless two extra comparator thresholds are added in the last pipeline stage, the signal range is reduced slightly due to the redundant ranges, which are not used.

Errors occur only at the comparator thresholds. Therefore, without affecting the ADC performance, shifting ADC thresholds by half the ideal comparator interval gives some advantages [9]. First, the comparator threshold can be placed away from the midpoint close to the signal offset so that unnecessary triggering of the comparator by noise or small signal can be avoided. Assume that the comparator thresholds are collectively shifted to the right (or to the left) by half a bit. The half-bit shifted residue would be as shown in Figure 5.24.

In this example, $V_{ref}/8$ corresponds to the half bit. Then, the lowest range of the input can use the full range down to the lower bound, but as shown by the dashed line, the upper bound is still not used. Therefore, as shown by the solid line, the upper range can be used if one comparator with the highest threshold is removed. The full ADC conversion range from 0 to V_{ref} can be used for the residue while saving one comparator, contrary to the previous case where the residue is confined in the middle half of the range.

An example of a 4b pipelined ADC made of three pipeline stages is shown in Figure 5.25. On the vertical axis, the signal and residue levels as well as the ADC comparator levels are marked to explain how the digital correction works in this example. The solid and dotted thick lines trace two separate residue paths that yield the same digital output codes regardless of the comparator thresholds.

This half-bit shift residue generation is valid for stages resolving any number of bits. Also note that due to the half-bit shift, the whole ADC thresholds move up by half a bit to

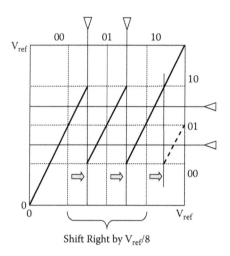

FIGURE 5.24
Half-bit shifted residue.

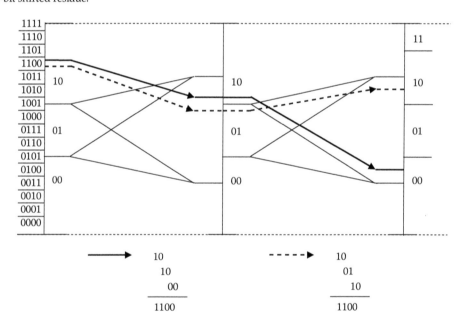

FIGURE 5.25
4b pipelined ADC with three 1.5b stages.

exhibit a systematic offset of half an LSB. This means that the top and bottom steps are 50% narrower and wider than the normal step, respectively. An additional comparator can be added to the last pipeline stage to recover this one extra level decision.

5.5.5 Generalized *N*-Bit Pipeline Stage

Digital error correction enables fast data conversion allowing the use of inaccurate but fast comparators. However, the DAC accuracy and the residue amplifier gain error remain as the fundamental limits in the pipelined ADC. The currently known ways to overcome

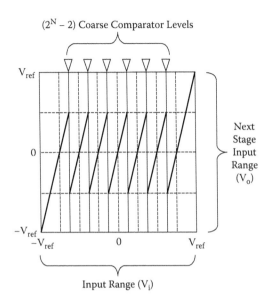

FIGURE 5.26
Differential N-bit (N = 3) residue.

these limits are self-calibration and oversampling techniques. In the general N-bit case of the half-bit shifted residue, the required number of comparators is $(2^N - 2)$, which is one less than the nominal $(2^N - 1)$. Because only two comparators are used instead of three, the 2b/stage case is often called a 1.5b/stage.

Figure 5.26 shows the residue plot of a general N-bit (N = 3) pipelined stage in a differential mode covering the full input and output range from $-V_{ref}$ to V_{ref}. In this half-bit shifted residue, the top digital code with all ones (111) is missing, as the highest-level comparator is saved after the shift. That is, the input ranges for the lowest and highest codes (000) and (110) are 50% wider than the rest, and each comparator level is equally spaced with the nominal $V_{ref}/2^N$ interval but biased higher by $V_{ref}/2^{N+1}$, which is equivalent to half a bit.

The residue plots of the multistage pipelined ADCs are shown in Figure 5.27. The residue output of each stage is the input to the next stage. The digital outputs from all stages can be added together with one bit overlapped for digital correction.

5.5.6 Trilevel Multiplying Digital-to-Analog Converter (MDAC)

Each pipelined stage consists of an N-bit thermometer-coded capacitor-array MDAC as shown in Figure 5.28. $(2^N - 1)$ unit capacitors produce 2^N DAC levels, and two-unit capacitors are in feedback for the gain of 2^{N-1}. Note that the total capacitance is increased by one more unit C to be $(2^N + 1)C$ to reduce the gain by half for digital correction. This one extra capacitance samples the ground while all other 2^N capacitors sample the input. During the residue amplification phase, this extra unit capacitor with ground sampled is connected in the feedback path together with the unit capacitor with input sampled, which effectively reduces the gain of the residue amplifier by half.

The differential signaling doubles the signal range from $-V_{ref}$ to V_{ref} compared to the single-ended signal range from 0 to V_{ref}. The ground connection goes to $-V_{ref}$, and the switching on the negative side is the mirror image of the positive side switching. When the digital bit is 1, the positive bottom plate is switched to V_{ref} while the negative bottom

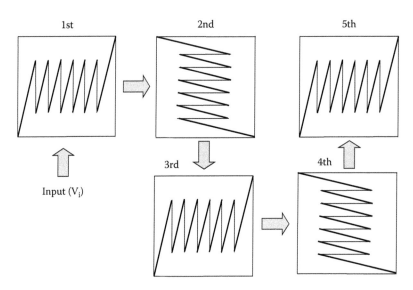

FIGURE 5.27
Residue plots for the pipelined stages.

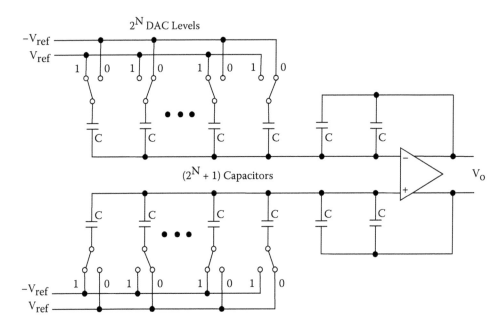

FIGURE 5.28
General N-bit differential capacitor-array MDAC.

plate goes to $-V_{ref}$. If the bit is 0, the positive and negative sides are switched to $-V_{ref}$ and V_{ref}, respectively. N-bit DAC outputs yield 2^N output levels. In pipelined stages, because the flash ADC is commonly used as a coarse ADC, its output is already thermometer coded. The coarse ADC output can directly drive the thermometer-coded DAC input.

Differential circuits need no grounding because only symmetric reference voltages of $\pm V_{ref}$ are used. However, the midpoint between V_{ref} and $-V_{ref}$ can be defined as the third reference level. This free midpoint reference is inherently accurate in differential signaling and can be obtained easily by just shorting the positive and negative inputs. If the midpoint ground is allowed, it is possible to use the three reference levels. This trilevel or 1.5b switching with an extra level greatly simplifies the capacitor-array MDAC. Conceptually, two unit capacitors can be combined into one unit, and their bottom plates can be switched together to three levels of $-V_{ref}$, 0, and V_{ref}, thereby reducing the total number of unit capacitors by half [10].

The simplified thermometer-coded differential MDAC modified with trilevel references is shown in Figure 5.29. The total number of unit capacitances is reduced to 2^{N-1}, which is now half the number ($2^N + 1$) needed for purely differential implementations. The disadvantage is the complexity in the switch control logic and the digital delay. In return for the complicated switching, half the capacitor array can be saved as shown in Figure 5.30 when using trilevel references.

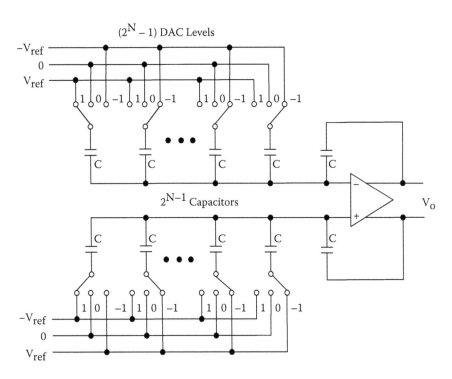

FIGURE 5.29
Simplified N-bit capacitor-array MDAC using trilevel references.

N	Number of Comps	Number of Caps	Number of DAC Levels
2	2	2	3
3	6	4	7
4	14	8	15
5	30	16	31
6	62	32	63
N	$2^N - 2$	2^{N-1}	$2^N - 1$

FIGURE 5.30
MDAC components using trilevel references.

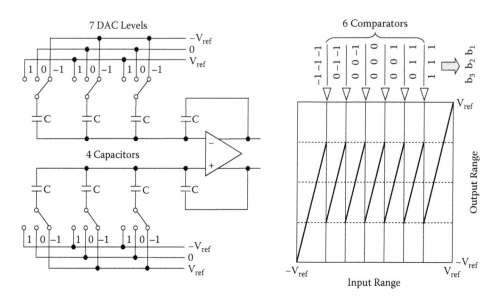

FIGURE 5.31
Example of 3b MDAC and its residue plot.

		Inputs for 3 DAC Caps		
Six Comparator Thresholds	**Coarse ADC Outputs**	V_{C1}	V_{C2}	V_{C3}
$5V_{ref}/8 < V_i$	110	V_{ref}	V_{ref}	V_{ref}
$3V_{ref}/8 < V_i < 5V_{ref}/8$	101	V_{ref}	V_{ref}	0
$V_{ref}/8 < V_i < 3V_{ref}/8$	100	V_{ref}	0	0
$-V_{ref}/8 < V_i < V_{ref}/8$	011	0	0	0
$-3V_{ref}/8 < V_i < -V_{ref}/8$	010	$-V_{ref}$	0	0
$-5V_{ref}/8 < V_i < -3V_{ref}/8$	001	$-V_{ref}$	$-V_{ref}$	0
$V_i < -5V_{ref}/8$	000	$-V_{ref}$	$-V_{ref}$	$-V_{ref}$

FIGURE 5.32
ADC thresholds, digital outputs, and DAC inputs for 3 bits.

A 3b MDAC example and its residue plot are shown in Figure 5.31. It is made of four unit capacitors to make seven DAC levels. The six thresholds are placed as sketched. Note that because it has only seven DAC outputs, one less than eight, the highest digital code 111 is not used.

These comparator thresholds are listed in Figure 5.32 along with the coarse digital outputs. The thermometer-coded DAC inputs corresponding to these digital outputs are shown in the last column. The reference voltages V_{C1}, V_{C2}, and V_{C3} that drive three capacitor bottom plates are thermometer coded using the trilevel references. These three levels are coded as a digital number b_3, b_2, and b_1 in Figure 5.31. They increase from the lowest −1, −1, and −1 to the highest 1, 1, and 1, where 1 corresponds to V_{ref}. At the midpoint, it is 0, 0, and 0.

As shown, the trilevel reference scheme works when the comparator thresholds are shifted up by half bit. Although only one comparator and one reference level are saved, the difference is more notable when $N = 2$. This degenerate case is named a 1.5b/stage MDAC

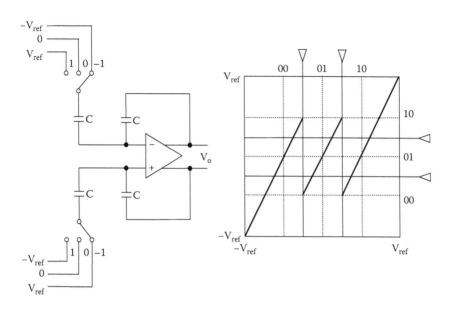

FIGURE 5.33
MDAC for degenerate case of $N = 2$.

Two Comparator Thresholds	Coarse ADC Outputs	One Capacitor DAC Input
$V_{ref}/4 < V_i$	10	V_{ref}
$-V_{ref}/4 < V_i < V_{ref}/4$	01	0
$V_i < -V_{ref}/4$	00	$-V_{ref}$

FIGURE 5.34
ADC thresholds, digital outputs, and DAC inputs for 1.5 bits.

as shown in Figure 5.33. Only two capacitors are used for the trilevel DAC. Three 2b digital outputs and the trilevel DAC inputs are shown in Figure 5.34. This stage performs the following function:

$$V_o = 2V_i - b_i V_{ref},\qquad(5.8)$$

where the thermometer coded one bit b_i can be −1, 0, and 1. This is the simplest residue amplifier for the pipelined ADC resolving only 1b per stage [11]. Rather than a full bit, one more comparator and reference level are added for redundancy.

5.5.7 Capacitor Matching

In CMOS switched-capacitor circuits, all capacitors are made of a unit capacitor for best matching. It is typically a square geometry with a side dimension of less than 10 μ. The value of the total capacitance is set by the kT/C noise requirement. Depending on the number of bits per MDAC stage, the capacitor matching affects the ADC resolution differently.

Figure 5.35 compares two 10b pipelined ADC examples resolving 2b and 5b in the first stage. The first-stage capacitor-array MDAC makes residues using 2- and 16-unit capacitors, respectively. Assuming that each unit capacitor matching accuracy is $\Delta C/C$, the gain errors of two MDACs due to the capacitor mismatch increase by $2^{1/2}$ and $16^{1/2}$, respectively.

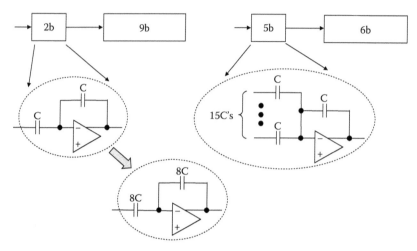

FIGURE 5.35
Capacitor matching considerations for 2b versus 5b per stage.

Because the later stages require 9b and 6b to further resolve, the capacitor-matching requirement is as follows:

$$N = 2: \quad \frac{\Delta C}{C} < \frac{1}{\sqrt{2} \times 2^9}.$$

$$N = 5: \quad \frac{\Delta C}{C} < \frac{1}{\sqrt{16} \times 2^6}.$$

(5.9)

Note that resolving more bits in the first stage reduces the matching requirement. The former case of $N = 2$ requires $2^{3/2}$-times more accurate capacitor matching than the latter case of $N = 5$. However, if 2^3-times larger capacitors are used in the 2b MDAC as shown, the capacitor-matching requirement stays the same because the total number of the unit capacitors becomes 16 in both cases. In general, resolving fewer bits in the first stage makes the MDAC simple and gives the speed advantage because the feedback factor is set higher and the loading capacitance becomes smaller. The down side is that it requires higher capacitor matching and opamp settling accuracies.

5.5.8 Opamp Gain Requirement

Like all feedback systems, MDAC settles with a time constant corresponding to the unity loop-gain frequency, which is affected by the feedback factor.

Figure 5.36 shows the capacitive feedback network during the residue amplification phase. In the capacitor-array MDAC, the feedback factor is greatly affected by both the parasitic capacitance C_p at the summing node and the opamp input capacitance C_i. C_p includes the parasitics of the top-plate initialization switch. C_i is the input capacitance of the differential pair and can be approximated as

$$C_i \approx C_{gs} + (1 + a_o) C_{gd},$$

(5.10)

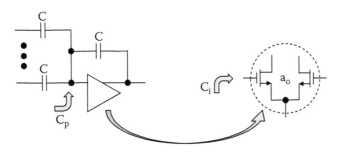

FIGURE 5.36
MDAC with a feedback network.

where C_{gs} and C_{gd} are the gate-source and gate-drain capacitances of the input devices, and a_o is the direct current (DC) gain of the input stage. The feedback factor can be defined as follows:

$$f = \frac{C}{2^{N-1}C + C_p + C_i} \tag{5.11}$$

This feedback factor f is a strong function of the number of bits to resolve and greatly affects the closed-loop DC gain and bandwidth. The standard flip-around bottom-plate S/H is the special case of Equation (5.11) when N is set to 1. Then, the feedback factor is ideally 1, which is the highest with the unity-gain feedback. However, it can be far smaller than 1 due to the sizable C_p and C_i. It implies the bandwidth requirement will be more stringent in the S/H or when N is low.

The DC loop gain $a_o f$ contributes to the gain error of about $-1/a_o f$, which should be smaller than one LSB of the later stage. For example, in a 10b pipelined ADC, the DC loop gain of the first N-bit MDAC stage should meet the following condition:

$$a_o f > 2^{10-(N-1)} = 2^{11-N}, \tag{5.12}$$

because the first stage resolves only $(N - 1)$ bits and uses 1b for redundancy. In the examples of 2b and 5b first stages, Equation (5.11) implies that the loop gain in the 5b case can be far lower than in the 2b case. The opamp open-loop DC gain requirement can be derived.

$$N = 2: \quad a_o > \frac{2^9}{f} = 2^{10}\left(1 + \frac{C_p + C_i}{2C}\right).$$

$$N = 5: \quad a_o > \frac{2^6}{f} = 2^{10}\left(1 + \frac{C_p + C_i}{16C}\right). \tag{5.13}$$

Note that unlike the loop-gain requirement, the opamp open-loop DC gain requirement is about the same for the two cases if both C_p and C_i are much smaller than the unit capacitor C. However, in most high-speed pipelined ADCs, the opamp input devices are made very large to get high g_m for high gain and bandwidth. As a result, C_i reduces the overall feedback factor f significantly. Resolving fewer bits in the first stage, the opamp DC gain requirement is more stringent as a result.

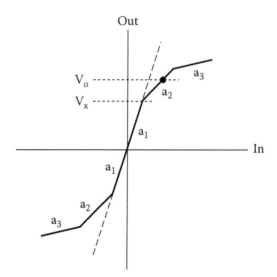

FIGURE 5.37
Opamp gain nonlinearity.

As shown, the finite loop-gain error of the MDAC also affects the overall ADC performance. However, it matters little in the S/H performance as long as the S/H function is linear. The nonlinearity of the open-loop DC transfer function is a critical factor in both S/H and MDAC designs.

Consider the piecewise nonlinear opamp transfer function as shown in Figure 5.37, where there are three gain slopes of $a_1 > a_2 > a_3$.

If the output V_o is allowed to swing higher than the high-gain boundary V_x, then the following error occurs in the output:

$$\Delta V_o = \left(V_o - V_x\right)\left(\frac{1}{a_2} - \frac{1}{a_1}\right). \tag{5.14}$$

For this nonlinearity error to stay smaller than 1 LSB of the later stage, the low gain a_2 should meet the requirement of Equation (5.12) to be safe in the MDAC design. However, in the S/H design, because the gain error is not critical, the nonlinearity requirement can be met with a gain slightly lower than required.

5.5.9 Opamp Bandwidth Requirement

During the residue amplification phase, the MDAC output needs to settle to the final value in time with required accuracy. Responding to the large step input, the MDAC needs to start to slew and then settle. However, most pipelined ADCs find high-frequency applications at conversion rates well over high tens of MS/s in low-voltage CMOS. Opamps are often biased with high currents, and capacitor sizes are made small enough to operate at high sampling rates. As a result, it is reasonable to expect that the MDAC settles with an exponential time constant without slewing.

Figure 5.38 shows the opamp compensation scheme in the Bode gain plot for different MDACs. The feedback factor f varies depending on the number of bits to resolve. For larger N, the opamp can be compensated so that the dominant pole frequency ω_{p1} can be

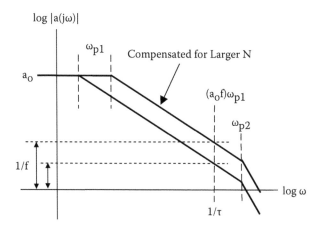

FIGURE 5.38
Opamp compensation for N-bit multiplying MDAC.

higher to maximize the closed-loop bandwidth because f is smaller. If single-stage opamps such as telescopic or folded-cascode opamps are used, the nondominant second pole ω_{p2} is set almost independently, and the unity-gain bandwidth can be maximized. If the pole-splitting Miller compensation is used for two-stage opamps, the second pole is also closely related to the compensation capacitance.

In the opamp open-loop frequency response, the unity loop-gain frequency ω_k is higher than ω_{p1} by the loop gain $a_o f$.

$$\omega_k \approx \left(a_o f\right)\omega_{p1} = 1/\tau. \tag{5.15}$$

Unless slew limited, this time constant τ and the resolution of the later stage set the upper bound of the sampling rate of the pipelined ADC. For a 10b pipelined ADC, the first N-bit MDAC requires a bandwidth of $1/\tau$ be wide enough to meet the following condition because it should settle accurately within half the sampling period $T/2$:

$$e^{-\frac{T}{2}\left(\frac{1}{\tau}\right)} < \frac{1}{2^{10-(N-1)}}. \tag{5.16}$$

For example, it takes about 6.9τ to settle exponentially with 0.1% accuracy, and settling should be completed within $T/2$. Also for proper settling, the nondominant second pole ω_{p2} should be placed higher so that the phase margin (PM) at the unity loop-gain frequency can be more than 60°.

$$PM = 90^o - \tan^{-1}\frac{\left(a_o f\right)\omega_{p1}}{\omega_{p2}} > 60^o. \tag{5.17}$$

MDAC settling is affected by PM and zeros in the transfer function. The transient response to the step input suffers from excessive ringing if PM is less than 60°. Even with a good PM, zero in the transfer function creates overshoot and can also cause slow settling with a pole-zero doublet.

5.5.10 Noise Considerations

The ADC resolution is limited by either noise or nonlinearity. The in-band thermal noise is fundamental, and it gets worse in wideband systems. Three types of thermal noises to consider are the sampled kT/C noise, opamp thermal noise, and opamp output noise aliased during sampling. Thermal noise should be handled associated with its noise bandwidth. As discussed, the kT/C noise is the switch thermal noise band-limited by the wideband sampling network and added directly to the signal. Because of the exponential increase of the sampling capacitor size, even the trade-off between speed and resolution is prohibitively difficult to make in low-voltage scaled CMOS. As a result, an SNR over 80 dB is rarely achievable with 1.8 V supply and at sampling rates over 100 MS/s.

On the other hand, the opamp input thermal noise is manageable because it is band-limited by the unity loop-gain frequency ω_k. The input-referred opamp thermal noise is dominated by the thermal noise of the input differential pair as discussed in Chapter 3. Therefore, assuming one-pole roll-off in the opamp frequency response, the average input-referred opamp noise can be approximated in the same way as the kT/C noise is derived.

$$v^2 = 2 \times 4kT \frac{2}{3g_{m1}} \times f \frac{g_{m1}}{2\pi C} \times \frac{\pi}{2} = 2 \times \frac{kT}{C} \times \frac{2}{3} f, \qquad (5.18)$$

where g_{m1} is the transconductance of the first-stage differential pair, and C is either the output loading of the single-stage opamp or the compensation capacitor of the two-stage opamp. Note that Equation (5.18) gives similar dependence on the capacitance like the kT/C noise, but the noise is weighted by the factor of $(2/3)f$ due mainly to the band limiting by the MDAC feedback. In most MDAC stages with a small feedback factor, the opamp thermal noise is negligible compared to the kT/C noise. However, the opamp thermal noise may contribute more in the S/H stage because the feedback factor f is high due to the unity-gain feedback.

In most bottom-plate sampling S/H and MDAC, the bandwidth of the sampling network is designed to be much wider than the clock frequency, and the opamp output noise can be aliased into the signal band multiple times. However, opamps used in the capacitor-array MDAC usually have high-impedance output nodes, and their output noise decreases as the frequency response rolls off at high frequencies. Opamps designed with low-impedance buffer amplifiers can have the wideband output thermal noise, but its level is low due to the high g_m of the output buffer. As a result, the aliasing effect is usually not as significant in most switched-capacitor circuits as expected.

5.5.11 Optimum Number of Bits per Stage

A few trade-offs should be made in the pipelined ADC design. How to set the number of bits per stage depends heavily on the system requirements, such as resolution, speed, and power. As summarized in Figure 5.39, resolving more bits per stage helps to alleviate the requirements for capacitor matching, DC loop gain, and loop bandwidth, while higher comparator resolution and complexity are required. If the goal is the speed, resolving one bit per stage ($N = 2$) is the best choice because the capacitive loading is minimum, and the feedback factor is maximized. On the other hand, if the goal is the resolution, resolving more bits like ($N = 5$ or 6) helps to achieve a higher accuracy of the amplified residue. Optimizing for both speed and resolution would lead to the typical number close to 2 to 3 bits ($N = 3$ or 4).

N	2	3	4	5	6
Capacitor Matching			⇒		
DC Loop Gain			⇒		
Loop Bandwidth			⇒		
Comparator Resolution			⇐		
Complexity			⇐		

FIGURE 5.39
Capacitor-array MDAC design.

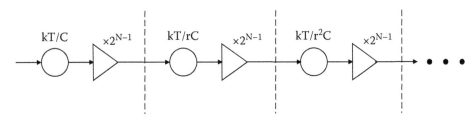

FIGURE 5.40
Pipelined stages scaled by *r*.

5.5.12 Scaling Pipelined ADC

In pipelined ADCs, each stage resolves a few bits and feeds the next stage with an accurate residue. The accuracy requirement of the residue is progressively reduced in later stages. The noise of the pipelined ADC is dominated by the kT/C noise, but the contribution from the later stages becomes less significant. So, both capacitor size and power consumption can be scaled down in later pipelined stages.

Figure 5.40 shows the noise contribution of each pipelined stage, which resolves the same number of bits, but capacitor is scaled by the fixed ratio *r*. Therefore, the interstage gain is still 2^{N-1}, but only the kT/C noise increases by $1/r$ if $r < 1$. All noises can be referred to the input after dividing it by the interstage gain, and can be added together to get the total input-referred kT/C noise.

$$v^2 = \frac{kT}{C}\left\{1 + \frac{1}{r\left(2^{N-1}\right)^2} + \frac{1}{r^2\left(2^{N-1}\right)^4} + \cdots\right\} \approx \frac{kT}{C}\left\{\frac{1}{1 - \frac{1}{r\left(2^{N-1}\right)^2}}\right\}. \tag{5.19}$$

When there is no scaling ($r = 1$), the later stages increase the total noise by 1.25, 0.56, and 0.07 dB for $N = 2, 3,$ and 4, respectively. It is due to the interstage gains, and the contribution from the second stage is most significant. In the case of more aggressive scaling equivalent to the interstage gain factor ($r = 1/2^{N-1}$), the noise contributions from the later stages increase to 3, 1.25, and 0.56 dB, respectively, because the kT/C noise increases by $1/r$. The scaling factor can be arbitrarily set as long as the total noise requirement is met.

The other important implication of the scaling is that because the residue settling requirement is less demanding in the later stages, the opamp gain and bandwidth can also be scaled down aggressively. The g_m of the opamp input stage can be scaled down by the capacitor scale ratio r, and still the same unity loop-gain frequency can be kept. This implies that both the device size (W/L) and the bias current can be scaled by the same ratio r while keeping the same g_m/C ratio. Therefore, in pipelined ADCs, the most stringent accuracy and power requirements are on the input S/H and the first-stage MDAC. More aggressive scaling can result in significant area and power savings in the later stages. The scaling does not need to be by a constant ratio. In high-speed, high-resolution pipelined ADCs, the settling requirement of the first-stage MDAC is severe, and some irregular scaling is often used. For example, the second stage resolves the minimum number of bits to reduce the loading to the first-stage MDAC.

5.5.13 S/H-Free Pipelined ADC

The S/H is the most demanding element to design in the pipelined ADC, and it is also the most power consuming. It should be linear with full ADC resolution. Because it has no gain, both its kT/C and opamp noise are directly added to the total ADC noise. Therefore, building pipelined ADCs without the input S/H is a challenging but rewarding task. Digital correction works based on the assumption that the redundancy range should be able to cover the total sum of the three offset: charge injection and clock feed-through offset during sampling, MDAC opamp offset, and comparator offset. This is to correct the ADC error resulting from the total difference between the opamp and comparator offsets. The function of the S/H helps to reduce this offset difference because both the MDAC and comparator inputs are the DC inputs already sampled. Eliminating the S/H runs a risk of operating the MDAC and the comparator with the alternating current (AC) inputs.

Normal timing of the pipelined ADC is illustrated in Figure 5.41. Note that both ADC and MDAC sample the same DC voltage held in the S/H during ϕ_1, because the ADC pre-amplifier can amplify the seed signal before the input is sampled.

By removing the input S/H stage, the AC input is directly applied to the MDAC and comparator as shown in Figure 5.42. Ideally, both MDAC and comparator sampling networks should have the same time constant, but the sampling-time mismatch results from the difference in the signal delays of the two paths. This mismatch error is in effect identical to

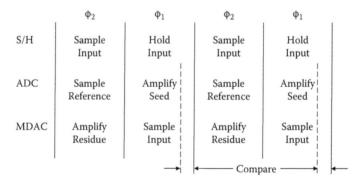

FIGURE 5.41
Functional timing for S/H, ADC, and MDAC.

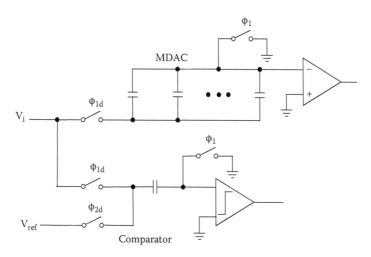

FIGURE 5.42
S/H-free pipelined ADC timing of MDAC and ADC.

the comparator offset, and affects the system performance more when the input frequency is high. That is, the comparator samples the signal with a scaled replica circuit similar to the amplifier sampling circuit so that the phase delays can be matched. This modified clock scheme imposes a trade-off between the sampling accuracy and the comparator offset. The longer the sampling time is, the more accurate is the seed sampling, but the preamplification time becomes shorter. That is, the input sampling period should be shortened slightly for preamplification. All RC networks or amplifiers exhibit group delays that are inversely proportional to their bandwidths. Therefore, it is difficult to match the comparator path delay with the exact simple RC delay of the MDAC path, in particular when the input frequency is high. However, digital correction allows some room for the ADC error as long as it is contained within the redundancy range [12]. This opens up a new possibility of implementing low-power pipelined ADCs.

5.6 Folding ADC

The folding ADC is a derivative of the flash ADC, but fewer comparators are used because the input range can be divided into many subranges. Unlike the subranging ADC, preamplifiers for comparators are replaced by folding amplifiers. In the original arrangements, the folding ADC digitizes the folded signal with a flash ADC [13]. The folded signal is in concept equivalent to the residue of the subranging, multistep, or pipelined ADC. The signal folding is done solely in the analog domain. Although multistep and pipeline ADCs shift the amplified residue subranges to fit into the next ADC range using coarse decisions, folding ADCs use analog folding amplifiers to contain the residue within the next ADC range as shown in Figure 5.43.

The analog folding is inaccurate and nonlinear, and the full residue range is not usable. Practical folding ADCs just rely on the zero-crossings of the folded outputs like the flash ADC. Due to the folding nature, the digitized code of the folding amplifier output repeats

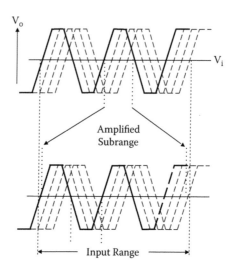

FIGURE 5.43
Amplified residue folding concept.

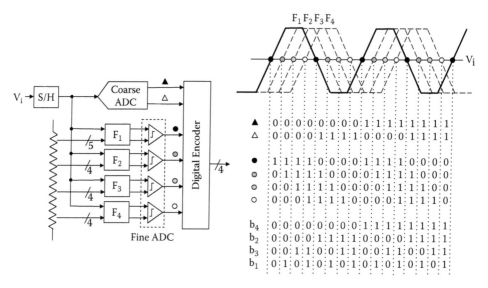

FIGURE 5.44
A 4b folding ADC example.

over the whole input range; therefore, a coarse coding is also required as in subranging-type ADCs. As a result, the same comparator can be used repeatedly as many times as the signal is folded.

A 4b folding ADC example is shown in Figure 5.44. Four folding amplifiers, F_1 through F_4, are placed in parallel to produce four folded signals as sketched on the right side as the input is swept. It is similar to the residue plot of the pipelined ADC. Four comparators check the outputs of the four folders for zero-crossings. Note that the outputs of the four fine comparators marked with circles show a repetitive pattern, and eight different codes can be obtained. Because there are two identical fine code patterns, an extra comparator is needed to distinguish them. However, if this coarse comparator zero-crossing is misaligned with that of the fine quantizer, DNL errors will result. The same digital correction concept as used in the multistep or pipelined ADC can be employed to correct the coarse ADC error. In this example, one redundant bit is used with two extra comparators used in the coarse ADC as marked with triangles.

5.6.1 Accuracy Considerations

How many times the folded ranges repeat represents the degree of folding (DF), and the number of folders (NF) used increases the number of zero-crossings. Similarly, the resolution of the folding ADC can be further improved using the interpolation concept. When the two neighboring folded outputs are used to interpolate the intermediate outputs between them, the number of zero-crossing points will increase accordingly. The upper bound of the degree of interpolation (DI) is set by the comparator resolution, the gain of the folders, the linearity of the folded signals, and the interpolation accuracy. That is, DI also increases the ADC resolution by $\log_2(DI)$ bits. Because the total number of quantization steps is the product of these three numbers, which may not be a binary number, the number of bits of the folding ADC is

$$N = \log_2(DF \times NF \times DI). \tag{5.20}$$

In the example shown in Figure 5.44, the total number of bits resolved is four, because DF = 4, NF = 4, and DI = 1.

Figure 5.45 shows a standard folder made of differential pairs. Note that three input differential pairs are needed to get DF = 3, and the folder needs the same number of reference voltages as DF. Their output currents are summed alternately at the output, and the output covers the full range three times as shown on the right side when the input is swept across the full input range, thereby making three zero-crossings. Therefore, the folding process increases the internal signal bandwidth by DF because the folding amplifiers should cover the full range DF times per single input sweep, and the folding ADC performance is also limited by the folding amplifier bandwidth.

If differential pairs and load resistances are matched, and all tail currents are assumed to be the same ($I_1 = I_2 = I_3 = I$), the zero-crossings occur exactly at V_{ref1}, V_{ref2}, and V_{ref3}, with the maximum output swing of ±IR. Having several folded signals instead of one and detecting only their zero-crossings gives unusual advantages in designing high-speed ADCs. The

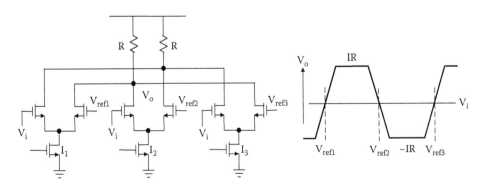

FIGURE 5.45
Folding amplifier with DF equal to 3.

folding amplifier requires neither linear output nor accurate settling, but the offset of the folders becomes the most critical design parameter. In the folding ADC, only the zero-crossing accuracy of the folded signals matters, but in the flash ADC their absolute values do not matter. As the zero-crossing accuracy of the comparator is of paramount interest, the offset of the preamplifier should be sampled, canceled, or averaged out. The offset averaging is the spatial averaging of the neighboring preamplifier offsets as discussed in Chapter 4.

Both the differential pair offset and the tail current mismatch contribute directly to the accuracy of the zero-crossings. The tail current mismatch effect is shown in Figure 5.46 with three different tail currents.

In the folder design, the transconductance (g_m) of the differential pair is optimized so that the folder output is the triangle waveform when the input is swept linearly.

$$g_m \approx \frac{2I}{\Delta V_{ref}}, \tag{5.21}$$

where ΔV_{ref} is the reference step such as $(V_{ref2} - V_{ref1})$. Then the tail current mismatch amount divided by g_m becomes the input zero-crossing error of the folder. The maximum INL error can be derived using Equation (5.21) as follows:

$$INL \approx \frac{\Delta I}{g_m} = \left(\frac{\Delta I}{I}\right)\frac{\Delta V_{ref}}{2}, \tag{5.22}$$

where ΔI is the tail current mismatch. If a total number of N codes are to be resolved with smaller than half LSB error within the ΔV_{ref} step, the tail current matching requirement becomes, from Equation (5.22),

$$\frac{\Delta I}{I} \approx INL \times \frac{2}{\Delta V_{ref}} < \frac{\Delta V_{ref}}{2N} \times \frac{2}{\Delta V_{ref}} = \frac{1}{N}. \tag{5.23}$$

That is, to resolve $N = 50$ more levels within the ΔV_{ref} step, the current matching accuracy should be less than 2%.

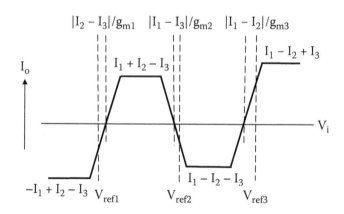

FIGURE 5.46
Folder output current with tail current mismatches.

5.6.2 Cascaded Folding

It is still costly to create as many zero-crossings as in the flash ADC by folding alone, because the number of preamplifiers stays the same. One way to produce more zero-crossings is to interpolate intermediate levels between the outputs of two neighboring folders. Interpolation works well in the folding ADC and helps to get extra quantization levels. However, the linearity performance tends to be sacrificed as the nonlinear range of the folding amplifier is used for interpolation. Another way is to increase the DF so that fewer comparators can be used repeatedly to cover the full input range. Cascading results in the large reduction in the number of comparators. However, the folding amplifier's gain should also be set high to maintain the comparator resolution. To alleviate this constraint in meeting both high gain and bandwidth requirements simultaneously, the folding stages should be either cascaded [14] or pipelined [15].

To cascade folders, a four-quadrant analog multiplier such as a Gilbert multiplier is required as shown in Figure 5.47. In the example of four folders, the DF is improved by multiplying two alternate folder outputs like F_1*F_3 and F_2*F_4. However, this multiplying element is difficult to implement in CMOS without using level shifters, and it introduces additional offset error and distortion. It is possible to cascade the same stages if the number of folders are multiples of odd numbers as shown in Figure 5.48.

If odd multiples of the folded signals are available, folder output currents can be added or subtracted alternately to achieve higher DF. It is possible to use the same stages for cascading, but the number of zero-crossings is no longer binary.

Although cascading can achieve high DF without increasing the folder bandwidth requirement, it is similar to the flash ADC in concept as shown in Figure 5.49.

For example, two folders need to settle simultaneously during the same clock phase as illustrated on the right side, and the conversion speed is still limited by the settling of the second folder. That is, the DF in the cascaded folding case is not as high as that of the single folder, but it takes a longer time for all cascaded folders to finish settling within half the clock period. The bandwidth requirement due to the DF becomes worse by cascading more stages.

This settling time constraint can be relieved if the pipelining concept is employed as shown in Figure 5.50. Like the residue in the pipelined ADC, the folded signal can be sampled, and pipelined. In this case, there are many folded signals, which are considered

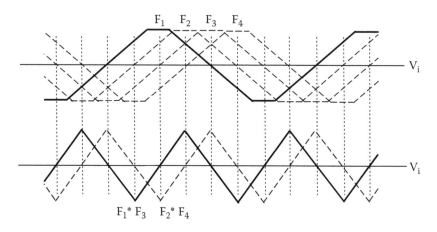

FIGURE 5.47
Cascaded folding using analog multiplier.

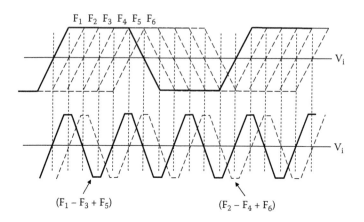

FIGURE 5.48
Cascaded folding without using analog multiplier.

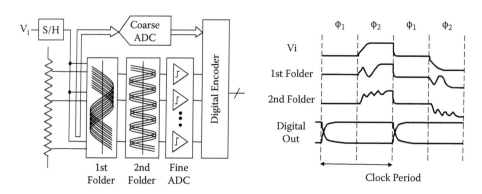

FIGURE 5.49
Cascaded folding ADC system.

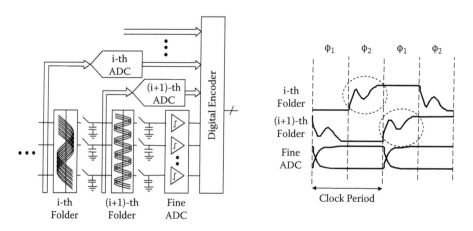

FIGURE 5.50
Pipelined folding ADC system.

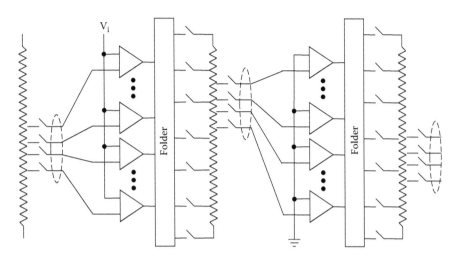

FIGURE 5.51
Generalized pipelined folding ADC with subranging and interpolation.

as multiple residues. If distributed interstage S/Hs are used to pipeline multiple residues between folders and the fine ADC, the settling time constraint is greatly relieved because each folder stage and the fine ADC have a full clock phase to complete settling, thereby achieving high DF without any speed degradation. However, in this multiple-residue system, the distributed S/Hs contribute more offset errors and add more complexity to already massively parallel systems.

Figure 5.51 sketches a generalized folding ADC with all other imaginable ADC concepts applied. This multiple-residue system is interpolated, subranged, folded, and pipelined. Compared to the single residue system like the pipelined ADC, it is faster and requires lower matching accuracy, lower amplifier gain, and less accurate settling, but it suffers greatly from offset errors and circuit complexity. The folding ADC can replace the flash ADC for high-speed but only low-resolution applications.

5.7 Other ADCs

Since early 1980, the demand for low-cost ADCs has been growing for mixed-mode embedded applications, and ADC designs have changed drastically as CMOS becomes ubiquitous. Many variations in ADC architectures have emerged for various reasons to meet specific goals at the system level, such as small chip area, low power, low voltage, and high sampling rate. Those changes are noticeable at various levels from architectures to circuits. However, in doing that, some performance at the system level has always been compromised.

5.7.1 Algorithmic ADC

When used in the multistep architecture, the interstage S/H offers a desirable feature of being able to pipeline multiple stages. Therefore, the same functional blocks can be

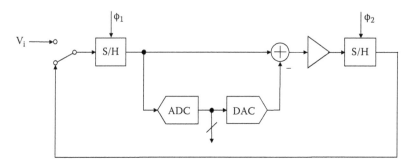

FIGURE 5.52
Algorithmic, cyclic, or recursive ADC architecture.

replicated to raise its throughput rate though the overall latency period increases by the number of stages in the pipeline. However, instead of replicating hardware like pipeline, using one residue amplifier repeatedly can lead to the great savings of the chip area and power as shown in Figure 5.52. It is also similar in concept to the SAR ADC, except for using the residue amplifier in place of the comparator. The difference is an additional half-clock delay in the loop, which can be implemented using a simple switch and capacitor S/H rather than an opamp. Unlike SAR, it can resolve more bits than one, which reduces the number of cycles. The operation is the same as in the pipelined ADC using nonoverlapping two-phase clocks.

The end result is that the high throughput rate of the pipeline is directly traded for small chip area and low power. Such a converter is called algorithmic, cyclic, or recursive ADC. The functional blocks used for the algorithmic ADC are identical to the ones used in the pipelined ADC, and thus all performance enhancement techniques such as digital correction and calibration developed for the pipelined ADC can also be applied.

5.7.2 Time-Interleaving ADC

Another logical variation in the ADC architecture is time interleaving, which raises the throughput rate by operating ADCs in parallel. Unlike the algorithmic ADC that sacrifices the throughput rate for small chip area, the time-interleaved ADC takes quite the opposite approach to replicate more hardware for higher throughput rate. In analogy, pipelining is a one-dimensional hardware replication, and time interleaving is a two-dimensional one.

This is the classical time-varying N-path system. In principle, any types of ADCs can be time-interleaved as shown in Figure 5.53 using N-phase clocks, where the throughput rate increases by the number of parallel paths [16]. However, each path still operates at the N-times lower reduced sampling rate of f_s/N. Although it improves the throughput rate significantly, it suffers from all ADC-related problems such as offset, gain, sample time, bandwidth, and even INL errors arising from mismatches among the multiplexed paths. They appear as errors that are periodic with a frequency of f_s/N, because the input signal always goes through each channel periodically at an interval of N/f_s. That is, the ADC sees the same error of each channel per every N samples repeatedly. As a result, the channel mismatches produce either fixed tones or modulated folded image bands at multiples of f_s/N in the output spectrum. Because the ADC transfer function is time varying, all non-idealities including DC parameters in the single-path ADC are modulated. The constant

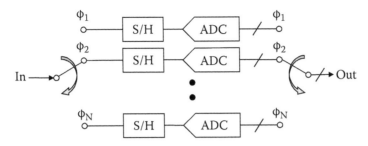

FIGURE 5.53
Time-interleaved parallel ADC architectures.

offset mismatches make a fixed tone, but the rest of the mismatches vary with the signal. Therefore, unless each channel ADC is ideal, channel mismatches degrade the noise performance of the time-interleaved system.

The SNR of the N-path system can be defined as in the single path system with an average channel transfer function. All the mismatch errors are modulated, and the system SNR approaches the value approximated by counting the mismatches just as amplitude and jitter noises in the single-path system. Because each channel subsamples at N times lower rates, the spectrum folding and aliasing complicate the noise analysis. Assuming that the variances of the $(N-1)$ mismatch errors reduced by N are summed, the SNR with a sinusoidal input tone of f_o can be estimated as

$$SNR \approx \frac{\sigma_{signal}^2}{\sigma_{noise}^2 + \frac{(N-1)^2}{N^2} \times \left\{ \sigma_{offset}^2 + \sigma_{gain}^2 + \left(2\pi f_o \sigma_{time} \right)^2 + \sigma_{BW}^2 + \sigma_{INL}^2 \right\}}, \quad (5.24)$$

where the mismatch error powers are represented by their amplitude and time variances. If the number N of parallel paths increases, it is the same SNR definition as in the single-path system. The only difference is that now even the DC static errors are modulated by f_s/N and its mutiples. For small N, it is slightly better by a factor of $N^2/(N-1)^2$ because the error powers shift out of the Nyquist band due to the subsampling. Of course, the spectral density will be shaped by the frequency response, but it is still reasonable to assume that the system bandwidth is much wider than the Nyquist band, and the signal is assumed to be band limited, meeting the Nyquist criteria.

Figure 5.54 explains three representative errors occurring in the time domain in a two-path example assuming that each channel has perfect DNL/INL, and its bandwidth is wide enough to ignore the bandwidth-related error. If one ADC sampling at ϕ_2 is assumed to be ideal, another ADC sampling at ϕ_1 converts samples with wrong offset, gain, and timing errors. The ideal sampling point is marked as darkened squares, but the actual sampling occurs at the darkened circles. The difference between the two points marked by the dot and square is the time-domain error that occurs at every sampling point ϕ_1.

The offset difference of the two ADCs becomes the peak-peak error, and it appears as a tone in the output spectrum at half the sampling frequency. Because the offset error is constant, it can be canceled by calibration, or notched out by filtering. However, the other two gain and timing errors are not constant but are heavily dependent on the signal. As shown, the gain error is proportional to the signal, and the error spectrum shows

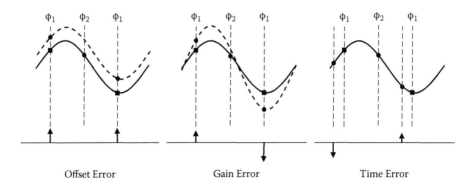

FIGURE 5.54
Three errors in two-path time interleaving.

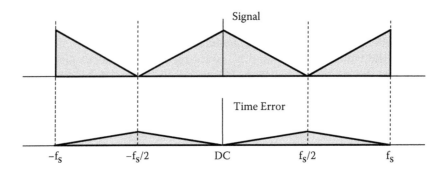

FIGURE 5.55
Two-path sample-time error spectrum.

up inside the Nyquist band after being modulated by half the sampling frequency. The constant average gains of the two ADCs can also be trimmed or calibrated to match each other. However, unless ADC transfer functions are perfectly linear and monotonic, even the static DNL and INL mismatches between them appear in the output spectrum as signal-dependent time-varying spurious tones, which in effect raises the quantization noise floor.

Last, the timing error in sampling is the most troublesome. The gain error is magnitude dependent, but the timing error is frequency dependent. It has the same effect as the sampling clock jitter and gets worse at high frequencies. The sample-time error spectrum is like a small signal leak, whose spectrum is frequency shifted by half the sampling frequency as shown in Figure 5.55. It is difficult to generate multiphase clocks with precise delays on the same order as the clock jitter, and inaccurate clocking raises the noise floor. Unlike the offset and gain errors, it is not simple to trim or calibrate the timing error. The bandwidth mismatches arise from the timing-constant mismatches of the RC sampling networks and create problems in very high-frequency sampling.

To avoid the sample-time error, one Nyquist-rate S/H can be shared by all time-interleaved stages as shown in Figure 5.56. Then, because one S/H is used, there exists no sample-time error related to the multipath operation. However, the speed and accuracy requirements of the S/H are prohibitively demanding as in the Nyquist-rate ADC, and the parallelism quickly loses some of its advantages because the S/H is the most critical element in all ADC designs.

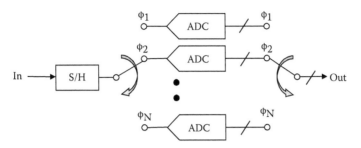

FIGURE 5.56
Time-interleaved ADC with Nyquist-rate S/H.

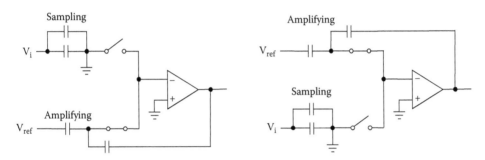

FIGURE 5.57
Opamp-sharing concept.

5.7.3 Opamp-Sharing ADC

The power consumed in wideband opamps often limits the power consumption of the pipelined ADC, especially when the stringent linearity and resolution requirements have to be met. Various approaches have been proposed in the literature with the aim to reduce the power consumption. Removing the input S/H by sampling the input directly on the first MDAC allows some power reduction as discussed. New ADC architectures can lead to power savings, but creative design solutions also contribute to the power reduction. Most low-power pipelined ADCs achieving good FOM save the power consumed in opamps by operating them dynamically.

Figure 5.57 shows the opamp-sharing concept to make two MDACs with one opamp [17]. In pipelined ADCs, the opamp is just reset during the sampling phase and is used with 50% duty for residue amplification during the other phase. The opamp sharing refers to the use of one opamp with a 100% duty cycle for two pipelined stages operating out of phase. This is done by disconnecting the summing node from the sampling capacitors during the sampling phase when the opamp is not used, and by attaching it to another stage, which is in the amplification phase. One opamp makes residue amplifiers for two different stages during two different clock phases. Therefore, the number of opamps is reduced by half, directly resulting in both power and area savings.

However, this violates the basic rule of the switched-capacitor circuit. Usually the capacitor top plates are switched to ground only for initialization. Because the top plates are switched back and forth, some performance degradation is expected. Switching them during the amplification phase increases the top-plate parasitics and allows more power supply coupling into the sensitive opamp summing nodes. The fundamental drawback of this method is that the summing node is never reset because the opamp is active in both phases. It also implies that the settling is slower because it does not start from the reset zero but from the previous output, and the input-referred offset is affected by the value of the previous residue and the $1/f$ noise.

This can be overcome by using a two-stage opamp in which the high-power second stage is shared by two pipelined stages, but the two first-stage preamplifiers powered on all the time provide two input summing nodes. Such a partial opamp sharing scheme requires the same number of preamplifiers as in a conventional design and has no significant power and area advantage. If the power of the preamplifiers is turned on and off, the summing node can still be reset when it is not in the signal path. However, it suffers from the problems associated with the turn-on delay when preamplifiers are turned back on. In addition, there is an area overhead for the additional preamplifier.

A simplified two-input preamplifier is shown in Figure 5.58, where each summing node can be reset during the sampling phase. The advantage is that two preamplifiers can be merged into a single bias branch, allowing the two amplifiers to share the same bias current resulting in both area and power savings. The two switches are turned on and off alternately using nonoverlapping clocks, ϕ_1 and ϕ_2. Note that this amplifier can be used as a single-stage opamp by itself. However, the switching transient problem still exists, and the input capacitance changes when the bias current is turned back on.

A modified version of this scheme is shown in Figure 5.59 [18]. A stacked P-channel metal-oxide semiconductor (PMOS) input differential pair is used instead of adding a parallel N-channel MOS (NMOS) input pair.

This arrangement allows the bias current to be reused, and the summing node can be reset with no additional turn-on delay. The switches are moved to the gate of the input devices, thus connecting the gate to the common-mode bias voltages V_{ipo}, V_{ipn} instead of switching the bias current. Connecting the inactive pair of transistors to the bias voltage allows these to act like an active load to the other.

Note that these switches disconnecting the opamp inputs are not necessary because the opamp input nodes are capacitive and can be initialized to any input common-mode bias level. Therefore, the summing node can be reset in every clock cycle because there

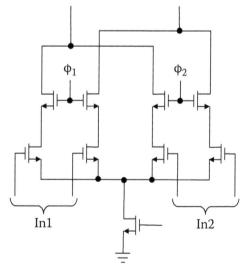

FIGURE 5.58
Switched preamplifier inputs.

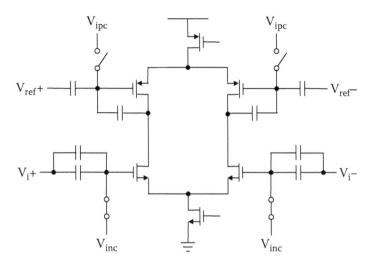

FIGURE 5.59
Opamp current reuse concept.

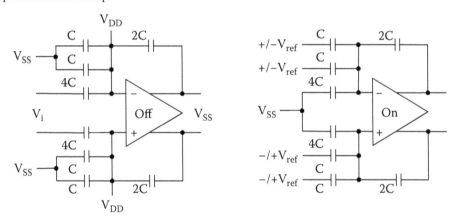

FIGURE 5.60
Switched-opamp trilevel MDAC.

are two separate input differential pairs that operate in two different clock phases. The input capacitance stays constant because all transistors are always operated in saturation. The disadvantage is that an additional tail current source has to be stacked in the branch, reducing the signal range slightly.

5.7.4 Dynamic Low-Voltage Low-Power Design

Another straightforward power-saving strategy is to turn off opamps completely during the clock phase when the opamps are inactive. One switched-opamp MDAC is shown in Figure 5.60 in two different phases for sampling and amplification as discussed in Chapter 4 [19].

The main reason to switch opamps is to operate ADC at low voltages as low as 1 V, because normal opamps are not operable. Because the signal swing is limited, there is no signal ground in the middle of the supply. The only available voltages are the high and low

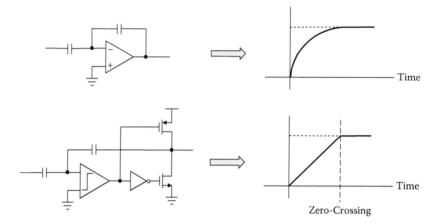

FIGURE 5.61
MDAC versus comparator-based MDAC.

supply voltages, V_{DD} and V_{SS}. As a result, to generate the trilevel MDAC output, eight unit capacitors are needed to amplify by the gain of 2, to set the input common-mode voltage, and to synthesize the middle level of the trilevel DAC.

This switched-opamp tactic also suffers from a turn-on delay when the opamp power is turned back on. There is a considerable time delay until the bias is stabilized, because the bias condition also settles with the same time constant as the signal does. Furthermore, the opamp input capacitance changes when the opamp is turned on and off. This makes the parasitic at the summing nodes affect the accuracy of the amplified residue.

Normal opamps need high supply and high current, and all power-saving tactics have been focused on how to reduce the opamp power. Implementing systems with open-loop amplifiers or comparators only without using closed-loop opamps can lower power consumption [20].

Figure 5.61 compares the settling behaviors of the regular opamp-based residue amplifier and the comparator-based one. The former settles exponentially as the opamp feedback reduces the error at the summing node, while the latter output is charged up with current source like slewing, and the current source is cut off instantly when the comparator detects the zero-crossing. For this to work, the comparator should have a well-defined offset, very high gain, and wide open-loop bandwidth. The differential sensing of the zero-crossings may reduce the aperture time error, and the open-loop system relies on the absolute comparator switching accuracy. The nanometer CMOS technology offers high switching accuracy with short gate delay, but the time-related error still gets worse for high-frequency inputs.

All low-voltage design compromises performance. The difficulty can be found in both S/H and amplifier designs. Using low-threshold devices simplifies most designs, and clock boosting properly done is the next best option. Switched-opamp or current-mode design sacrifices the performance both in accuracy and linearity, but may be useful for battery operation with medium to low performance.

5.7.5 Time-Domain ADC

In low-voltage scaled CMOS, the signal swing of the opamp is severely limited, and alternative ADC architectures without using opamps and even DACs have been revisited.

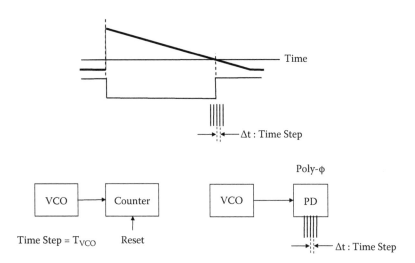

FIGURE 5.62
Three time-domain ADC concepts.

One way is to rely on the time resolution of the digital circuits. The obvious advantage of the time-domain ADC, which is also called the time-to-digital converter (TDC) in digital phase-locked loop (PLL), is to use mostly digital circuitry and require no large signal swing. As a result, it can offer a significantly low figure of merit (FOM), which has become a new performance standard. However, even the time-domain ADC requires that some functions such as voltage-to-current, voltage-to-time, and voltage-to-frequency converters should be linear. Such a trade-off between speed and performance can be made with oversampling.

In time-domain ADC, the unit time resolution step makes the quantization step, and the step mismatch limits linearity as shown in the three examples of Figure 5.62. Now noise appears in the form of jitter, frequency noise, or phase noise. The simple example is to use a counter as in the slope-type ADC discussed earlier, which is a serial ADC that makes decisions for all incremental time steps. Like the slope-type ADC, the frequency counter type relies entirely on the voltage-to-frequency converter, and the resolution improves as the ratio of the VCO and the counter frequencies increase. Either the ramp generator or VCO should be linear. The counter type can be monotonic, but the time to make all decisions limits speed. The polyphase phase detector (PD) can be used as a multilevel time quantizer similar to the flash ADC in the voltage domain. Then, the phase or time interval of the polyphase clock becomes the quantization step.

Figure 5.63 shows the basic concept of the polyphase time or phase quantization. The input reference phase is sliced with oversampling polyphase VCO outputs. The sliced bit outputs are the snapshot of the data showing where the zero-crossing is. The distance of the 0-1 or 1-0 transition from the reference center is the digital output. The same digital PD concept can be used for digital loop filtering in the digital PLL or clock and data recovery [21]. This digital polyphase PD is fast like flash, but it is difficult to generate multiple polyphases. Coarse accuracy in the decision can be enhanced by oversampling and feedback for high resolution as shown in Figure 5.64 [22].

VCO integrates its phase continuously, and the integrator can be used as a first-order loop filter for $\Delta\Sigma$ modulators. The VCO nonlinearity is suppressed using the loop filter $H(s)$. The ring oscillator provides polyphase clocks for the D flip-flop array that performs

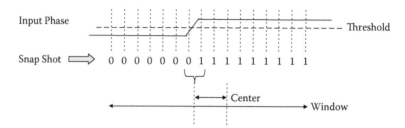

FIGURE 5.63
Flash-type polyphase time quantization.

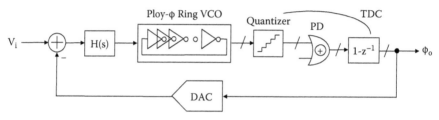

FIGURE 5.64
VCO-based ΔΣ ADC.

as a multibit quantizer. Due to the integration of the VCO phase, the operation of $(1 - z^{-1})$ is performed to get the phase corresponding to the input. The quantizer, PD, and the differentiator make a TDC as shown. This is similar to the CT ΔΣ modulator design. It needs a local feedback loop for stability. The DAC nonlinearity can be shuffled, but the loop filter nonlinearity ultimately limits the resolution. The VCO nonlinearity can also be calibrated to enhance the resolution using multilevel dithers [23]. Although the ring oscillator–type VCO is calibrated, the voltage-to-current converter still needs high voltage and good linearity. Quantizing in the time domain using low-voltage digital circuitry is attractive when using low-voltage CMOS.

5.8 Stand-Alone DACs

When DAC is used as a subblock of ADC, DAC need only to settle accurately within a given time interval. As a stand-alone device, its output transient behavior limited by slew, clock jitter, settling, glitch, and so forth, is of paramount importance. The multibit digital input word should be synchronously applied to the DAC with a precise timing accuracy, and the switching of the DAC elements should be timed accurately. Three most common DAC topologies are those using a resistor string, a capacitor array, and current sources. The current-steering DAC has ubiquitously been used as a high-speed DAC, while the resistor string and capacitor array are used mainly as ADC subblocks. Capacitor DACs need output buffers to operate.

5.8.1 Resistor-String DAC

The simplest voltage divider is a resistor string. Reference levels are generated with 2^N identical resistors connected in series between V_{ref} and 0. Switches connecting the divided

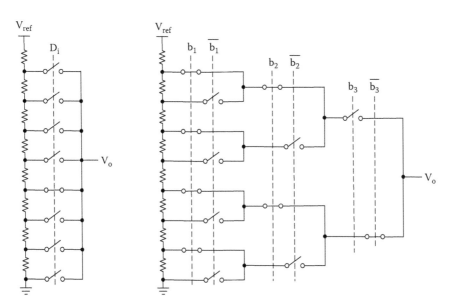

FIGURE 5.65
3b example of a resistor-string DAC.

reference voltages to the output can be either 1-out-of-2^N decoded or binary tree decoded as shown in Figure 5.65.

The resistor-string DAC is used widely as a reference divider, which is an integral part of the flash ADC. Resistor-string DACs are inherently monotonic and exhibit good DNL. However, they suffer from poor INL and have a critical drawback that the output impedance depends on the digital code. The latter causes a code-dependent settling time when charging the capacitive load of the output bus. This code-dependent settling has no effect on the reference divider performance when used as ADC subblocks, but would degrade the dynamic performance significantly when used as a stand-alone DAC. This nonuniform settling problem can be alleviated by adding low-resistance parallel resistors and by compensating the MOS switch overdrive voltages.

In BJT technology, the most common resistors are thin-film resistors made of tanatalum, Ni-Cr, or Cr-Sio, which exhibit very low voltage and temperature coefficients. In CMOS, diffusion, well, or undoped poly layers are used for resistors. The use of large geometry and careful layout are effective in improving the matching. Large geometry reduces the random edge effect, and layout using a common centroid or geometric averaging can reduce the process gradient effect. However, the resistor string layout is very difficult due to the contact resistances at the tap and termination points, and typical matching of resistors in integrated circuits is still limited to 8 to 10b level due to the mobility and resistor thickness variation. Differential-resistor DACs with large feature sizes are known to exhibit a high matching accuracy at the 12 to 14b level.

Figure 5.66 shows two ways to improve the INL of resistor-string DACs. It is possible to force the tap points to specified voltages by connecting low-impedance sources. The more taps are adjusted, the better the INL gets. An additional benefit of this method is the reduced RC time constant due to the voltage sources at the taps. Instead of using voltage sources, the required voltage can be obtained by low parallel resistance trimming or calibration. This electrical trimming is equivalent to other irreversible laser trimming, Zener

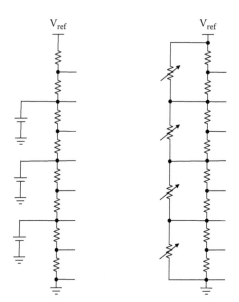

FIGURE 5.66
INL improvements for resistor-string DAC.

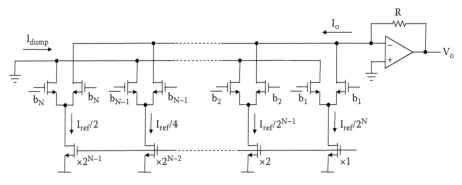

FIGURE 5.67
Binary-weighted current-steering DAC.

zapping, and trimming using programmable read-only memory (PROM). While being trimmed, the resistor string can be biased with a constant current, and the tap voltage can be sensed. Because electrical trimming is reversible, the long-term stability of the trimmed resistor is not a major concern.

5.8.2 Current-Steering DAC

The most popular stand-alone DACs commonly in use are current-steering DACs as shown in Figure 5.67. It is made of an array of switched binary-weighted current sources and the current summing network. In BJT, the binary weighting is achieved using transistors and emitter resistors. In CMOS, only transistors biased with large V_{gs} can be used. Current source matching is about at the 8b level with 0.2 ~ 0.5% matching achieved using 10 ~ 20 μ

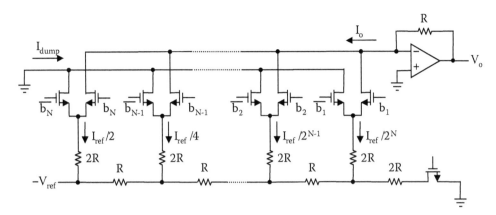

FIGURE 5.68
Binary-weighted R-2R DAC.

feature sizes. The output current can be summed using a wideband transresistance amplifier, but in most high-speed DAC applications, a simple resistor load converts the summed current into a voltage. Because the low load resistance is used for maximum speed, the DAC power increases, and the output swing is limited accordingly. Although the ratioed current DAC is simple and fast, it is difficult to implement a high-resolution DAC since the DAC element values are widely spread.

The same binary-weighted current sources can be generated using an R-2R ladder as shown in Figure 5.68. The large resistance ratio problem in the binary ratio can be alleviated this way. The R-2R array is made of series resistors of R and shunt resistors of 2R. The top of each shunt resistor 2R is switched using a single-pole double-throw switch, which connects the resistor either to ground or to the virtual ground of the current summing network. The operation of the R-2R ladder is based on the binary division of current as it flows down the ladder. At any junction of the series resistor R, the resistance looking to the right side is 2R. Therefore, the input resistance at the junction becomes R, and the current splits into two equal parts at the junction because it sees the equal resistances 2R in both current paths. The result is the binary-ratio currents flowing into each shunt resistor in the ladder. The advantage of the R-2R method is that only two resistance values are used, thereby greatly simplifying the task of matching or trimming and temperature tracking. In addition, relatively low resistance values can be used for high-speed applications.

5.8.3 Segmented DAC for Monotonicity

Although simple, it is difficult to make high-resolution DACs using either weighted-current DAC or R-2R DAC only. For large DAC arrays, controlling thermometer-coded DAC will be prohibitively complicated. However, the MSB of the DAC should be thermometer coded to warrant monotonicity. Applying the same multistep conversion concept as used in (R + C), (C + R), and (C + C) DAC examples, current steering DACs can also be implemented in multiple steps. The segmented two-step DAC approach as shown in Figure 5.69 can combine the features of both thermometer-coded and binary-ratio DACs. The LSB segment DAC can be made using a simple current divider. Binary-ratio DAC or R-2R DAC can be used as a current divider to make a sub-DAC. That is, the MSB M bits are selected by a thermometer code, but one of the MSB currents corresponding to the next segment of the thermometer

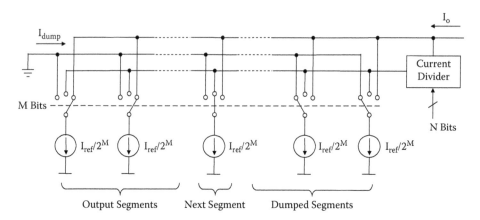

FIGURE 5.69
Next-segment multilevel DAC.

code is subdivided in the current divider to generate fine LSBs. To further reduce the number of DAC elements, the segmented DAC can be implemented using more steps.

If one fixed MSB segment is subdivided into LSBs, matching among MSB segments creates a nonmonotonicity problem. However, if the next segment is subdivided instead of the fixed segment, the segmented DAC can maintain monotonicity regardless of the MSB matching. The next-segment DAC performs the same function as the residue amplifier in the multistep subranging ADC. Unless the next-segment approach is used to make a segmented DAC with a total $(M + N)$ bits, the MSB DAC should have a resolution of $(M + N)$ bits, and the LSB DAC requires N-bit resolution. However, using the next-segment approach, making the MSB DAC with 2^M identical elements can guarantee monotonicity although INL is still limited by the MSB matching.

References

1. J. L. McCreary and P. R. Gray, "All-MOS charge redistribution analog-to-digital conversion techniques, I," *IEEE J. Solid-State Circuits*, vol. SC-10, pp. 371–379, December 1975.
2. B. Fotouhi and D. A. Hodges, "High-resolution A/D conversion in MOS/LSI," *IEEE J. Solid-State Circuits*, vol. SC-14, pp. 920–926, December 1979.
3. H. S. Lee, D. A. Hodges, and P. R. Gray, "A self-calibrating 15 bit CMOS A/D converter," *IEEE J. Solid-State Circuits*, vol. SC-19, pp. 813–819, December 1984.
4. K. B. Ohri and M. J. Callahan, "Integrated PCM Codec," *IEEE J. Solid-State Circuits*, vol. SC-14, pp. 38–46, February 1979.
5. A. G. F. Dingwall and V. Zazzu, "An 8-MHz CMOS subranging 8-bit A/D converter," *IEEE J. Solid-State Circuits*, vol. SC-20, pp. 1138–1143, December 1985.
6. B. S. Song, S. H. Lee, and M. F. Tompsett, "A 10b 15MHz CMOS recycling two-step A/D converter," *IEEE J. Solid-State Circuits*, vol. SC-25, pp. 1328–1338, December 1990.
7. H. Schmid, *Electronic Analog/Digital Conversion Techniques*, New York: Van Nostrand, pp. 323–325, 1970.
8. S. H. Lewis and P. R. Gray, "A pipelined 5-Msample/s 9-bit analog-to-digital converter," *IEEE J. Solid-State Circuits*, vol. SC-22, pp. 954–961, December 1987.

9. S. H. Lewis, H. S. Fetterman, G. F. Gross Jr., R. Ramachandran, and T. R. Viswanathan, "A 10-b 20-Msample/s analog-to-digital converter," *IEEE J. Solid-State Circuits*, vol. SC-27, pp. 351–358, March 1992.

10. P. C. Yu, S. Shehata, A. Joharapurkar, P. Chugh, A. R. Bugeja, X. Du, S. U. Kwak, Y. Papantonopoulous, and T. Kuyel, "A 14 b 40 MSample/s pipelined ADC with DFCA," *Dig. Tech. Papers, IEEE Int. Solid-State Circuits Conf.*, pp. 136–137, February 2001.

11. B. S. Song, M. F. Tompsett, and K. R. Lakshmikumar, "A 12-Bit 1-Msample/s capacitor error averaging pipelined A/D converter," *IEEE J. Solid-State Circuits*, vol. SC-23, pp. 1324–1333, December 1988.

12. I. Mehr and L. Singer, "A 55-mW, 10-bit, 40-Msample/s Nyquist-rate CMOS ADC," *IEEE J. Solid-State Circuits*, vol. SC-35, no. 3, pp. 318–325, March 2000.

13. R. J. van de Plassche and P. Baltus, "An 8-bit 100-MHz full-Nyquist analog-to-digital converter," *IEEE J. Solid-State Circuits*, vol. SC-23, pp. 1334–1344, December 1988.

14. P. Vorenkamp and R. Roovers, "A 12-b, 60-MSamples cascaded folding and interpolation ADC," *IEEE J. Solid-State Circuits*, vol. SC-32, pp. 1876–1886, December 1997.

15. M. J. Choe, B. S. Song, and K. Bacrania, "An 8b 100MHz pipelined folding CMOS ADC," *IEEE J. Solid-State Circuits*, vol. SC-36, pp. 184–194, December 1990.

16. C. S. G. Conroy, D. W. Cline, and P. R. Gray, "An 8-b 85-MS/s parallel pipeline A/D converter in 1-μm CMOS," *IEEE J. Solid-State Circuits*, vol. SC-28, pp. 447–454, April 1993.

17. K. Nagaraj, H. S. Fetterman, J. Andjar, S. H. Lewis, and R. G. Renninger, "A 250-mW, 8-b, 52-Msamples/s parallel-pipelined A/D converter with reduced number of amplifiers," *IEEE J. Solid-State Circuits*, vol. SC-32, pp. 312–320, March 1997.

18. S. T. Ryu, B. S. Song, and K. Bacrania, "A 10-bit 50-MS/s pipelined ADC with opamp current reuse," *IEEE J. Solid-State Circuits*, vol. SC-42, pp. 475–485, December 1990.

19. M. Waltari and K. Halonen, "1-V 9-bit pipelined switched-opamp ADC," *IEEE J. Solid-State Circuits*, vol. SC-36, pp. 129–134, January 2001.

20. L. Brooks and H. S. Lee, "A zero-crossing-based 8-bit 200MS/s pipelined ADC," *IEEE J. Solid-State Circuits*, vol. SC-42, pp. 2677–2687, December 2007.

21. B. Kim, D. N. Helman, and P. R. Gray, "A 30-MHz hybrid analog/digital clock recovery circuit in 2-μm CMOS," *IEEE J. Solid-State Circuits*, vol. SC-25, pp. 1385–1394, December 1990.

22. M. Park and M. Perrott, "A 0.13μm CMOS 78dB SNDR 87mW 20MHz BW CT ΔΣ ADC with VCO-based integrator and quantizer," *Dig. Tech. Papers, IEEE Int. Solid-State Circuits Conf.*, pp. 170–171, February 2009.

23. G. Taylor and I. Galton, "A mostly digital variable-rate continuous-time ADC ΔΣ modulator," *Dig. Tech. Papers, IEEE Int. Solid-State Circuits Conf.*, pp. 298–299, February 2010.

6

Oversampling Data Converters

The Nyquist-rate analog-to-digital converter (ADC) has no choice but to trade high power for low quantization noise. Because it operates in open loop, its performance is also heavily influenced by all nonideal circuit imperfections, in particular, given by inaccurate references, digital-to-analog converters (DACs), and comparators. When it is operated at oversampling rates, the in-band quantization noise is lowered by the oversampling ratio as the quantization noise is spread evenly over the Nyquist band. If it is placed inside a feedback loop, its quantization noise can be even further suppressed by the loop gain. This oversampling quantized feedback system offers a unique advantage over the Nyquist-rate system, and the system based on this noise-shaping property is called the $\Delta\Sigma$ modulator [1,2]. However, its pulse-density modulated output needs digital decimation and filtering, and the modulator together with the digital processor makes an oversampling ADC.

6.1 Concept of Quantizer in Feedback

An $\Delta\Sigma$ modulator is basically an active feedback filter made of a high-gain loop filter that performs either as a low-pass filter (LPF) or band-pass filter (BPF). With a nonlinear quantizer placed inside a loop after the high-gain filter, the quantization noise is reduced by the gain of the loop filter when it is referred to the input. Therefore, the $\Delta\Sigma$ modulator design boils down to the active filter design.

6.1.1 Active Filter by Feedback

Active filters can be made using integrators and resonators. Assume an active filter has an open-loop transfer function $H(s)$ with a high gain at direct current (DC) or at a resonant frequency, respectively, as conceptually sketched in Figure 6.1 without proper scales.

As shown, simply applying negative feedback around the high-gain filter, the same LPF or BPF is obtained but with a wider bandwidth, because the feedback gain is traded for bandwidth. The closed-loop gain is defined as the signal transfer function (STF):

$$STF = \frac{H(s)}{1+H(s)}. \tag{6.1}$$

However, because the quantizer is inserted after the filter, its transfer function makes either high-pass filter (HPF) or band-reject filter (BRF), and is defined as noise transfer function (NTF).

$$NTF = \frac{1}{1+H(s)}. \tag{6.2}$$

For the simple integrator of $H(s) = 1/s$, the STF and NTF are $1/(1+s)$ and $s/(1+s)$, respectively, as expected from Equations (6.1) and (6.2). This implies that the quantization noise is shaped by the NTF.

The ideal filters for noise shaping are, of course, the brick-wall filters with infinite gains as shown in Figure 6.2. However, in reality, they can be approximated using only integrators or resonators as shown on the right.

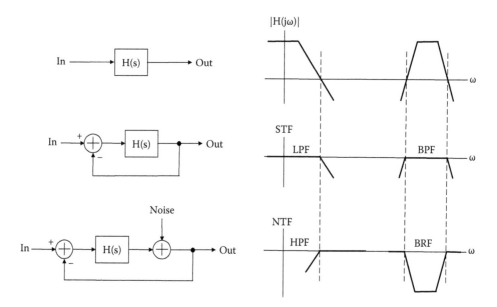

FIGURE 6.1
High-gain filter, feedback filter, and quantizer inside a loop.

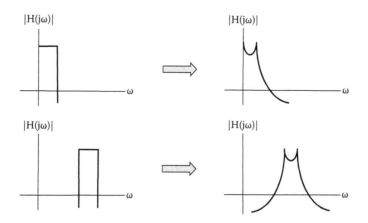

FIGURE 6.2
Ideal versus real low-pass filter (LPF) and band-pass filter (BPF).

6.1.2 Loop Stability

It is not possible to analyze the stability of the $\Delta\Sigma$ modulator accurately using a small-signal linear model due to the nonlinear quantizer inside the feedback loop. Only transients can be simulated. However, all nonlinear feedback systems such as $\Delta\Sigma$ modulator and phase-locked loop are well understood using linear small-signal analysis. The loop LPF and BPF are usually made of high-order filters with substantial delays. The loop stability is affected by all delay elements in the loop, including the quantizer's half clock delay. For stability, zeros are commonly inserted to warrant phase margin (PM) at the unity loop-gain frequency.

To achieve high gain like the brick-wall filter, poles can be placed at low frequencies as shown in Figure 6.3, where one integrator pole is placed at DC, and two resonator poles are placed at the signal band edge. The number of poles determines how steep the noise shaping is. The more poles there are, the steeper is the noise shaping. However, for example, these three poles make an almost 270° phase shift at the unity loop-gain frequency, which becomes the closed-loop bandwidth (BW). The three-pole feedback would be unstable. For the system to be stable with enough PM, at least two zeros should be inserted at frequencies lower than the unity loop-gain frequency. The STF is just a closed-loop LPF with a cut-off frequency of BW, and the feedback loop gain leads to the NTF, by which the quantization noise is suppressed.

6.1.3 Quantization Noise Shaping

The $\Delta\Sigma$ modulator along with spectral densities at various points is shown in Figure 6.4. The digital output spectrum repeats per sampling frequency and is made of the low-frequency input signal and the shaped high-frequency quantization noise. After the digital output is converted back into the analog domain using a DAC, the output is subtracted from the input. Then only the small residual signal and the quantization noise remain within the narrow signal band, but the shaped quantization noise stays high above the signal band. The function of the loop filter is to pass the signal band but attenuate the out-of-band quantization noise. The uniform quantization noise is added back again so that its low-frequency in-band noise can be suppressed by feedback.

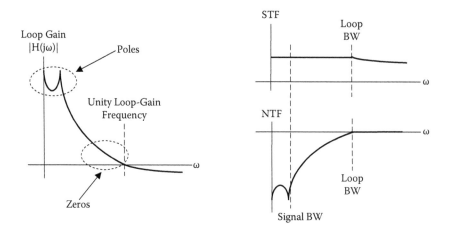

FIGURE 6.3
Loop gain, signal transfer function (STF), and noise transfer function (NTF).

Using a small-signal model, the following relation can be derived from the definitions of the STF in Equation (6.1) and the NTF in Equation (6.2):

$$Y(s) = \frac{H(s)}{1+H(s)} X(s) + \frac{1}{1+H(s)} Q(s) = STF \times X(s) + NTF \times Q(s). \tag{6.3}$$

Inside the signal band, if the loop filter gain is assumed to be very high, Equation (6.3) can be approximated as follows:

$$Y(s) \approx X(s) + \frac{Q(s)}{H(s)}. \tag{6.4}$$

This implies that the input is passing without attenuation, but the quantization noise is suppressed by the loop filter gain within the signal bandwidth. For example, if the loop filter is just an integrator, $H(s) = 1/s$, the noise shaping can be approximated just by s.

Figure 6.5 shows the quantization noise spectral density before and after being shaped by the NTF. The in-band quantization noise is uniformly spread over the Nyquist band with the following power spectral density:

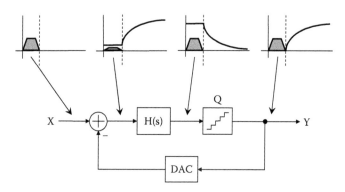

FIGURE 6.4
Spectrum shaped in the ΔΣ modulator.

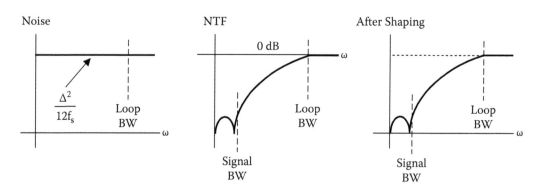

FIGURE 6.5
Shaped quantization noise spectral density.

$$\frac{q^2}{\Delta f} = \frac{\Delta^2}{12} \times \frac{1}{f_s},$$ (6.5)

where Δ is the quantization step, and f_s is the sampling rate of the quantizer. There are three effective ways to reduce the quantization noise inside the signal band. They are to raise the oversampling rate, to increase the order of the loop filter, and to use multiple bits in the quantizer. Two design issues remain to be resolved. One is the loop stability of the high-order modulator, and the other is the linearity of the feedback DAC.

6.1.4 Loop Filter and Bandwidth Requirements

The loop filter is the most critical element in the $\Delta\Sigma$ modulator design. The loop filter can be implemented using either active or passive filters as shown in Figure 6.6.

It is far easier to meet the high-gain requirement using active filters because high-gain filters are implemented using integrators and resonators with poles placed at DC and on the imaginary axis. Due to the high gain, the noise of the later stages is reduced by the filter gain when referred to the input. The drawbacks of active filters are nonlinearity and power consumption. On the other hand, passive filters are linear and consume no power, but the filter design is not flexible because they only make negative real poles and DC zeros unless inductors are used. Because they have no gain, the noise of the later stages is also directly referred to the input. Thus, passive filters are not useful for most practical modulator designs.

DC poles can be made cascading multiple integrator stages. The feedback loop with up to two integrator DC poles can be stabilized easily by inserting one feed-forward zero, but stabilizing high-order cascaded stages is challenging because it requires that more zeros be inserted and that integrator gains be scaled. If loop filters are made using only integrators, the loop filter gain at the signal band edge is far lower than the DC gain. Resonators can implement non-DC imaginary poles that keep the loop gain high over the signal band. However, because one resonator contributes two poles, the modulator can become unstable.

The integrator unity-gain bandwidth is another critical parameter in the $\Delta\Sigma$ modulator design. It is set by the unity-gain frequency of the integrator. Setting it too high may help to increase the in-band filter gain, but the system performance is sacrificed because the feedback system becomes more nonlinear and even unstable. Because the $\Delta\Sigma$ modulator is a discrete-time (DT) feedback system, both DT and continuous-time (CT) integrators can be used. In the switched-capacitor DT integrator, a certain amount of quantized feedback

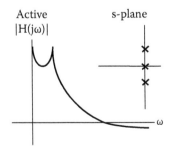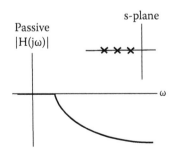

FIGURE 6.6
Active versus passive loop filter.

is periodically made at the clock interval, which is equivalent to the function of the CT resistance-capacitance (RC) integrator as shown in Figure 6.7.

The plain DT integrator has a z-domain transfer function of $z^{-1}/(1 - z^{-1})$, and its CT frequency response can be derived by substituting z by $\exp(j\omega T)$.

$$H(j\omega) = \frac{1}{e^{j\omega T} - 1} = \frac{1}{\left\{1 + j\omega T + \frac{(j\omega T)^2}{2!} + \frac{(j\omega T)^3}{3!} + \cdots\right\} - 1} \approx \frac{1}{j\omega T}, \tag{6.6}$$

where T is the sampling period, and the oversampling condition of $\omega T \ll 1$ is used for approximation. The NTF becomes just $(1 - z^{-1})$, which is the delay and subtraction equivalent to the differentiator transfer function of $j2\pi fT$ or $j2\pi f/f_s$ in the frequency domain. Therefore, the unity-gain frequency f_{unity} can be related to the sampling frequency in both continuous-time and discrete-time integrators.

$$f_{unity} = \frac{1}{2\pi RC} = \frac{f_s C_s}{2\pi C} = \frac{1}{2\pi T} = \frac{f_s}{2\pi}. \tag{6.7}$$

This is the optimum loop bandwidth, and all integrator unity-gain frequencies shown in Figure 6.8 should be scaled relative to Equation (6.7).

From Equation (6.7), because the integrator gain in the signal band is greater than 1, $f_{BW} < f_{unity}$, and the following oversampling condition should be met:

$$M = \frac{f_s}{2 f_{BW}} = \frac{\pi f_{unity}}{f_{BW}} > \pi, \tag{6.8}$$

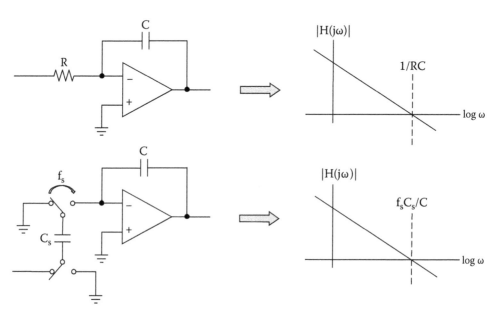

FIGURE 6.7
Continuous-time and discrete-time integrators.

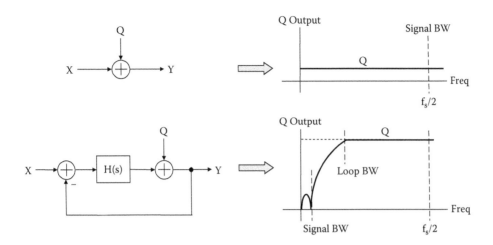

FIGURE 6.8
Quantization noise: Nyquist-rate versus oversampling quantizer.

where M is an oversampling ratio. If M is lower than π, the system can be more nonlinear and unstable. As a result, it is well known that empirically, M lower than about four does not offer any oversampling advantage over the Nyquist-rate sampling. The similar conclusion applies to PLL, and its loop bandwidth should be commonly set to be lower than $1/8$ of the clock or reference frequency.

6.2 $\Delta\Sigma$ Modulator

The quantization noise spectral densities of the pulse-code modulation (PCM) and $\Delta\Sigma$ modulation are compared in Figure 6.8.

The normal PCM coder (Nyquist-rate ADC) adds the quantization noise (Q) to the signal (X) to get the quantized (digitized) output Y, and the quantization noise is additive, and uniformly spread over the Nyquist band. On the other hand, the $\Delta\Sigma$ modulator modifies the quantization noise spectrum by the NTF. Therefore, the quantization noise can be reduced in the narrow signal band where the loop gain is high. There are two such types of coders (ADCs) that benefit from oversampling. One is the predictive coder called *delta modulation* that shapes the signal by quantizing the error. The drawback of the delta modulation is that the decoder needs an integrator to reconstruct the original signal. As the integrator is basically DC unstable, reconstructing the signal using an integrator is challenging. The other is the noise-shaping coder that shapes the noise instead and reconstructs the signal just using a DAC as shown in Figure 6.9. The signal subtraction at the input is Δ, and the high-gain LPF or integrator is Σ. Thus, this coding system is called $\Delta\Sigma$ modulation.

Because only the signal is quantized, the digital output is pulse-density modulated, often described as controlled oscillation. Because the output is digital, to complete the feedback loop a DAC is needed to convert the digital output back into the analog domain. That is, the DAC output should oscillate at high rates to contain the signal within the DAC range, and the quantization noise inside the signal band is reduced by feedback. The high-frequency noise in the pulse-density modulated output spectrum is removed using digital decimation and low-pass filtering.

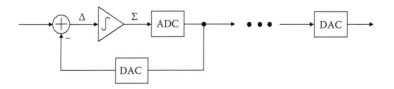

FIGURE 6.9
Noise-shaping coder and decoder chain.

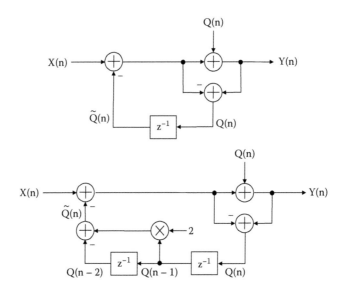

FIGURE 6.10
Error estimation in first- and second-order modulators.

6.2.1 Quantization Error Estimation

The $\Delta\Sigma$ modulator concept is similar to the improbable idea that ideal noise-free quantization is feasible if the quantization error is known in advance and subtracted before it occurs. However, because it is impossible to know what the error will be before it actually occurs, the next best choice is to estimate the quantization error based on the previous quantization errors that already occurred. This is based on an assumption that the quantization error changes little within a few sampling periods. It is partly true if the input is oversampled.

Figure 6.10 explains the error estimation concept for the first- and second-order $\Delta\Sigma$ modulators. Assume that $Q(n)$ is the quantization error that will occur at time n. The simplest estimate of $Q(n)$ is the quantization error $Q(n-1)$ that occurred in the previous time slot $(n-1)$:

$$\tilde{Q}(n) = Q(n-1). \tag{6.9}$$

That is, this first-order estimate of the quantization error can be subtracted from the input to reduce the quantization noise. The flaw of the first-order estimate is the assumption that the quantization error stays constant, and the difference between the actual error and the estimate remains. A better estimate can be made if the quantization error is assumed to vary linearly. It can be estimated by extrapolating the errors that occurred in the two previous consecutive samples.

$$\tilde{Q}(n) = Q(n-1) + \{Q(n-1) - Q(n-2)\} = 2Q(n-1) - Q(n-2). \tag{6.10}$$

The fact is that more elaborate higher-order estimates are more accurate.

6.2.2 Quantization Noise Shaping

Figure 6.10 shows that the subtraction of the estimated error is equivalent to the feedback loop integrating the error. Therefore, if the delay in the feedback is moved into the forward signal path, the error subtraction concept can be transformed into the first-, second-, and third-order $\Delta\Sigma$ modulators as shown in Figure 6.11.

The first-order modulator is stable because it has only one integrator inside the loop. The second-order modulator is basically a cascade of two integrators. Although two integrators are inside the loop, it is also stable because the feed-forward path with the gain of two inserts a zero at half the unity loop-gain frequency of the integrator. This way, even a fourth-order modulator can be made with the feedback coefficients chosen to be 4, 6, and 4, but modulators of third order or higher would be unstable unless the integrator gains are scaled down more aggressively.

The transfer function of the first-order modulator can be derived as follows:

$$Y(z) = z^{-1} X(z) + (1 - z^{-1})Q(z). \tag{6.11}$$

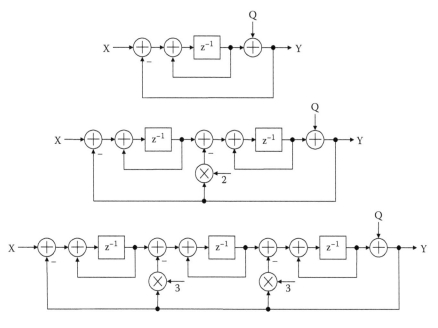

FIGURE 6.11
Models of first-, second-, and third-order $\Delta\Sigma$ modulators.

If $f_s \gg 2f_{BW}$ due to oversampling, $(1 - z^{-1})$ is approximated as $j2\pi(f/f_s)$ in the frequency domain. The in-band quantization noise after noise shaped can be obtained from Equation (6.5) as

$$In-band\ Noise = \int_{-f_{BW}}^{f_{BW}} \frac{\Delta^2}{12 f_s} \left(\frac{2\pi f}{f_s} \right)^2 df = \frac{\Delta^2}{12} \times \frac{\pi^2}{3} \left(\frac{1}{M} \right)^3, \tag{6.12}$$

where Δ is the quantization step. This implies that the quantization noise is suppressed by 9 dB per every 2× oversampling. Note that just oversampling reduces the in-band quantization noise by 3 dB, but an extra 6 dB suppression results from the noise shaping given by the integrator gain. Although it is the simplest, the first-order modulator suffers from its fixed pattern noise for DC inputs. For example, if the input X is a constant 0.25 for the normalized input range of 1, the output digital bit stream Y from the 1b first-order modulator will be 00010001000100 $\cdots$. This is also why it is called *pulse-density modulation*, and the average pulse density represents the input magnitude. That is, the density of 1 is about 25% of the total, and this digital bit stream has a strong tone due to the fixed periodic pattern of 0001. Higher-order modulators are preferred as they are free of the fixed pattern noise.

Similarly, the second-order modulator has a transfer function that shapes the quantization noise by the second-order shaping function of $(1 - z^{-1})^2$.

$$Y(z) = z^{-2} X(z) + (1 - z^{-1})^2 Q(z). \tag{6.13}$$

Under the same assumption as in the first-order modulator case, we get the following:

$$In-band\ Noise = \int_{-f_{BW}}^{f_{BW}} \frac{\Delta^2}{12 f_s} \left(\frac{2\pi f}{f_s} \right)^4 df = \frac{\Delta^2}{12} \times \frac{\pi^4}{5} \left(\frac{1}{M} \right)^5. \tag{6.14}$$

The in-band quantization noise is suppressed by 15 dB per every 2× oversampling, and the fixed pattern noise prominent in the first-order modulator is greatly reduced.

6.2.3 Signal-to-Quantization Noise

Signal-to-quantization noise ratio (SQNR) is a figure of merit to describe the performance of the $\Delta\Sigma$ modulator. Assuming that the signal range V_{ref} is divided into K quantization levels, SQNR can be shown as a function of the oversampling ratio M for different orders from Equations (6.12) and (6.14).

$$SQNR = \frac{\left(\dfrac{V_{ref}}{2\sqrt{2}} \right)^2}{\dfrac{1}{12} \left(\dfrac{V_{ref}}{K-1} \right)^2 \times \dfrac{\pi^{2L}}{(2L+1)} \left(\dfrac{1}{M} \right)^{2L+1}} = \frac{3}{2}(K-1)^2 \times \frac{(2L+1)}{\pi^{2L}} M^{2L+1}, \tag{6.15}$$

where L is the order of the modulator. Note that the first term in Equation (6.15) is the SNR of an ideal quantizer with K quantization levels, and the second term is an improvement factor by oversampling. The value of K is usually $(2^N - 1)$ for the N-bit quantizer because the additional top level is commonly used for the feedback in the $\Delta\Sigma$ modulator. Therefore, the quantization step size is reduced to $V_{ref}/(K - 1)$ for multilevel quantizers. Because SQNR increases at 9 dB and 15 dB per 2× oversampling, respectively, the effective number of bits (ENOB) can be estimated as

$$
\begin{aligned}
ENOB &= 1.5\log_2 M - 0.86\,(Bits) \quad \text{for } L = 1. \\
&= 2.5\log_2 M - 2.14\,(Bits) \quad \text{for } L = 2.
\end{aligned}
\tag{6.16}
$$

If a single-bit modulator is sampled at ×256 oversampling rate, the ENOBs are about 11 and 17.8b for first and second-order modulators, respectively. The second-order modulator suppresses the quantization noise far more sharply at low frequencies as expected. Figure 6.12 shows the SQNR versus M for different orders given by Equation (6.15).

6.2.4 Stability and Integrator Overload

The stability condition based on the linear small-signal analysis is not sufficient for nonlinear systems. High-order modulators can become unstable if integrator outputs are overloaded by large quantization errors. If integrators are overloaded or clipped, the feedback loop is momentarily disabled. Therefore, even in nonlinear systems, it is of paramount interest to

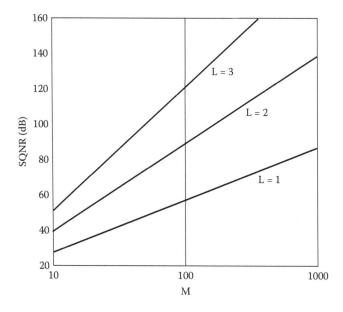

FIGURE 6.12
SQNR versus M for different orders.

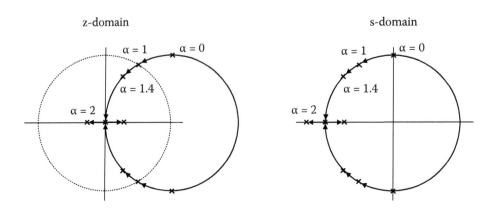

FIGURE 6.13
Second-order loop poles in *z* and *s* planes.

operate all integrators within their linear ranges so that any outputs of the integrators can be contained within the linear range of the integrators, and the averaged feedback error can decay exponentially. For any linear system to be stable, either pole should be placed in the open left-half complex *s*-plane or PM > 0 at the unity loop-gain frequency.

As shown in Figure 6.11, the second-order loop is stable due to the feed-forward zero, which is realized with the gain-of-2 path bypassing the first integrator. Without it, the loop gets unstable due to the two poles inside the loop. Assuming the feed-forward path gain is α, the complex conjugate poles can be found both in the *z* and *s* planes as shown in Figure 6.13.

If α > 1, the loop gets stable. If α = $2^{1/2}$, the poles are approximately at the 45° angle, and the pass-band becomes the maximally flat Butterworth response. If α > 2, the poles become negative real poles.

It is always easier to check PM for stability. Three loop gains for the first-, second-, and third-order modulators shown in Figure 6.11 are given as follows after the discrete-time frequency response of the integrator as approximated by 1/s, and the unity-gain frequency is normalized to 1 rad/s.

$$T(s) = \frac{1}{s}, \quad \frac{2s+1}{s^2}, \quad \frac{3s^2+3s+1}{s^3}. \tag{6.17}$$

Figure 6.14 shows the loop filter and its Bode plots of the second-order modulator. The loop is made of two integrators with one integrator bypassed when its gain falls below 2. Then from Equation (6.17), the PM can be simply estimated as $\tan^{-1}2 = 63.5°$; therefore, it is stable. However, the PM of the third-order modulator is negative due to three poles inside the loop, and it is unstable. In practice, there are more loop delays to consider other than the phase delay of the loop filter. Each DT integrator has one clock delay, and the comparator has also one clock delay. These delays can be made insignificant for DT ΔΣ modulators with high oversampling ratios. However, in broadband CT ΔΣ modulators, the oversampling ratio approaches a single digit, and even the half-clock delay in the comparator is becoming critical for loop stability.

Unless integrator outputs are heavily clipped, even this nonlinear feedback system behaves well like any other small-signal linear systems. However, once overloaded, the

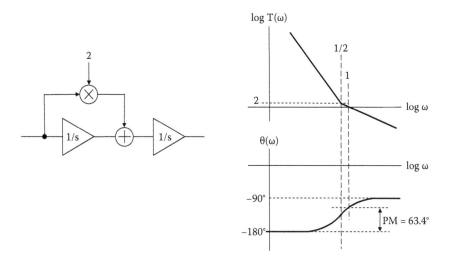

FIGURE 6.14
Loop gain and bode plots of the second-order modulator.

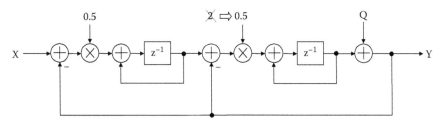

FIGURE 6.15
Second-order modulator scaled for dynamic range (DR).

quantizer output cannot correct large errors quickly by feedback. Therefore, any ΔΣ modulators become unstable regardless of their orders if the quantizer gain becomes lower than unity. For this reason, most ΔΣ modulators' performance degrades with the full-range input applied, and the performance of the ΔΣ modulator is always measured, typically with a −3 dB signal of the full range. To warrant reliable and stable operation, it is critical to reduce the integrator gain and to keep all integrators within their linear ranges. Multibit quantizers also help to avoid the integrator overload condition.

Figure 6.15 shows one example of integrator gain scaling to limit the integrator output swing and to make room for larger quantization errors [3]. The gain of 2 in the second-order modulator structure shown in Figure 6.11 is moved into each integrator path after scaling the first integrator gain by half. Then the gain in the second integrator should be 2. Even this gain of 2 can be scaled to 0.5 if a 1b quantizer is used, because the 1b quantizer only detects the polarity of the signal with an infinite gain. Whether the integrator gain is 2 or 0.5 does not matter for the 1b quantizer. Even with multibit quantizers, the integrator scaling helps to stabilize the loop.

6.3 High-Order Architectures

For most high-resolution applications such as digital audio, the dynamic range requirement is extremely high, and high-order modulators are more suitable to meet such a demand. There are various high-order modulator architectures that can be configured.

6.3.1 Direct Multiloop Feedback

Directly cascading multiple stages as shown in Figure 6.11 is the multiloop feedback approach. However, cascading more stages than two is not straightforward due to the stability concern. High-order modulators can be stabilized if the gains of integrators are scaled aggressively.

Figure 6.16 shows an example of the generalized multiloop feedback approach with the following STF and NTF:

$$Y(z) = \frac{z^{-4}}{\left(1-z^{-1}\right)^4} \times \frac{X(z)}{1+T(z)} + \frac{Q(z)}{1+T(z)}.$$

$$T(z) = \frac{Az^{-1}}{\left(1-z^{-1}\right)} + \frac{Bz^{-2}}{\left(1-z^{-1}\right)^2} + \frac{Cz^{-3}}{\left(1-z^{-1}\right)^3} + \frac{Dz^{-4}}{\left(1-z^{-1}\right)^4},$$

(6.18)

where $T(z)$ is the loop gain. The NTF is fourth-order, shaped with four DC poles, and the z-domain transfer function can be derived as follows: To get the fourth-order noise shaping and make the all-pass STF, A, B, C, and D can be set to 4, 6, 4, and 1, respectively.

The values of four coefficients A, B, C, and D affect the STF, and to ensure the stability, they should be scaled so that the poles of Equation (6.18) can be placed inside of the unit circle in the z-domain, or the loop gain should have a PM at its unity loop-gain frequency. For lower loop gain, the integrator gains should be scaled to be at least one order low. For more phase margin, earlier integrators should be bypassed well below their unity-gain frequencies, which is in effect equivalent to inserting zeros at very low frequencies. Without the low loop gain and feed-forward zeros, higher-order modulators would be unstable. Note that the feedback is from the digital output to multiple analog inputs. The major drawback of the multiloop feedback is that many feedback DACs are required to complete the feedback loop.

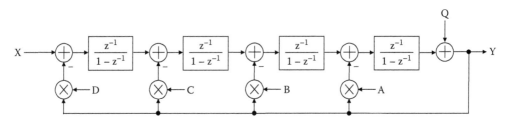

FIGURE 6.16
Direct feedback fourth-order modulator example.

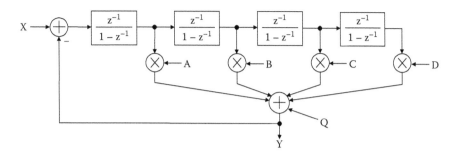

FIGURE 6.17
Single-loop high-order modulator.

6.3.2 Single-Loop Feedback

The single-loop feedback needs only one feedback DAC because the integrator outputs are weighted and summed for feedback as shown in Figure 6.17 [4].

The same STF and NTF as in the multiloop feedback can be derived as follows:

$$Y(z) = \frac{T(z)X(z)}{1+T(z)} + \frac{Q(z)}{1+T(z)}.$$

$$T(z) = \frac{Az^{-1}}{\left(1-z^{-1}\right)} + \frac{Bz^{-2}}{\left(1-z^{-1}\right)^2} + \frac{Cz^{-3}}{\left(1-z^{-1}\right)^3} + \frac{Dz^{-4}}{\left(1-z^{-1}\right)^4}. \tag{6.19}$$

Note that the loop gain $T(z)$ and NTF stay the same as in Equation (6.18), but in this case, $T(z)$ also affects the STF. Therefore, as in the multiloop feedback case, the modulator is also conditionally stable due to the cascaded integrators. Also all high-order loops are somewhat sensitive to the high-Q poles. Therefore, although unlikely, it is necessary to reset integrators when overloaded.

Zeros can also be inserted in the STF, but the choice of the coefficients is very limited only to shape the STF. All NTF zeros inside the signal band are at DC because all integrator poles make NTF zeros at DC. However, to suppress the noise at the pass-band edge, it is necessary to add complex conjugate poles on the imaginary axis. Resonators made of two integrators connected in negative feedback can make non-DC poles, which is translated into non-DC zeros after feedback. These non-DC NTF zeros achieve a high level of the noise suppression at the signal band edge. A fourth-order CT modulator model with two DC poles and one resonator is shown in Figure 6.18.

A resonator is formed with two integrators connected inside the local loop. Then the modulator has the following transfer function in the s-domain [4]:

$$Y(s) = \frac{\left\{a_1s^3 + a_2s^2 + \left(a_3 + a_1b_4\right)s + \left(a_4 + a_2b_4\right)\right\}X(s) + s^2\left(s^2 + b_4\right)Q(s)}{s^4 + a_1s^3 + \left(a_2 + b_4\right)s^2 + \left(a_3 + a_1b_4\right)s + \left(a_4 + a_2b_4\right)}. \tag{6.20}$$

The NTF in Equation (6.20) has two zeros at DC and two complex-conjugate zeros on the imaginary axis at $\pm j(b_4)^{1/2}$, which can be placed inside the signal band as in the simulated example shown in Figure 6.19.

Note that two non-DC complex zeros notch out the noise at the band edge to evenly distribute the quantization noise in the signal band. With mismatched-shaped multibit DACs, this has been the dominant oversampling $\Delta\Sigma$ modulator design for digital audio.

Figure 6.20 shows a way to avoid the overloading of the integrators by feed-forwarding the signal from the input to the quantizer input directly [5]. This relieves the integrator

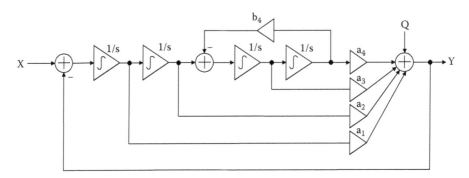

FIGURE 6.18
A single-loop fourth-order modulator with resonator inside.

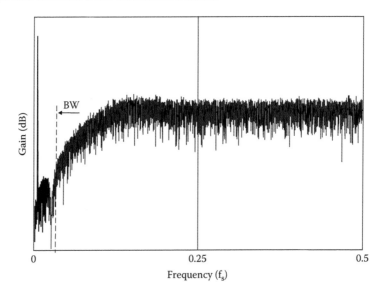

FIGURE 6.19
Noise shaping with a resonator in the loop.

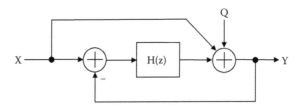

FIGURE 6.20
Signal feed-forwarding architecture.

linearity and dynamic range requirements as the signal is routed to the quantizer input through a separate path. As a result, the first-stage integrator output swing is greatly reduced because integrators filter only the feedback error.

6.3.3 Cascaded Modulators

An alternative way to ensure stability with high-order noise shaping is to cascade stable low-order modulators [6]. The cascaded modulator based on the multistage noise shaping (MASH) structure is similar in concept to the pipelined Nyquist-rate ADC. Each stage resolves a few bits and feeds a residue to the next stage.

Figure 6.21 shows that two second-order modulators can be cascaded to make a fourth-order modulator. The quantization noise Q_1, which is the residue of the first stage, is quantized again by the later stage. For each stage, the following is true:

$$Y_1 = z^{-2}X + \left(1 - z^{-1}\right)^2 Q_1.$$

$$Y_2 = z^{-2}Q_1 + \left(1 - z^{-1}\right)^2 Q_2. \tag{6.21}$$

After multiplying Y_1 by z^{-2} and Y_2 by $(1 - z^{-1})^2$, Q_1 can be removed, and Y is related to X and Q_2:

$$Y = z^{-2}Y_1 - \left(1 - z^{-1}\right)^2 Y_2 = z^{-4}X + \left(1 - z^{-1}\right)^4 Q_2. \tag{6.22}$$

That is, the first-stage quantization noise does not appear in the output, and the fourth-order shaping of Q_2 is achieved. As in the pipelined ADC, the later stage resolution can be lower because the first stage made coarse decisions. Because two stages are independent low-order stages, the stability is guaranteed. However, if the first-stage integrator is not ideal, the Q_1 in Y_1 is different from the Q_1 in Y_2, and the mismatched quantization noise will leak to the output. It is similar to the reference mismatch in the pipelined stages. For

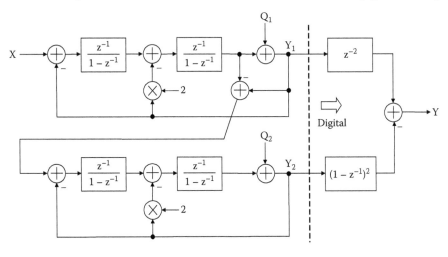

FIGURE 6.21
Cascaded high-order modulator.

this cascading to be meaningful, this two-path mismatch error should be smaller than the resolution of the second stage.

6.4 Discrete-Time (DT) versus Continuous-Time (CT) Modulators

Active loop filters can be implemented in the DT or CT domain. The CT modulator design is quite different from that of the DT version and so is the system performance. In general, the problems of the CT modulator become the very advantages of the DT modulator, and vice versa. Note that the location of the S/H in the dotted circle in both DT and CT modulators is as shown in Figure 6.22.

In the CT modulator, the S/H can be placed just before the quantizer, and it should be at the input of the filter in the DT modulator. As a result, compared to the DT modulator, the CT modulator exhibits two main advantages. One is the anti-aliasing requirement, and the other is the sampling clock jitter. The DT modulator needs to band-limit the input with a continuous-time anti-aliasing filter. This additional filtering requirement becomes very stringent if the oversampling ratio is low. On the other hand, the CT loop filter performs the anti-aliasing filter function. The clock jitter in the DT modulator should be very low because the input should be sampled with high accuracy. However, the sampling jitter is not much of an issue in the CT modulator because the sampling is just for the low-resolution quantizer after the loop filter. However, the CT modulator suffers from other accuracy problems.

Very accurate filter coefficients can be implemented using DT switched-capacitor filters. However, filter coefficients of CT RC or G_m-C filters vary widely due to RC and C/G_m variations. CT loop filters need coefficients tuning or calibration. Furthermore, in the DT modulator, the shape of the DAC output is not important as long as the filter output settles during the half clock period. However, in the CT modulator, the DAC pulse shape affects the loop filter transfer function, and incorrect DAC pulse design can lead to intersymbol interference (ISI) and result in nonlinear settling of the feedback DAC. The clock jitter also affects the feedback DAC pulse timing. Last, in the DT modulator, the comparator meta-stability

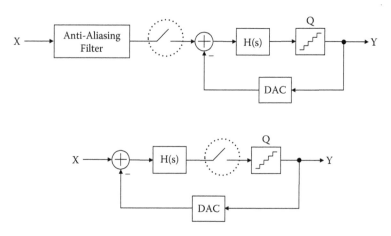

FIGURE 6.22
DT versus CT modulators.

Discrete Time (DT)	Continuous Time (CT)
Switched-capacitor filter	Continuous-time filter
Accurate coefficients	Inaccurate coefficients
Speed limited by settling	Speed limited by comparator
Sampled signal value is important	CT signal waveform is important
Jitter affects the input sampling	Jitter affects the DAC output
Need anti-aliasing filter	Built-in anti-alias filtering

FIGURE 6.23
Comparison of DT and CT modulators.

or hysteresis is not a problem, while in the CT modulator, it adds the signal-dependant loop delay that degrades the performance. The comparator meta-stability can cause the nonuniform quantizer delay.

The features of the DT and CT modulators are summarized in Figure 6.23. DT modulators are mainly used for premium high-resolution quantizers at low sampling rates, such as in digital voice-band and audio signal processing. CT modulators find many applications in medium-resolution baseband or intermediate-frequency (IF) quantizers at high sampling rates such as in wireless receivers. In particular, the lack of anti-aliasing requirement is a great feature in the radio frequency (RF) baseband quantizer design. It can lead to substantial chip area and power savings at the system level of digital receivers that need to operate with strong blockers.

Because sampling occurs at the quantizer input, large sampling errors are allowed, and any sampling-related nonideal errors such as jitter, aperture, and nonlinearity are reduced by the loop filter gain when referred to the input. Furthermore, because filters are implemented in continuous time, the bandwidth requirement of operational amplifiers (opamps) is greatly reduced, and no high-voltage switches are necessary, either. As a result, the CT modulator works at lower supply voltages and consumes lower power than the DT version. Because the speed of the CT modulator is limited by the speed of the comparator, it can also achieve high sampling rates with low oversampling ratios for any given process technology, compared to the DT modulator.

6.5 Discrete-Time Modulator Design

Due to the nature of the quantized feedback, $\Delta\Sigma$ modulators are easier to implement in DT than in CT. The CT modulator suffers from the quantization noise feedback error resulting from the modulation by the clock jitter, but the DT modulator is insensitive to the clock jitter as the signal is already in the sampled-data domain. Therefore, the DT modulator design mainly focuses only on improving the linearity performance of the first-stage integrator and the feedback DAC.

6.5.1 Switched-Capacitor Integrators

The first-order switched-capacitor modulator implemented differentially with a 1b quantizer and a DAC is shown in Figure 6.24. This is the simplest switched-capacitor implementation using the nonoverlapping two-phase bottom-plate sampling principle.

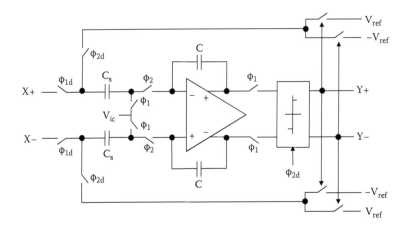

FIGURE 6.24
First-order switched-capacitor $\Delta\Sigma$ modulator.

Both the integrator and the 1b comparator sample their inputs at ϕ_1. During ϕ_2, the integrator integrates, and the comparator latches to produce either digital 1 or 0. These digital bits decide whether V_{ref} is subtracted from or added to the input. From Equation (6.7), the unity-gain frequency of the integrator should be set to f_s (rad/sec) or $f_s/2\pi$ (Hz). To make the same integrator using a switched-capacitor integrator, the C_s and C values can be chosen so that the following relation can be met:

$$\frac{f_s C_s}{2\pi C} = \frac{f_s}{2\pi}. \tag{6.23}$$

Therefore, nominally, $C_s = C$ unless the integrator gain is intentionally scaled to be lower for stability reason.

As long as the amplifier finishes settling within the required accuracy and the quantizer makes decisions within the half clock period, there are no major issues in this 1b system. Because the opamp is inside the feedback loop, the finite DC gain of the opamp is not critical, and the DC output transfer function of the opamp is linearized by the loop gain. However, the opamp should still be linear enough to meet the linearity requirement at the integrator output.

In most high-order modulators, the quantization noise can be easily shaped to be far lower than the required, but the system noise performance is often limited by either the opamp thermal noise or the kT/C noise of the input stage. The former is more or less a trade-off issue between power and dynamic range (DR), but the latter is fundamental due to the power and supply voltage constraints. Assuming the kT/C noise is dominant, the DR of the first-order differential modulator can be defined as

$$DR = \frac{\left(\dfrac{V_{ref}}{\sqrt{2}}\right)^2}{\dfrac{2kT}{C_s} \times \dfrac{1}{M}}. \tag{6.24}$$

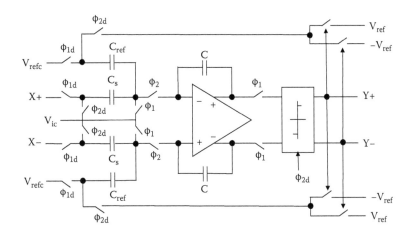

FIGURE 6.25
Separation of input and reference paths.

As expected, the oversampling ratio M helps to improve the DR of the system because the kT/C noise is spread over the frequency range M-times wider than the signal bandwidth. The advantage of oversampling is that the sampling capacitor looks M times larger in effect than in the Nyquist-rate sampling case.

To achieve extremely high resolution above 16b, there are additional issues to consider. One is the input-dependent loading to V_{ref}, and the other is the input common-mode change. To keep the reference loading constant, the input and reference paths can be separated as shown in Figure 6.25.

Because the common-mode voltages for the reference V_{refc} and input V_{ic} are separated, the reference feedback is independent of the input common mode. In some implementations, the references are not injected through the bottom plates as shown but through the top plates. It is because the reference is constant, and the charge injection and feed-through effect can be assumed to also be constant. However, the input common-mode change is still there and should be rejected by the common-mode rejection ratio (CMRR) of the differential integrator.

One disadvantage in this arrangement is that the additional kT/C noise contributed by the reference sampling capacitor C_{ref} and the signal range are scaled by the ratio of two capacitors. Therefore, DR degrades as

$$DR = \frac{\left(\dfrac{V_{ref}}{\sqrt{2}} \times \dfrac{C_{ref}}{C_s}\right)^2}{\dfrac{2kT}{\left(C_s + C_{ref}\right)} \times \dfrac{1}{M}}. \tag{6.25}$$

The SQNR given by the quantization noise can be arbitrarily lowered by properly choosing design parameters such as the order of the modulator, the number of bits, and the oversampling ratio. However, the DR given by the kT/C and opamp thermal noise can be traded for only with the power consumption if the technology is given.

Figure 6.26 shows the straightforward implementation of the second-order modulator shown in Figure 6.15 with integrator gains scaled by half [3]. Note that the gain of half is

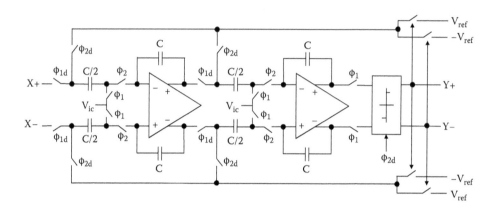

FIGURE 6.26
Second-order modulator with gain-scaled integrators.

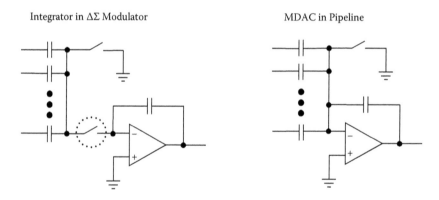

FIGURE 6.27
Multibit integrator versus multiplying digital-to-analog converter (MDAC).

obtained by the capacitor ratio. However, due to the reduction of the loop gain, the SQNR is reduced by 12dB compared to the one that is not scaled.

6.5.2 Multibit Integrator versus Multiplying Digital-to-Analog Converter

As shown in Equation (6.15), the quantization noise is inversely proportional to the number of quantizer levels and the order of the modulator. Therefore, using the multibit integrator and feedback DAC helps to implement high-resolution modulators.

Figure 6.27 compares the multibit integrator in switched-capacitor $\Delta\Sigma$ modulators and the capacitor-array multiplying digital-to-analog converter (MDAC) in pipelined ADCs. They are basically identical except for the switch inside the dotted circle. In the MDAC, all top plates connected to the summing node are reset in one clock phase so that the residue can be amplified in the next clock phase. However, in the integrator, the top plates are never reset or initialized. They are disconnected from the summing node for sampling in one clock phase and connected to the summing node in the next clock phase so that the charges on the capacitor array can be integrated onto the feedback capacitor. If the feedback DAC has K levels, for example, the top plates of all K unit capacitors are switched into

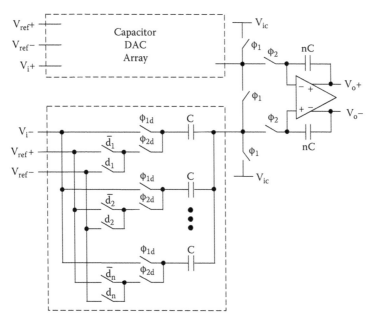

FIGURE 6.28
Multibit differential DAC and integrator.

the summing node. Therefore, the multibit differential MDAC with a thermometer-coded input $(d_n d_{n-1} \dots d_2 d_1)$ can be modified to make a multibit integrator as shown in Figure 6.28, where only the negative-side capacitor array is shown for simplicity.

The top-plate switching inevitably degrades the performance of switched-capacitor circuits due to the increased parasitic and the noise coupling into the summing node. However, due to the high oversampling and feedback, the performance degradation by coupling is M-times smaller and minimal within the narrow signal bandwidth. The integrator is inside the feedback loop, and the opamp gain and linearity requirements for the integrator are greatly relieved in the $\Delta\Sigma$ modulator. Note that the opamp in the MDAC of the pipelined ADC requires very high gain and linearity due to the open-loop operation. As a result, even linear settling is allowed in the integrator, while absolute settling accuracy is required in the MDAC. Even large offset is tolerable in the integrator due to feedback, but in the MDAC, the offset should be far smaller than the redundant digital correction range.

6.5.3 Multilevel Feedback Digital-to-Analog Converters

In the DT modulator, because nonidealities of the later stages are noise shaped, only the input sampling limits both noise and linearity performance. The quantization error is reduced by the loop filter gain in the signal band. Therefore, the first-stage integrator and the feedback DAC are the most critical elements in the $\Delta\Sigma$ modulator design. The 1b two-level DAC is an ideal DAC with no mismatch error, as is the trilevel 1.5b differential DAC. However, to reduce the quantization step further, 3 or 4b quantizers are commonly used in the modulator design. Although SQNR can be improved using multibit quantizers, the feedback DAC accuracy is limited by the DAC mismatch error. In switched-capacitor implementations, the DAC is combined with the integrator. If the integrator settles with the required accuracy, the DAC error results mostly from the static DAC mismatch error. There are diverse DAC linearization techniques called dynamic element matching that improve the linearity of the feedback DAC. They are methods such as shuffling, randomization, or mismatch shaping.

Figure 6.29 shows the most straightforward shuffling algorithm [7]. The thermometer-coded eight-level DAC, for example, can be shuffled using a 3b digital number $(b_2b_1b_0)$. By choosing the DAC elements as shuffled, each DAC element has an equal probability to be selected to contribute to any of the eight DAC output levels. As a result, the mismatch errors of the DAC elements are averaged over time. This is in effect to modulate and evenly spread the DAC nonlinearity error power over the oversampling bandwidth. The actual DAC nonlinearity is therefore reduced by the oversampling ratio inside the narrow signal band. Although the DAC nonlinearity error spectrum is whitened at an oversampling rate, periodic fixed tones arise in this simple shuffling.

A simple fix for the fixed tone is to select DAC elements using data sequentially as shown in Figure 6.30 [8]. With this data-weighted averaging (DWA), the mismatch error is first-order shaped, and fixed tones can be suppressed. However, the mismatch shaping still depends on the signal. Bidirectional DWA that switches the direction of rotation at every sample point can be applied to further reduce fixed tones. In oversampling systems, just pushing the nonlinearity error spectrum out of band is very effective, but employing more elaborate pseudo-random (PN) sequences in the shuffling or mismatch shaping can lead to a way to correlate out and even subtract the residual in-band shaped noise digitally.

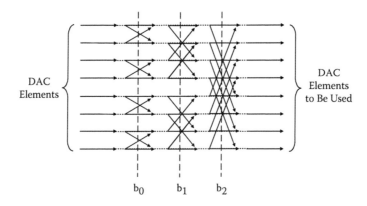

FIGURE 6.29
Multibit digital-to-analog converter (DAC) linearization by shuffling.

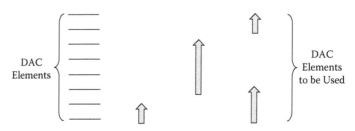

Data Sequence ...., $D(n-1) = 2$, $D(n) = 4$, $D(n+1) = 5$, ...

FIGURE 6.30
DAC linearization by data-weighted averaging.

6.5.4 Design Considerations

Like the pipelined ADC, the accuracy requirements of later modulator stages are less demanding. By cascading modulators, very high-order noise shaping is possible using low-order or even 1 or 1.5b modulators as long as good gain matching between the analog and digital cancellation filters is achieved. Low-order modulators can take large input signals without overloading integrators, but the quantization noise level is high. In most modulator designs, because NTF is designed to reject in-band quantization noise that is not low enough, the noise performance is often limited by conventional noise sources such as kT/C, $1/f$, opamp thermal noise, and jitter. However, due to oversampling, all random white noises except for the $1/f$ noise are reduced by the oversampling ratio. The $1/f$ noise affects only the low end of the spectrum. Large P-channel metal-oxide semiconductor (PMOS) input devices are known to exhibit lower $1/f$ noise than N-channel MOS (NMOS) devices. Offset cancellation techniques such as correlated double sampling and chopper stabilization are also effective to suppress the $1/f$ noise. The band-limited opamp thermal noise is about the same order as the kT/C noise but is reduced by the feedback factor. Because the dominant pole is commonly set by g_m/C_s or g_m/C_c, where C_s and C_c are the sampling and compensation capacitors, respectively, they should be sized large accordingly.

Other than noise, active devices contribute to nonlinearity. Finite opamp gain is not important in the feedback system, but its nonlinearity matters. The linearity of the $\Delta\Sigma$ modulator is also limited by the nonlinearity of the switch on-resistance and the charge injection and feed-through error of the input sampling network as in the Nyquist-rate S/H. Figure 6.31 shows an example of the input sampling network for low-frequency applications. The principle of high-linearity sampling is the same as in the Nyquist input sampling. It is the bottom-plate sampling with top-plate switches turned off early as discussed in Chapter 4. To reduce the kickback effect when sampling, the large input bypass cap is added. For high linearity, the input switch resistance should be made low and linear. To improve linearity, even a series degeneration resistor R_s is added to further reduce the signal across the switch. The input source-side impedance should be low, and the bottom-plate switch on-resistance should be sized far smaller than the top-plate switch on-resistance. In addition, it is important to properly ratio NMOS and PMOS for linearity. Too large of a switch increases nonlinear S/D junction capacitance that also limits the sampling linearity. Using low-threshold devices and boosting switch overdrive voltages helps to reduce the on-resistance and to linearize the switch on-resistance. Capacitors also contribute to the nonlinearity. Metal-to-metal capacitors exhibit a second-order voltage coefficient of about 10 ppm/V². However, in most low-frequency differential modulator designs, the

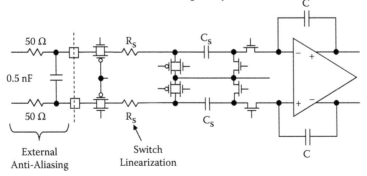

FIGURE 6.31
Input sampling for high linearity.

capacitor nonlinearity effect is almost negligible because typical third-order distortion can be measured to be below 110 dB with several Volt-level signals.

For DC input, it is well known that low-order modulators exhibit discrete fixed idle tones depending on the DC level and modulator topology. Tones are strong near $f_s/2$, and folded tones also show up after being modulated by $f_s/2$. Random dithering effectively removes the fixed idle tones by whitening the dither spectrum, but large dithers in the 1b case can degrade SNR and even cause overload. However, if dither is modulated using PN sequence, it can be recovered by the same PN correlation and subtracted digitally. Fixed idle tones would get far smaller if multibit quantizers or higher-order modulators are used. Any intentional DC offset also helps to move them out of the signal band. Integrator settling time has the same effect as the finite gain, and the unity loop-gain frequency should be wide enough for the integrator to settle accurately. For 10b settling, the bandwidth should be wider than $(6.9/\pi)f_s$. However, because limited slew rate causes nonlinearity in settling, it should be set far higher than $2f_sV_{ref}$. The thermal noise of the V_{ref} source is band limited, but its $1/f$ noise is dominant. The low-frequency V_{ref} noise is less significant than the quantization noise. However, the finite opamp gain affects the cascaded high-order modulator performance severely with data-dependent V_{ref} coupling. Higher V_{ref} provides higher SQNR, but the nonlinearity and the integrator overload issues set the upper bound of the V_{ref} range.

6.5.5 Broadband Modulators

Due to the difficulty in suppressing the in-band quantization noise with practical low oversampling ratios, higher-order loops with multibit quantizers and DACs are common in high-performance broadband modulator designs based on high-order direct multiloop feedback, single-loop feedback, and cascading approaches. Also for high DAC linearity, either dynamic element matching or calibration with PN correlation method can be used. However, most high-order modulators are easily overloaded with large signals, and the feedback becomes nonlinear. As a result, typically, a few dB of DR is lost, and the loop stability is more of a serious concern.

The current design trend is toward high speed (Nyquist bandwidth > 5 MHz) and high resolution (SNR > 90 dB, total harmonic distortion (THD) > 100 dB, and spurious-free dynamic range (SFDR) > 110 dB). Although it is possible at a high oversampling ratio of 128 in the audio band, it is difficult to achieve such high noise suppression at an oversampling ratio as low as 8 with high-order feedback alone. The multibit cascaded MASH architecture has been widely used for broadband modulators operating with very low oversampling ratios [9]. As expected, high-performance modulators commonly suffer more from idle tones in the low-level performance, and are limited by kT/C, $1/f$, and V_{ref} noises. Dithering helps to eliminate low-level idle tones, and environmental noise coupling through substrate and power lines should be tightly controlled. The first-stage integrator and feedback DAC are the most critical elements in the modulator design, and the quantizaion noise leakage still plagues the cascaded modulator designs.

The same switched-capacitor integrator as in Figure 6.24 is shown again with a finite opamp DC gain of a_o in Figure 6.32. The transfer function is derived with the parameters as

$$H(z) \approx \frac{C_s}{C}\left(1 - \alpha - \beta \times \frac{z^{-1}}{1 - z^{-1}}\right)\frac{z^{-1}}{1 - z^{-1}}, \quad \alpha = \frac{C_s + C}{a_oC}, \quad \beta = \frac{C_s}{a_oC}. \tag{6.26}$$

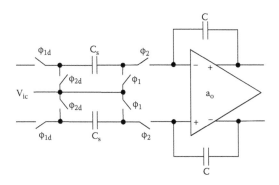

FIGURE 6.32
Switched-capacitor integrator with finite opamp DC gain.

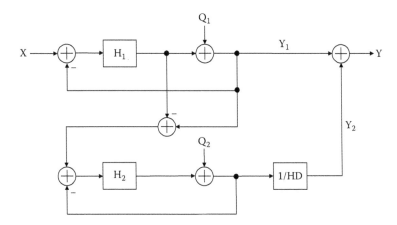

FIGURE 6.33
Noise cancellation in cascaded modulators.

In practice, the gain error α results from the finite opamp DC gain, capacitor mismatch, and imperfect settling error, and the finite gain contributes to the pole error β. The integrator becomes a leaky integrator due to incomplete charge transfer. That is, the integrator pole is not on DC, and the NTF is slightly modified as a result. Note that the pole error is integrated twice while the gain error is integrated once. Therefore, if first-order stages are cascaded to make a second-order modulator, the first-stage gain error leaks to the output after second-order noise shaping of $(1 - z^{-1})^2$, but its pole error leaks to the output only after first-order noise shaping of $(1 - z^{-1})$.

Figure 6.33 shows the noise cancellation mechanism in cascaded modulators. If there is a gain mismatch between filter stages, the first-stage quantization noise Q_1 is not canceled completely and leaks to the output, although the second-stage noise is higher-order shaped.

$$\frac{1}{1+H_1} \neq \frac{H_2}{1+H_2} \times \frac{1}{H_D}. \tag{6.27}$$

The matching between the analog filter H_1 and the digital filter H_D matters most if the analog filter gain is high ($H_2 \gg 1$). In switched-capacitor implementations, capacitors are very well matched, and only high opamp DC gain warrants proper noise cancellation. However, this filter gain matching is still critical in the CT cascaded modulator design to be discussed later.

Robust high-order modulator designs prevent integrators from being overloaded. The most straightforward way is the scaling for attenuation as shown in Figure 6.34. It degrades the system noise performance as the modulator noise is amplified by the same gain factor G, but it is required for single-loop high-order modulators. The same gain scaling can be applied to the interstage gain in the cascaded modulator. It is obviously better than the direct input scaling, because later stage modulators tend to be overloaded. After scaled, the second stage is not overloaded, but the first stage can be overloaded. However, imprecise gain factor can increase the leak of the first-stage quantization noise. Another way to prevent overloading is the local feedback as shown in the lower left side of Figure 6.34. A local quantizer resolving only a few bits (3 ~ 4b) can be added to each integrator to complete the local feedback. This limits the signal swing of the integrator output, and the local quantizer error can be canceled as in the cascaded modulator.

From Equation (6.15), a fourth-order modulator with a 4b quantizer gives about 78 dB of SQNR with a low 8× oversampling ratio if all poles are placed at DC. That is, 30 more dB of the loop gain is required to implement a broadband $\Delta\Sigma$ modulator that achieves a dynamic range over 100 dB with such a low oversampling ratio. It is unlikely that any single-stage modulators achieve more than 100 dB noise shaping within that short span of frequency. A logical solution is to use cascaded topologies as shown in Figure 6.35 [9].

Note that the interstage gain scaling can be used advantageously to add an extra gain in this case. The first stage is usually made of a second-order modulator as it reduces the accuracy required in the next cascaded stage. The 4b DAC should be linearized using any dynamic element matching techniques. The interstage gains of four and eight give an extra loop gain of 30 dB so that the total SQNR can be 108 dB. The first-stage gain and pole errors leak to the output after fourth and third-order noise shaping. If properly canceled, the third-stage quantization noise Q_3 comes out after being shaped by the fourth-order shaping function of $(1 - z^{-1})^4$.

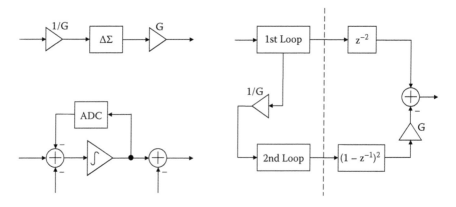

FIGURE 6.34
Integrator overload protection.

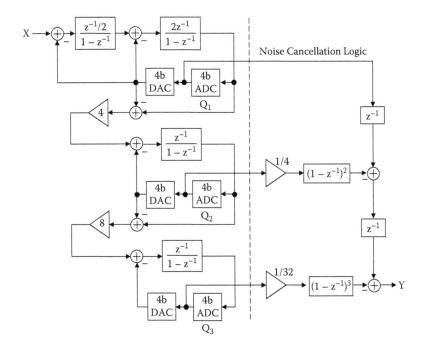

FIGURE 6.35
Broadband cascaded modulator example.

The parallel time-interleaved or pipelined concept can also be applied to make Nyquist-rate $\Delta\Sigma$ modulators, but the same sets of problems such as path mismatch and clock skew still exist. Therefore, in parallel implementations, at least one input S/H should work at the Nyquist rate. In the pipelined approach, an accurate analog delay line enables time-interleaving because multiple modulators can work at different sampling points simultaneously. Analog delay lines such as in a charge-coupled device (CCD) can be used as delay lines, but high accuracy is difficult to attain.

To sum up, switched-capacitor modulator designs can be refined following the following rules. The front-end integrator design is more critical than the later-stage designs. The kT/C noise and the nonlinearity error of the first stage set the fundamental limits. The input S/H should use the bottom-plate sampling with the bottom-plate switch linearized. Opamp gain should be high enough to avoid the data-dependent V_{ref} effect. Integrator gain is not critical as long as the transfer function stays linear. Integrator gain can be scaled, or low-order stages can be cascaded to avoid the integrator overload. Opamp settling is not critical, but its slew rate should be high enough. For practical oversampling ratio, multibit quantization is necessary, and multibit feedback DAC should be dynamically matched or calibrated.

6.6 Band-Pass Modulator Design

If the loop filter has a high gain BPF centered at a non-DC frequency, its NTF becomes a BRF. In principle, the band-pass (BP) modulator design is no different from the low-pass

(LP) modulator design except for the BP loop filter that replaces the LP loop filter. However, in the BP $\Delta\Sigma$ modulator, the sampling frequency to the signal bandwidth is oversampling, but the sampling frequency to the signal frequency is not oversampling. Therefore, the BP modulator performance is more affected by analog nonidealities than the LP counterpart if the pass-band is at high frequencies. The BP modulator can also be implemented using a CT high-Q resonator such as an LC-tank circuit as a loop filter. In practice, the real difficulty of implementing high-frequency BP modulators is in the high-frequency feedback DAC design. It is even more difficult to design the CT feedback DAC. For this reason, BP modulators are not common in real applications. However, once implemented, CT BP modulators can find many useful applications in modern wireless receiver systems.

The standard LP-to-BP transform is to replace an integrator with a resonator as shown in Figure 6.36. If z^{-1} in the LPF is replaced by $-z^{-2}$, the BPF is obtained.

Figure 6.37 shows a fourth-order BP modulator example. The order of the BPF is twice that of the LPF. If the center frequency of the BPF is moved to DC, its transfer function is identical to its LP counterpart. Because only the positive-frequency bandwidth is counted,

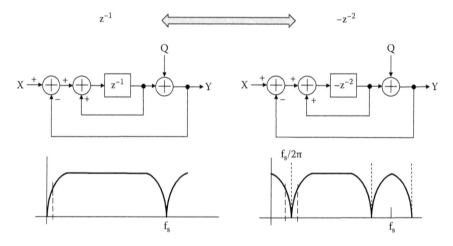

FIGURE 6.36
LP-to-BP transform and their noise transfer functions (NTFs).

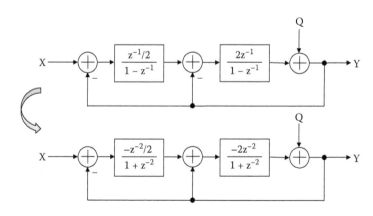

FIGURE 6.37
Second-order low-pass (LP) and fourth-order band-pass (BP) $\Delta\Sigma$ modulators.

the bandwidth of the BPF is also twice that of the LP bandwidth. Basically, the BP modulator design is the same as the switched-capacitor BPF design.

Because the unity-gain frequency of the integrator is $f_s/2\pi$, the resonator has a pole at the same frequency. The second-order BP transfer function can be obtained from the first-order LPF as follows:

$$LPF : H(z) = \frac{z^{-1}}{1-z^{-1}}, \qquad BPF : H(z) = \frac{-z^{-2}}{1+z^{-2}}. \tag{6.28}$$

The resonator has poles at $\pm j$, which are translated into $\pm f_s/2\pi$. This implies that the notch at DC due to the integrator is moved to $f_s/2\pi$. Of course, the loop gain can be adjusted to center the notch to any frequencies.

Rather than directly implementing BPF, the BP function can be implemented using two LPFs as shown in Figure 6.38. This two-path modulator system is from the same concept as the direct-conversion front-end in wireless receivers [10]. It is a complex down/up conversion using a complex carrier $e^{j\omega t}$, which has in-phase (I) and quadrature (Q) orthogonal carriers. In the specific case of the pass-band placed at $f_s/4$, the sine and cosine carriers can be made using trilevel sinusoidal carriers. This opens up a possibility to use one integrator for the two-path integrators. Because the pass-band moves to DC, two LP integrators can be used. These two I/Q down/up mixings are transparent due to the trigonometric identity of $\sin^2 + \cos^2 = 1$.

In reality, it is not necessary to up-convert the baseband I/Q signals back to the BP signal. Modern digital signal processors take the baseband signals and perform all filtering, equalization, and demodulation functions. Even inaccurate I/Q paths can create the in-band image, but they can be rejected in the digital domain [11]. However, it is still very challenging to directly quantize the BP spectrum with strong blockers because it requires very high SFDR. Unfortunately, there exists no advantage of oversampling when designing high-frequency BP modulators.

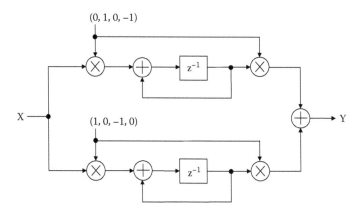

FIGURE 6.38
Two-path BP modulator.

6.7 Continuous-Time Modulator Design

In recent years, the way RF systems are designed has changed drastically as digital CMOS scaling continues toward the nanometer range. Modern single-chip multimode RF receivers with low or zero IF are based on a simple direct-conversion architecture and use wide dynamic-range I/Q ADCs for digital channel filtering and signal processing. For such applications, ADCs with wide SFDR are required to quantize the weak desired channel while providing enough headroom for strong adjacent and alternate channel blockers. One of the key components that makes advanced digital processing possible in the system-on-chip (SoC) environment is the oversampling CT $\Delta\Sigma$ modulator.

6.7.1 Continuous-Time $\Delta\Sigma$ Modulator

Sampling and quantizing any broadband spectrum are based on an assumption that the aliased band is filtered out using CT filters before sampling. DT modulators require high-order anti-aliasing filters, and unless operated with high oversampling ratio (OSR), the requirements for the input sampling linearity and clock jitter are very demanding for broadband ADCs. The requirements for the CT modulator are quite different from those for the DT modulator. CT-DSMs need no anti-aliasing filters and are insensitive to the sampling linearity and clock jitter, because the sampling is done after the CT loop filter. Furthermore, because no highly linear input sampling is required, they are suitable for scaled CMOS implementations with low supply voltages. Although the DT modulator offers premium performance at high oversampling rates, the CT modulator performs better in the low-voltage broadband environment with low oversampling ratios. CT modulators of third to fourth order with a 3 ~ 4b quantizer can achieve a DR greater than 80 dB with a very low oversampling ratio of 8 ~ 12. In particular, the low-power consumption at 10s of MS/s Nyquist rate with power supplies as low as 1 V is favorably compared to the performance of the Nyquist-rate pipelined ADC.

However, the down side is that the linearity of the front-end CT integrator is critical, and the same linearity and jitter problems remain in the DAC feedback path. CT RC integrators can meet the high linearity requirement easily with the feedback loop gain, but the opamp bandwidth should be made very wide to provide a good input virtual ground for the integrator. Although CT G_m-C integrators are simple and consume low power, transistor-based G_m cells are sensitive to process, voltage, and temperature (PVT) variations, and their performance depends heavily on the input linear range that is somewhat limited. The time constant of active filters as it is set by the absolute value of RC or C/G_m may vary by as much as ±20% in CMOS. Such a large variation of a key parameter either makes high-order single-loop modulators unstable or lowers DR. As a result, most modulators are conservatively designed to take this large time constant variation into account, and their performance can be compromised with somewhat lower peak DR. The common master–slave approach has been used to tune the time constant with a tuning accuracy on the order of a few percent. For more precise time constant tuning, elaborate tuning methods such as by pulse injection and dithering should be used.

6.7.2 Built-In Anti-Aliasing and Blocker Filtering

One distinct advantage resulting from sampling after the CT loop filter and just before the quantizer is the anti-aliasing function of the loop filter. That is, the loop filter performs dual functions. The CT loop transfer function shapes the quantization noise and filters out the aliased band around f_s, as shown in Figure 6.39.

At low oversampling rates, the anti-aliasing requirement demands prohibitively high power to make high-order CT anti-alias filters. In the CT modulator, the in-band quantization noise is suppressed by the loop gain, which should be intentionally set high. Therefore, the aliased band is suppressed more than the in-band quantization noise. This feature of CT filtering can also be advantageously used to suppress some blockers in RF system designs. In the wireless environment, the desired channel to receive is hidden under 40 ~ 60 dB stronger neighboring channels called blockers as shown in Figure 6.40, where the CT STF is superposed on a typical RF blocker mask.

In conventional RF receivers, blockers are prefiltered using passive filters such as surface acoustic wave (SAW) filters, but in modern RF direct-conversion receivers, no blocker filters are used, and very high SFDR quantizers are used instead. If CT $\Delta\Sigma$ modulators are used to quantize the baseband I/Q channels, their STFs can be shaped to perform a certain degree of channel and blocker filtering. In addition to filtering the aliased band, the high-order anti-alias filtering also helps to attenuate far-away blocker channels, thereby alleviating the quantizer DR requirement and reducing the third-order

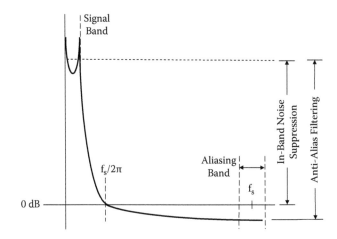

FIGURE 6.39
Noise shaping and anti-aliasing by CT loop filter.

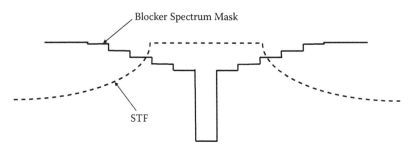

FIGURE 6.40
Blocker filtering by CT loop filter.

intermodulation (IM3) requirement of the RF front end. Furthermore, STFs can be modified with a zero inserted on the specific blocker channels for suppression. These features of anti-aliasing and blocker filtering simplify the RF receiver design and save power and area at the system level.

6.7.3 DAC Pulse Position and Pulse Width Jitters

Although CT implementations have no input sampling nonlinearity and jitter issues, the clock jitter affects the feedback DAC performance instead. The feedback DAC is still a DT sampled-data circuit, and its timing inaccuracy results in the same error equivalent to the sampling error at the input.

There exist two clock jitters to consider in the feedback DAC design as shown in Figure 6.41. One is the pulse position jitter, and the other is the pulse width jitter. The former results from the sampling edge uncertainty, but this random jitter error is averaged out in the integrator while the latter pulse width jitter directly modulates the magnitude of the DAC. What is worse is that the jitter error is not proportional to the signal but to the magnitude of the DAC steps. Therefore, the jitter error of the 1b DAC is the largest.

The DAC pulse shape also matters. Finite and unequal rise and fall times of the DAC pulse make the injected charge depend on the past DAC symbols, called intersymbol interference (ISI). Pulse shaping examples of 1b no-return-to-zero (NRZ) and return-to-zero (RZ) DACs using differential pairs are compared in Figure 6.42.

Note that symmetric crossover of the DAC output helps to mitigate the ISI effect. However, there is no ISI if the DAC output does not change. Therefore, RZ DAC pulses exhibit lower ISI than NRZ DAC pulses. RZ DAC is more linear, independent of the digital pattern, and also insensitive to the pulse width jitter but more susceptible to the pulse position jitter than NRZ pulses, because they make twice as many pulse transitions.

In the worst 1b DAC case, either the pulse width or position jitter energy is highest with the largest DAC step as compared to the sampling jitter in Figure 6.43. Assume the current DAC quantization step is $\pm I_{\text{dac}}$. The RMS jitter charge transferred to the integrator is the DAC step multiplied by the RMS jitter during the period T. Assume the jitter noise is white

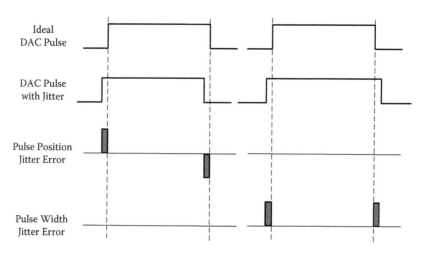

FIGURE 6.41
DAC pulse position and pulse width jitters.

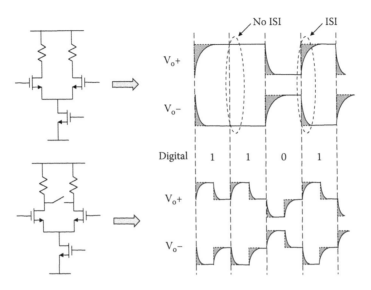

FIGURE 6.42
Pulse shaping examples of 1b no-return-to-zero (NRZ) and return-to-zero (RZ) DACs.

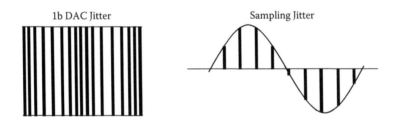

FIGURE 6.43
1b DAC jitter versus sampling jitter error.

and evenly spread over the sampling bandwidth. The maximum SNR within the signal band f_{BW} due to the clock jitter can be derived as follows [12]:

$$SNR = 10 \times \log \left\{ \frac{\dfrac{(I_{dac} \times T)^2}{4}}{(I_{dac} \times \Delta t_{rms})^2 \times \dfrac{2f_{BW}}{f_s}} \right\} = 10 \times \log \left(\frac{1}{16 M f_{BW}^2 \Delta t_{rms}^2} \right),$$ (6.29)

where M is an oversampling ratio of $f_s/2f_{BW}$, and Δt_{rms} is the DAC clock jitter. The factor of 4 in the power of the signal RMS charge results from getting an RMS signal with the maximum −3 dB amplitude.

The DAC jitter noise is independent of the input frequency, and the SNR using 1b DAC is $4M/\pi^2$ times worse than the worst-case Nyquist-rate sampling jitter noise given by Equation (4.18). Unlike the sampling jitter noise, the oversampling ratio M helps to lower the DAC clock jitter sensitivity. For $M = 16$, the sampling clock jitter is about the same level as the

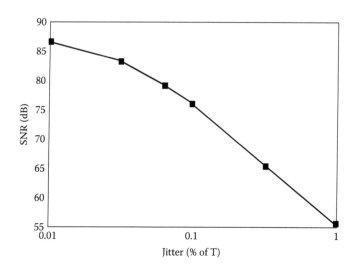

FIGURE 6.44
Signal-to-noise ratio (SNR) versus jitter for third-order 16× modulator with 4b quantizer.

clock jitter of a 3b DAC, because the DAC pulse jitter has a constant magnitude while the sampling jitter varies with the signal. Therefore, it is safe to assume that the pulse-width jitter is not critical when using multibit DACs with 3b or higher.

Figure 6.44 shows the simulated plot of SNR versus jitter percentage of the sampling period. To get the 14b resolution over 20 MHz signal band, a third-order 16× oversampling modulator with a 4b quantizer should be clocked at 640 MS/s with an RMS jitter lower than 1ps. The pulse width jitter only affects the CT modulator when the standard current DAC (I-DAC) is used, because the CT integrator is constantly charged during the whole clock period. Two-level 1b or trilevel DACs are simple and linear but suffer more from high quantization noise, settling error, and jitter. Therefore, CT modulators have been implemented using multibit feedback NRZ DACs. Multibit DACs also offer lower quantization noise and better linearity in settling than the 1b or trilevel DACs.

6.7.4 Current DAC versus Switched-Capacitor DAC

Voltage output DAC needs a voltage-to-current converter or a G_m cell to produce the feedback current for the integrator. The standard current-steering I-DAC works well for CT integrators. I-DAC is a straightforward thermometer-coded current array switched by differential pairs. By switching currents in and out of the opamp summing node, NRZ currents are injected into the integrating capacitor. The DAC element mismatch in multibit I-DACs leads to the injection of unequal unit charges into the loop as in the multibit DT modulator. Therefore, the multibit current-steering I-DAC requires either dynamic element matching or calibration to eliminate its static nonlinearity. On the other hand, the two-level 1b DAC is inherently linear, and it can be used for feedback if the pulse width jitter can be suppressed. The jitter effect can be alleviated if the 1b DAC is converted into a charge DAC at high oversampling rates.

If a switched-capacitor DAC (SC-DAC) is used, the DAC output is shaped into an impulse current or charge packet with a duty cycle shorter than the clock period, and the pulse width jitter contributes far less to the noise [12]. Differential trilevel DAC is also linear like the 1b DAC but with a 6-dB smaller quantization error.

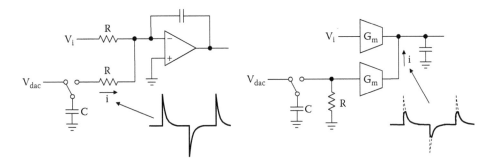

FIGURE 6.45
1b switched-capacitor digital-to-analog converter (SC-DAC) for RC and G_m-C integrators.

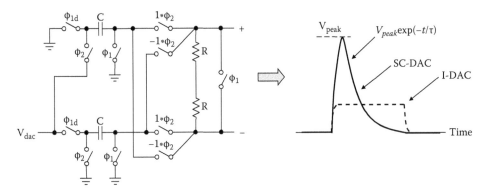

FIGURE 6.46
Differential 1b SC-DAC concept.

Examples of a 1b SC-DAC used for RC and G_m-C integrators are shown in Figure 6.45. The pulse shaping is by discharging the voltage sampled on the capacitor through the resistance so that the capacitor voltage can exponentially decay with a time constant of RC. The input resistor performs as a voltage-to-current converter in the RC integrator, while the additional G_m cell does the same in the G_m-C integrator. With RC integrators, the switched-capacitor DAC can be very linear if the integrator performs well as a current summing circuit. On the other hand, in the G_m-C integrator, the limited linear range of the G_m cell clips the impulse peak and limits the signal swing and linearity.

A 1b differential SC-DAC shaping circuit is shown in Figure 6.46. The DAC voltage V_{dac} is sampled on one capacitor during ϕ_1, and the other capacitor samples ground. Two capacitors are switched onto two resistors during the next clock phase ϕ_2. At this time, the bottom of one capacitor that sampled ground is raised to V_{dac}, while the bottom of the other capacitor is grounded. This is to make sure that the injected $V_{dac}+$ and $V_{dac}-$ to the SC-DAC are identical. Note that the polarity of the discharge path is swapped depending on the DAC digital input. This can create an impulse voltage with an exponentially decaying tail. The accuracy of the feedback SC-DAC depends on many factors such as the parasitic capacitance C_p at the input node, the switch on-resistance R_{on}, the R and C values, and the RC integrator or G_m-C cell nonlinearity. The unit DAC element should be shaped so that the total unit charge to be integrated can be the same for both the SC-DAC and I-DAC.

The peak voltage V_{peak} affected by the parasitics C_p and the time constant τ of the decaying slope are approximated as follows if switch on-resistances are assumed to be small:

$$V_{peak} \approx V_{dac} \frac{C}{C+C_p}.$$

$$\tau \approx (R+R_{on})(C+C_p). \tag{6.30}$$

At the end of ϕ_2, the capacitor voltage is not fully discharged to the ground. To remove the residual charge, the time constant RC_p can be shortened by closing the reset switch during ϕ_1. Because the impulse charge is tapered out, the integrated charge is relatively insensitive to the pulse width jitter, but the pulse position jitter matters. If an ideal exponential decay is assumed during ϕ_2, the amount of the total charge to be integrated will be proportional to the product of the peak of the impulse voltage and the time constant.

$$Area \approx \int_0^\infty V_{peak} e^{-\frac{t}{\tau}} dt \approx V_{peak}\tau. \tag{6.31}$$

To make the total integrated charges equal in both SC-DAC and I-DAC cases, the peak voltage of the impulse should be far higher than the average voltage of the I-DAC. Due to this constraint, G_m-C integrators are acceptable as 1b SC-DACs, but it is difficult to design them with wider linear ranges for multibit feedback SC-DACs.

6.7.5 Integrators for SC-DAC

CT modulators can be realized with a single-loop topology with NTF zeros optimally placed over the band of interest or with multistage cascaded modulators made of stable low-order stages. In the opamp-based RC integrators, when the capacitor is switched to the resistor input, the current through the resistor decays exponentially with a time constant of $\tau = RC$, and the capacitor voltage is attenuated by $e^{-T/2\tau}$ at the end of the half clock period $T/2$. Therefore, the SNR due to the pulse width jitter error given by Equation (6.29) is improved approximately by the following amount compared to the standard NRZ I-DAC given by Equation (6.31).

$$\Delta SNR \approx 20\log\left|\frac{1-e^{T/2\tau}}{T/2\tau}\right|. \tag{6.32}$$

That is, if $T \gg \tau$, the effect of the pulse width jitter can be made insignificant, but the current pulse amplitude becomes too large. If $T/\tau = 4.4$, the jitter effect is reduced by 11.24 dB. This is true if the opamp input summing node is an ideal virtual ground, but in reality, the input node fluctuates whenever the impulse charge is dumped on it.

If the opamp has a unity-gain bandwidth of ω_{unity}, neglecting the low-frequency pole and zero and approximating the opamp open-loop gain as ω_{unity}/s, the voltage gain from the integrator input to the opamp input summing node is approximated as follows:

$$\frac{V_s}{V_i} \approx \frac{1}{(1+\omega_{unity}RC+sRC)}, \tag{6.33}$$

where C is the integrating capacitor. Note that Equation (6.33) implies that the opamp input summing node is not a virtual ground for high-frequency transient signals [13]. To minimize the gain at low frequencies, the opamp unity-gain bandwidth ω_{unity} should be far higher than $1/RC$.

Figure 6.47 shows a simulation example of the gain-bandwidth requirement for the signal bandwidth of 2 MHz with a trilevel DAC at 32× oversampling. According to the results, the opamp unity-gain bandwidth should be wider than 1 GHz to achieve an SNR > 70 dB at 128 MS/s. It is challenging to design such high-gain wideband opamps with low power. Therefore, the G_m-C integrator topology can be far simpler and consumes less power than the RC integrator in the 1b SC-DAC case. For multibit DACs at low oversampling ratios, the gain-bandwidth requirement is greatly reduced, but multibit SC-DACs are far more complicated to implement.

6.7.6 Quantizer Meta-Stability

Like jitter, quantizer meta-stability causes a loop delay in the CT modulator and leads to unequal injection of charge into the loop unless the outputs of the comparator are latched and delayed. The signal-dependent delay of the feedback latch output of the quantizer is shown in Figure 6.48.

Bottom-plate sampling comparators can minimize signal-dependent sampling jitter, but regardless of the sampling jitter, the latch delay alone introduces extra signal-dependent jitters. For a large seed input, the latching time is short, but for a very small seed input, even the positive-feedback latch is slow. Due to this meta-stability, the comparator cannot tell whether the output is high or low within a given time. The comparator

Gain-Bandwidth	Gain	Signal-to-Noise Ratio (SNR)
10 GHz	80 dB	85.7 dB
1 GHz	80 dB	72.3 dB
100 MHz	80 dB	59.9 dB

FIGURE 6.47
Operational amplifier (opamp) for 32×, 128 MS/s CT modulator with a trilevel DAC.

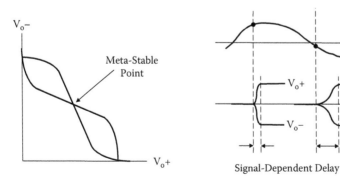

FIGURE 6.48
Latch transfer function and nonuniform pulse delay.

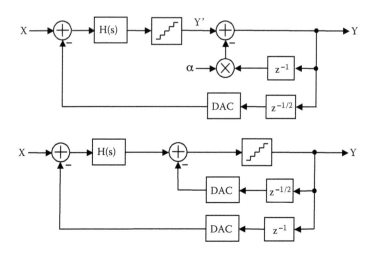

FIGURE 6.49
Local feedback for half-clock delay compensation.

gain can be increased further to reduce the meta-stability somewhat, but this signal-dependent jitter due to the latch delay can significantly degrade the performance of the CT modulator.

The common solution is to relatch the output after a half-clock delay. In broadband modulators with low oversampling ratios, this half-clock delay contributes to the loop phase delay significantly. At the unity loop-gain frequency, the half-clock period of $T/2$ $(1/2f_s)$ corresponds to an excess loop phase delay of

$$\Delta\phi = -360° \times \frac{f_s}{2\pi} \times \frac{1}{2f_s} \approx -29°. \tag{6.34}$$

The extra half-clock delay can be compensated for by using a variable coefficient local feedback loop as shown in Figure 6.49 [14,15].

This feedback loop can be either implemented in the digital or in the analog domain. In the latter, an additional DAC is needed to close the loop. The parameter α can be set to adjust the variable delay.

$$\frac{Y}{Y'}(z) = \frac{z^{1/2}}{z^{1/2} + \alpha \times z^{-1/2}}. \tag{6.35}$$

However, this digital feedback causes the gain to peak severely at half the clock frequency of $f_s/2$. Therefore, the NTF needs to be modified, and extra comparators should be added so that the output of the last-stage integrator can stay bounded in the linear range.

6.7.7 CT Modulator Architectures

CT modulator topology should be configured to properly address problems arising when the CT loop filter is used with the DT feedback DAC. Unique problems related to the CT modulator are the nonuniform quantizer delay, DAC pulse width jitter and rise/fall times,

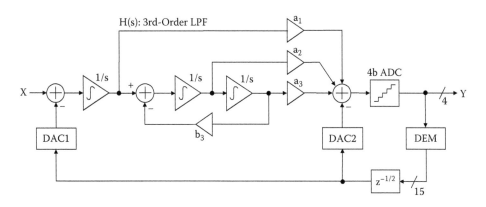

FIGURE 6.50
Third-order modulator with 4b quantizer.

and inaccurate analog filter time constant. CT modulators always include a fixed half-clock ($T/2$) delay of the quantizer output, and its excess loop delay can be compensated for by the local quantizer feedback. The multibit current-steering I-DAC or single-bit SC-DAC can be used to reduce the DAC pulse width jitter effect. Fast DAC switching with symmetric crossover mitigates the ISI problem. An inaccurate filter time constant can be trimmed or auto-tuned like other CT analog filters.

In addition, the same sets of problems related to DT modulators need to be addressed. The front-end integrator should be linear, and also for high DAC linearity, multibit DAC needs some dynamic mismatch averaging, shaping, shuffling, scrambling, or calibration unless 1b or trilevel DAC is used. To avoid some of the complexity in the CT implementation, it is possible to take a hybrid approach that uses a CT integrator at front and DT integrators in later stages. Such an arrangement benefits from the CT front-end that needs no anti-aliasing filter and eliminates the clock boosting for switches in low-voltage DT design, but the drawbacks are still the linearity of the front-end CT integrator and the feedback DAC accuracy.

Figure 6.50 shows an example of a third-order CT modulator with 4b quantizer, which is the same as the single-loop high-order modulator of Figure 6.18. As in the DT modulator, feed-forward paths improve stability and immunity to quantizer overload. A zero is added at $\pm jb_3^{1/2}$ inside the signal band. A local feedback to the quantizer in the DAC2 path is needed to cancel the half-clock delay of the quantizer. The loop filter transfer function is as follows:

$$H(s) = \frac{a_1 s^2 + a_2 s + a_3 + a_1 b_3}{s(s^2 + b_3)}. \tag{6.36}$$

Once this filter is enclosed in the feedback loop, the transfer function is slightly modified due to the local feedback path to the quantizer.

$$Y = \frac{H(s)}{2 + H(s)}X + \frac{1}{2 + H(s)}Q \approx X + \frac{Q}{H(s)}. \tag{6.37}$$

The local feedback for the delay compensation does not affect the overall stability. The small-signal stability requires that the poles of $1/\{2 + H(s)\}$ should be placed in the open left-half s-plane. One modification to make to this topology is the signal feed-forward from the input to the quantizer as shown in Figure 6.20. It helps to avoid the overload condition of integrators. However, the band to be aliased is only suppressed by the loop gain due to the direct path. Normally it is suppressed by the gain difference between the signal and aliased bands in the open-loop transfer function.

Zeros are inserted into the loop filter by bypassing integrators with feed-forward paths for phase margin (PM) at the unity loop-gain frequency, $f_s/2\pi$. Because the loop filter zeros should be placed at frequencies lower than the unity loop-gain frequency, the gains of the feedforward paths should be higher than unity. This implies that the coefficients should meet the following conditions:

$$a_1 > a_2 > a_3. \tag{6.38}$$

The feedback coefficient b_3 is just to set the zero frequency in the pass-band. When 16× oversampled with the in-band zero at 0.8* (Nyquist BW), an example of a third-order modulator with 4b quantizer can achieve SQNR greater than 80 dB with $a_1 = 5/2$, $a_2 = 1$, $a_3 = 5/8$, and $b_3 = (1/40)^2$. To mitigate integrator overloads, integrator gains can be scaled appropriately. In this case, if the three integrators are scaled as 2/s, 1/2s, and 1/2s, the coefficients can be scaled again as $a_1 = 5/4$, $a_2 = 1$, $a_3 = 5/4$, and $b_3 = (1/10)^2$.

The high gain in the feed-forward path results in the low integrator swing, and the high feed-forward gain makes the multiple-input summing circuit design very complicated. This is because values of $1/G_m$ or R get smaller, and integrators are heavily loaded, which inevitably slows the summing circuit. Figure 6.51 presents simulations showing STF, NTF, and the aliased band.

6.7.8 Integrator Design Considerations

For most CT modulators, the order of the modulator is higher than 3, and typically 3 ~ 4b DACs are required. Therefore, in-band noise suppression given by NTF is only about 60 ~ 70 dB. Such low DC gain can be easily obtained if three integrators are cascaded

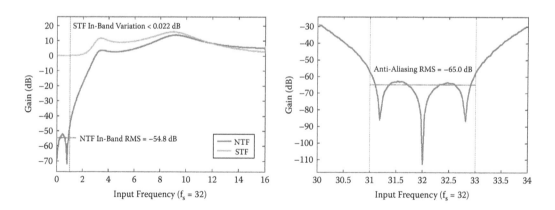

FIGURE 6.51
Simulated signal transfer function (STF), noise transfer function (NTF), and aliased band.

in the third-order loop filter. CT integrators can be implemented in the form of G_m-C or RC integrators. Three basic constraints in the integrator design are linearity, large signal swing, and high gain-bandwidth product. With high supply voltages, the first two requirements can be better met with DT integrators. However, at low voltages, only CT modulators can offer reasonable performance. Although the first stage still demands as high linearity as the Nyquist-rate ADC, the DC gain requirement is less demanding due to the feedback. Large output swing is desirable because large swing leads to the large input signal, which in turn improves SNR. Large resistors can be used to save power to the extent that their noise does not sacrifice SNR.

6.7.9 G_m-C Integrators

Although so many CT G_m cells have been developed for decades, the fundamental issue concerning their use has always been their nonlinearity and limited signal swing. Basically, it performs the voltage-to-current conversion function. Opamps for RC integrators just need high gain due to the feedback, but G_m cells for CT integrators operate in open loop. Therefore, designing G_m-C integrators for large signal range severely limits the choice of G_m cell topologies.

Three common G_m cells are shown in Figure 6.52. Its design is the same as the opamp input-stage design. The standard differential pair is the simplest, but the linear input range depends entirely on the overdrive voltage. Input devices can be linearized if operated in triode range without the tail current, but due to the difference in the input and output common-mode voltages, it is difficult to operate with low voltages. It is not differential but only makes a difference circuit. It is also difficult to bias, and its offset is high. To achieve higher linearity, the voltage-to-current converter can be used using the G_m-boosting concept. The input resistance performs the function, and the virtual ground is formed with shunt feedback. It is an example of the current-mode circuit, and its complexity approaches that of the voltage-mode circuit. The critical flaw of the current-mode circuit is the high noise, because the noise current is dominated not by the signal current noise but by the high bias current noise. Unfortunately, the bias current should be set to be a lot higher than the signal current for the current-mode circuit to work.

6.7.10 RC Integrators

The G_m-C integrator approach still uses lowest power if applied to low-resolution modulators with high oversampling ratios. For such applications, 1b or trilevel feedback DACs are common. However, the strict linearity requirement of the first stage is difficult to meet

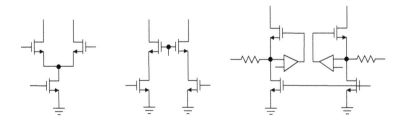

FIGURE 6.52
G_m cells.

with G_m cells with low-supply voltages. Therefore, most high-resolution CT modulators are made of a standard RC integrator as shown in Figure 6.53.

If the opamp is not ideal with a finite gain of a_o and bandwidth of ω_p, a pole is introduced in the integrator transfer function at the same frequency as the gain-bandwidth product $a_o\omega_p$ of the opamp as discussed in Chapter 1:

$$H(s) = \frac{a_o}{\left\{1 + \dfrac{s}{1/(1+a_o)RC}\right\}\left(1 + \dfrac{s}{a_o\omega_p}\right)}. \tag{6.39}$$

This pole not only changes the unity gain frequency of the integrator but also introduces some phase delay in the loop. Extra phase delay tends to make the loop unstable, and performance is sacrificed at a system level. High-frequency poles and zeros of the opamp also become the poles and zeros of the integrator.

For broadband modulators, the differential pair g_m happens to be large enough to drive resistive loads of later stages directly, and there is no need to add a low-impedance output buffer amplifier as shown in the single-stage opamp case in Figure 6.54. The unity gain bandwidth of the integrator is about $1/RC$. Because C shorts the input and output at high frequencies, the right-half-plane zero and the nondominant pole are placed at g_m/C and

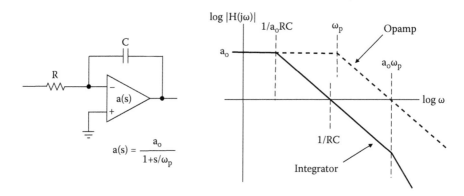

FIGURE 6.53
Opamp-based resistance-capacitance (RC) integrator and its Bode gain plot.

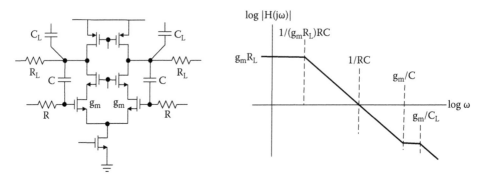

FIGURE 6.54
RC integrator with single-stage operational amplifier (opamp) and its Bode gain plot.

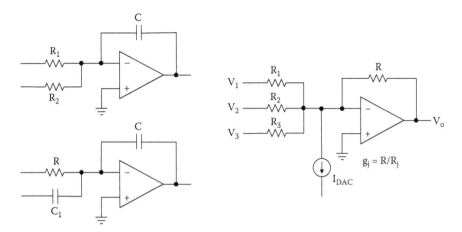

FIGURE 6.55
Multiple-input integrator and summer.

g_m/C_L, respectively, as in the Miller-compensated two-stage amplifier. An additional non-dominant pole is produced at the cascode node.

The goal is to maintain the phase delay of about −90° at the unity-gain frequency, and the right-half-plane zero is canceled by introducing the left-half-plane zero as in the standard opamp design. As shown, to increase the DC gain and also to push the high-frequency right-half-plane zero and the nondominant pole higher, g_m should be set very high though its upper limit is capped by technology.

Figure 6.55 shows three cases of a multiple-input situation that slightly modifies the gain and bandwidth. In the double input integrator case, its gain, bandwidth, and unity-gain frequency are derived as follows:

$$G = \frac{R_2}{R_1 + R_2} a_o, \quad B = \frac{1}{(1+a_o)(R_1 \| R_2)C}, \quad GB = \frac{a_o}{(1+a_o)R_1C}. \tag{6.40}$$

In the capacitive input coupling case, the integrator DC transfer function is not modified, but the nondominant high-frequency pole is moved down to $g_m/(C_1 + C_L)$, including the input capacitance C_1, and can degrade the integrator phase. Such extra phase delay can also be compensated together with the right-half-plane zero.

The multiple-input summer has been the bottleneck in most broadband CT modulator designs, and utmost care should be taken to achieve wide enough bandwidth. The three-input transfer function is modified as

$$V_o = \frac{g_1 V_1 + g_2 V_2 + g_3 V_3 + I_{DAC}R}{\left(1 + \dfrac{1 + g_1 + g_2 + g_3}{a_o}\right) + \dfrac{s}{\dfrac{a_o \omega_p}{1 + g_1 + g_2 + g_3}}}. \tag{6.41}$$

Note that the bandwidth of the summer is significantly lowered by the sum of the multiple-input gain factor $g_j = R/R_j$, $j = 1, 2,$ and 3 though the gain reduction is not a serious issue. Limited bandwidth also contributes to the extra loop delay that risks the loop stability.

The DC gain $g_m R_L$ of the single-stage opamp is limited to typically a little over 20 dB, and the output swing is somewhat limited. With the supply as low as 1.2 V, the possible signal swing approaches about 100 mV_{peak} range. Although it is still a usable signal range for small-signal applications such as in RF receivers, only two-stage opamps offer higher swing and DC gain. In the low-voltage environment with power supply approaching 1 V, modulators should have a wide bandwidth over 20 MHz and over 90 dB SFDR with 0.5 V_{peak} signal. The opamp gain of the first-stage integrator should be higher than 60 dB with input referred thermal noise below –90 dB.

The standard design of Miller-compensated two-stage opamps gives the following parameters to optimize:

$$G = g_{m1} R_{o1} g_{m2} \left(R_{o2} \,||\, R_L \right), \quad B = \frac{1}{R_{o1} \left\{ 1 + g_{m2} \left(R_{o2} \,||\, R_L \right) \right\} C_c}, \quad GB \approx \frac{g_{m1}}{C_c}, \tag{6.42}$$

where the subscripts 1 and 2 denote the first and second stages, respectively, and C_c is the Miller-compensation capacitor as discussed in Chapter 3. The advantage of the two-stage opamp is that g_{m2} is increased by the amount of the first-stage gain of $g_{m1} R_{o1}$. However, more phase delay results from the high-frequency pole and the right-half-plane zero.

$$\omega_{p2} \approx \frac{g_{m2}}{C_L}, \quad \omega_z \approx \frac{g_{m2}}{C_C}. \tag{6.43}$$

The nondominant pole ω_{p2} is lower than that of the single-stage opamp. The unity-gain bandwidth of the integrator is also narrower, and the right-half-plane zero both in the RC filter and inside the opamp should be canceled.

6.7.11 Feedback Path Design

A quantizer and a DAC are included in the feedback path of the CT modulator. Therefore, one constraint in the feedback loop design is that the total delay in the comparator plus the DAC feedback element selection logic should be shorter than half the clock period. In nanometer-range scaled CMOS, comparators can make decisions in 0.2 ~ 0.3 ns with offsets smaller than 10 mV. The delay in the DAC element selection logic is about of the same order. The comparator design for CT modulators is no different from that of other flash ADCs. As usual, the comparator should resolve half the least significant bit (LSB) with its offset kept smaller than that. Larger offsets would raise the noise floor and decrease SNR. However, the preamp delay contributes little to the loop delay as long as decisions are made within the fixed half-clock period. The input capacitance of each comparator directly loads the summing circuit. Such loading significantly slows the comparator and ultimately limits the sampling rate of the CT modulator.

In the third-order modulator of Figure 6.50, the current cell mismatch in DAC1 is the most critical, as its error is directly applied to the input. The nonlinearity of DAC1 can be considered as input noise and degrades the SNR. On the other hand, the nonlinearity of DAC2 is just added to the quantizer input, and therefore is noise-shaped as quantization noise. Although 10b linearity with 0.1% mismatch error can be achieved with carefully designed current cells, the multibit I-DAC requires dynamic element matching for improved linearity. As in the capacitor-array DAC, selecting current cells sequentially is equivalent to the first-order noise shaping because the mismatch errors are averaged to be

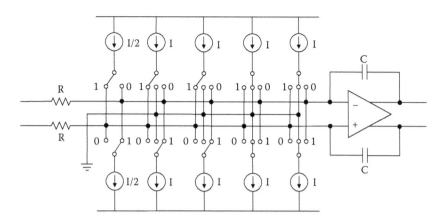

FIGURE 6.56
Differential multilevel standard current digital-to-analog converter (I-DAC).

zero. However, the periodical select pattern introduces aliasing tones in the signal band of interest. Because the nonlinearity error spreads over a wide sampled bandwidth, dynamic matching such as DWA lowers harmonic tones but slightly raises the noise floor. The parasitic capacitance of the DAC output should be minimized. The DAC1 output parasitic lowers the nondominant pole frequency of the first integrator, and the DAC2 parasitic lowers the pole frequency of the summing circuit.

An example of feedback I-DAC made of equal thermometer-coded currents is shown in Figure 6.56. For example, the currents are switched so that the output levels can be $(N + 0.5)^*2I$, where $N = -4, -3, ..., 2, 3$. Note that the DAC level is raised by $I/2$ to be symmetric. To help the common-mode settling at the integrator input, a small constant bias current can be left on all the time. The current cells are usually made of single transistors switched with differential pairs, and their noises can be referred to the input.

$$\frac{v^2}{\Delta f} = 4kT\left(\frac{2}{3}g_m\right) \times R^2 = 4kTR \times \left(\frac{2}{3}g_m R\right). \tag{6.44}$$

The actual noise in the Nyquist band is lower due to the oversampling ratio M. For lower noise, the current cells of DAC1 should have low g_m ($g_m \ll 1/R$) because the input-referred thermal noise is reduced by the factor $g_m R$. The current cell should have small W/L ratios with long channel length. The higher the overdrive voltage ($V_{GS} - V_{th}$), the better is the current matching. Typical current source matching of 4b I-DAC in nanometer CMOS is of 9 ~ 10b level, and dynamic element matching improves it above 14 ~ 15b level. At an oversampling ratio of 32, the 4b I-DAC only switches less than three current cells 99% of the time.

6.7.12 Filter Time-Constant Calibration

The transfer function $H(s)$ given by CT active filters is poorly defined as the time constant is set by the absolute value of RC or C/G_m that heavily depends on PVT variations. Inaccurate RC or C/G_m values result in wrong loop filters and make CT $\Delta\Sigma$ modulators suffer from instability or dynamic range loss. The effect of the inaccurate filter time constant on the modulator performance is more prominent in high-order, low-OSR broadband modulator designs.

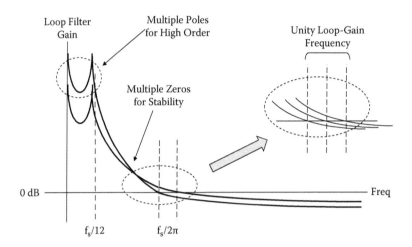

FIGURE 6.57
Unity loop-gain frequency sensitivity over process, voltage, and temperature (PVT).

For example, sketched in Figure 6.57 is the frequency response of a loop filter with low OSR and the sampling rate of f_s. The function of the loop filter is to get a very high gain in the signal band ($<f_s/12$) by placing multiple poles at DC and at the edge of the signal band. However, in order to warrant the closed-loop stability, the extra phase delay contributed by the low-frequency poles should be compensated for with as many zeros. The situation gets far worse as OSR is lowered for broad-banding CT modulators. Multiple zeros stop the steep decline of the loop gain before it becomes unity. This implies that the unity loop-gain frequency is very sensitive to the filter time-constant variation. When the loop gain approaches unity, its slope approaches the single-pole roll-off due to multiple zeros inserted to compensate for the delays of the low-frequency poles. As a result, the unity-gain frequency becomes very sensitive to PVT variations, and time constants of CT loop filters need auto-tuning.

How to calibrate C, R, or G_m components of active filters has been the classical analog filter tuning issue. They have been tuned either manually or by using the master–slave approach with a tuning accuracy of a few percent. Although the manual tuning is effective, the auto-tuning not only calibrates time constants over PVT, but also tracks their long-term drifts. In most CT filters, the master filter is tuned in the background, and the same tunable components in the slave filters are replicated after being scaled by ratios. Their tuning accuracy is limited to component mismatch in the master and slave filters.

Non-DC NTF zero can facilitate the auto-tuning task as shown in Figure 6.58 [16]. The NTF zero is detected based on sign-sign least-mean-square (LMS) algorithm. Because it only detects the error polarity and erases the error incrementally by feedback, it is called zero-forcing LMS algorithm. Assume that a fixed tone f_{zero} is injected into the input of the quantizer. It is embedded in the quantization noise and then filtered by the NTF. The poles of $H(s)$ become the zeros of the NTF that is given by $1/\{1 + H(s)\}$. If the injected tone is exactly at the resonant frequency of $H(s)$, it is nulled by the NTF. Otherwise, it survives after being filtered. This adaptive zero-forcing sign-sign LMS algorithm can detect and correct the time-constant error without interrupting the normal operation.

Figure 6.59 shows the simulated fast Fourier transform (FFT) spectrum of a third-order modulator with a NTF zero at DC and two on the imaginary axis. With no time constant error, the injected tone disappears as shown in the middle, but a residual tone remains in

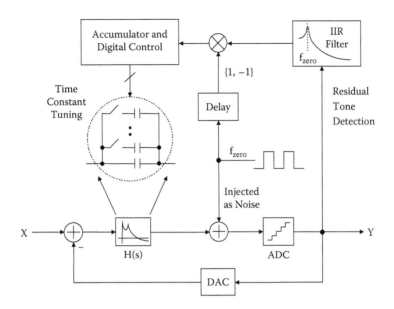

FIGURE 6.58
Auto-tuning by residual tone detection at noise transfer function (NTF) zero.

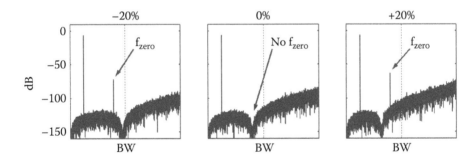

FIGURE 6.59
Tone injection to detect the NTF zero.

the other two cases. If the time constant is −20% shorter or 20% longer than that desired, the non-DC zero of the NTF is placed at a higher or lower frequency, respectively. In the former case, the system may have a poor phase margin and become unstable. In the latter case, the in-band quantization noise is elevated at the edge of the signal band, and the dynamic range is reduced. It peaks within a narrow range of time-constant values, and the modulator easily becomes unstable with large time-constant errors. Therefore, unless auto-tuned, CT filters should be designed for somewhat compromised performance to make sure that the system stays stable without regard to the time-constant variation. Selecting the right filter coefficients is a trade-off between the dynamic range and stability requirements, and it is difficult to meet both simultaneously unless they are very well controlled.

For LMS algorithm to work, only the sign of the time-constant error needs to be detected. Assume that NTF has three zeros at DC and non-DC zeros, $\pm\omega_z$ ($\pm f_z$).

$$NTF \propto s(s - \omega_z)(s + \omega_z). \tag{6.45}$$

Note that depending on whether f_z is higher or lower than f_{zero}, the vector of $(s - \omega_z)$ changes its polarity. Because this change results in the 180° phase shift of the residual injected tone, the sign of the time-constant error can be detected. Correlating the residual injected tone by the same tone produces the time-constant error power, and its sign changes at the NTF zero frequency. The time constant can be adjusted according to this sign change adaptively until the residual tone disappears in the digital output. A binary pulse used instead of a single tone can simplify the circuit implementation, but the modulator's out-of-band noise at its harmonic frequencies is also mixed down to DC in the correlation process. A digital infinite impulse response (IIR) filter with a resonant frequency centered at f_{zero} can prevent this harmonic mixing from happening as shown in Figure 6.58. The delay block is to match the phase shift that the injected f_{zero} experiences in the NTF and the IIR filter. A small delay error does not change the sign of the correlated output but only reduces its magnitude. A high degree of accuracy in the residual tone detection can be achieved as more correlated output samples are accumulated.

The accumulation time is affected by the tone magnitude. Large tones make the correlation output accumulate fast, but the residual injected tone and its harmonics may reduce the system DR as they use up some of the headroom reserved for the out-of-band quantization noise. The accumulation time also depends on the magnitude of the residual tone. In general, the larger is the time-constant error, the larger is the residual tone, and also the faster it accumulates. When the time constant is fully calibrated to be the desired value, the accumulator output stays close to zero as shown in Figure 6.60.

An example of a third-order modulator with the tone-injection for auto-tuning is shown in Figure 6.61. The NTF has one zero at DC and a pair of imaginary-axis zeros typically at 4/5 of the signal bandwidth. The sampling clock is divided by $2*M*(5/4)$ to generate the injected pulse of f_{zero} accurately. A small binary pulse is injected by switching an additional current into the summing node. It is done with two resistors and four switches as shown. There is no strict accuracy requirement on the injected pulse magnitude because it

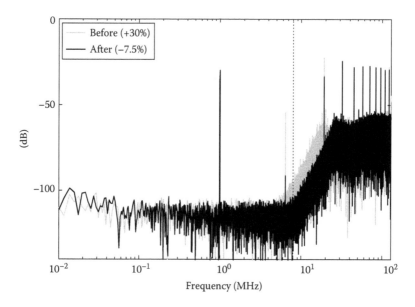

FIGURE 6.60
Auto-tuning effect before and after.

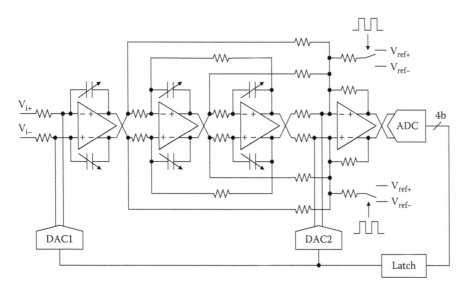

FIGURE 6.61
Pulse injection into third-order, 4b CT modulator.

will be suppressed by the NTF zero. Although unlikely, any constant input at f_{zero} with its phase also perfectly aligned with f_{zero} can interfere with the residual tone detection.

6.8 Interpolative Oversampling DAC

Ordinary Nyquist-rate DACs generate a discrete output level for every digital word applied to their inputs. Oversampling DACs achieve fine resolution by covering the signal range with a few widely spaced levels interpolating values between them. By rapidly oscillating between coarse output levels, the average output that corresponds to the applied digital code can be produced with reduced noise in the signal band.

Figure 6.62 shows the general architecture of the interpolative oversampling DAC scheme. A digital filter interpolates the sampled input values in order to raise the sampling rate to a frequency well above the Nyquist rate. The core of the oversampling technique is the digital truncator to shorten the input words. These short words are then converted into analog values at high oversampling rates so that the truncation error in the signal band can be satisfactorily low. The sampling rate up-conversion for this is usually done in steps using two or three up-sampling digital filters. The first filter is typically a ×2 or ×4 oversampling finite impulse response (FIR) to shape the signal band for the sampling rate up-conversion, and to equalize the pass-band droop resulting from the second sinx/x (SINC) filter for higher-rate oversampling.

6.8.1 ΔΣ Modulator as Digital Truncator

A digital noise-shaping ΔΣ modulator makes a digital truncator. As discussed in the ΔΣ modulator, for the Nth-order loop, the in-band noise falls by $(6N + 3)$ dB for every doubling of the sampling rate providing $(N + 0.5)$ extra bits of resolution. Because the advantage of oversampling begins to appear when the oversampling ratio of M is greater than 4, a practically achievable dynamic range by oversampling is about

$$DR > (6N + 3)(\log_2 M - 1). \tag{6.46}$$

For example, a second-order loop with ×256 oversampling can give a DR greater than 105 dB, but the same dynamic range can be achieved using a third-order loop with only ×64 oversampling. The DR is not a problem in the digital modulator, but in practice, it is often limited by the rear-end analog DAC and smoothing filters.

6.8.2 One-Bit or Multibit DAC

The rear end of the interpolative DAC is still an analog DAC. The same question as in the feedback DAC of the $\Delta\Sigma$ modulator arises in the rear-end DAC design. Because the digital processing in the interpolator and truncator are exact, the performance of the oversampling DAC owes its performance to the rear-end analog parts because the conversion of the truncated digital words into analog takes place in the rear-end DAC. The multibit system is always limited by the matching accuracy of the DAC, and the DAC error should be shaped and pushed out of the signal band so that it can be filtered out. Although analog techniques such as dynamic element matching and calibration can improve the performance of the multibit DAC, the single-bit DAC is always easier to implement without any concern about component matching.

It is true that any continuous-time filter can convert the 1b digital bitstream into a band-limited analog waveform, but it is difficult to reconstruct an ideal undistorted digital waveform without clock jitter and distortion. However, if the 1b bitstream is converted into a charge packet, a high linearity is guaranteed due to the uniformity of the charge packet as shown in Figure 6.63.

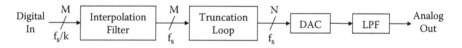

FIGURE 6.62
Interpolative oversampling DAC chain.

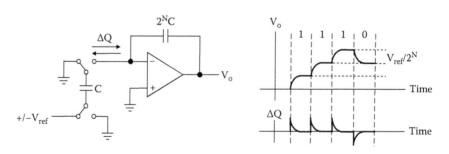

FIGURE 6.63
1b charge DAC.

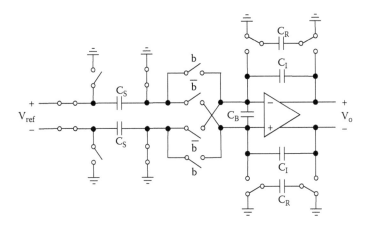

FIGURE 6.64
Bitstream 1b differential DAC.

Assume that V_{ref} or $-V_{\text{ref}}$ is sampled on the sampling capacitor C depending on the digital bitstream. The discrete-time integrator transfers the constant positive or negative charge sampled on C repeatedly on a large capacitor $2^N C$. Then, the integrator output changes incrementally by a small voltage step $V_{\text{ref}}/2^N C$, which is the same as one LSB step size. That is, it is equivalent to generating equally spaced small reference voltages. Once low-pass filtered by adding a resistor in parallel with the integrating capacitor, the output of the incremental DAC would be very linear and monotonic.

6.8.3 Monotonic Oversampling Bitstream DAC

This bitstream DAC in a switched-capacitor differential form is shown in Figure 6.64 with a one-pole roll-off low-pass filter integrated together.

It is differential and driven by two-phase nonoverlapping clocks. Although it is the constant V_{ref} sampling, the bottom-plate sampling would help to improve linearity. Note that the resistor for a lossy element is replaced by another switched capacitor C_R. Therefore, the bandwidth of this DAC/filter is set by $f_s C_R/C_I$, and the DC gain is set by C_S/C_R. The digital bitstream is converted into a charge packet by sampling the constant reference voltage on the bottom plate of the sampling capacitor C_S. Depending on the digital bit b, only the polarity of integration is reversed so that the corresponding positive or negative charge can be integrated.

The opamp DC gain requirement can be alleviated as long as the open-loop transfer function is linear within the signal swing range. To prevent the opamp from slewing when a large voltage is dumped onto the opamp summing node, the input can be shorted by a bypass capacitor C_B. The larger is the C_B, the smaller the voltage step gets. However, too large C_B would narrowband the feedback opamp, and it will take longer for settling as a result.

6.8.4 Postfiltering Requirement

The one-pole roll-off will substantially attenuate high-frequency components at f_s in the one-bit DAC case, but multibit DAC needs an elaborate continuous-time postfilter so that the charge packet can be smoothed out further. Unlike the $\Delta\Sigma$ modulator that filters out-of-band shaped noise digitally, the DAC output should be filtered only using an analog filter. Because the shaped noise is out-of-band, it does not affect the in-band performance directly, but the large high-frequency shaped noise tends to generate in-band intermodulation components and limits the DR of the system. Therefore, the shaped high-frequency noise needs to be filtered with a low-pass filter at least one order higher than the order of the modulator. It is challenging to meet this postfiltering requirement with on-chip analog filters. Analog filters for this are often implemented in continuous time using a cascade of Sallen-Key filters, but both switched-capacitor and continuous-time filters can be used. Another possibility is the hybrid implementation of an FIR filter using digital delays and an analog current summing network if current-steering DACs are used.

References

1. J. C. Candy, "A use of double integration in sigma delta modulation," *IEEE Trans. Comm.*, vol. COM-33. pp. 249–258, March 1985.
2. M. W. Hauser, P. J. Hurst, and R. W. Brodersen, "MOS ADC-filter combination that does not require precision analog components," *Dig. Tech. Papers, IEEE Int. Solid-State Circuits Conf.*, pp. 81–82, February 1985.
3. B. E. Boser and B. A. Wooley, "The design of sigma-delta modulation analog-to-digital converters," *IEEE J. Solid-State Circuits*, vol. SC-23, pp. 1298–1308, December 1988.
4. K. C. Chao, S. Nadeem, W. Lee, and C. G. Sodini, "A higher order topology for interpolative modulators for oversampling A/D converters," *IEEE Trans. Circuits and Systems*, vol. 37, pp. 309–318, March 1990.
5. K. Y. Nam, S. M. Lee, D. Su, and B. A. Wooley, "A low-voltage low-power sigma-delta modulator for broadband analog-to-digital conversion," *IEEE J. Solid-State Circuits*, vol. SC-40, pp. 1855–1864, September 2005.
6. T. Hayashi, Y. Inabe, K. Uchimura, and T. Kimura, "A multi-stage delta-sigma modulator without double integrator loop," *Dig. Tech. Papers, IEEE Int. Solid-State Circuits Conf.*, pp. 182–183, February 1986.
7. L. R. Carley, "A noise-shaping coder topology for 15+ bit converters," *IEEE J. Solid-State Circuits*, vol. SC-24, pp. 267–273, April 1989.
8. R. T. Baird and T. S. Fiez, "Linearity enhancement of multibit $\Delta\Sigma$ A/D and D/A converters using data weighted averaging," *IEEE Trans. Circuits and Systems II*, vol. 42, pp. 753–762, December 1995.
9. I. Fujimori, L. Longo, A. Hairapetian, K. Seiyama, S. Kosic, J. Cao, and S. L. Chan, "A 90-dB SNR 2.5-MHz output-rate ADC using cascaded multibit delta-sigma modulation at 8x oversampling ratio," *IEEE J. Solid-State Circuits*, vol. SC-35, pp. 1820–1828, December 2000.
10. B. S. Song, "A fourth-order bandpass delta-sigma modulator with reduced numbers of opamps," *IEEE J. Solid-State Circuits*, vol. SC-30, pp. 1309–1315, December 1995.
11. S. Lerstaveesin and B. S. Song, "A complex image rejection circuit with sign detection only," *IEEE J. Solid-State Circuits*, vol. SC-41, pp. 2693–2702, December 2006.
12. E. J. van der Zwan and E. C. Dijkmans, "A 0.2mW CMOS $\Sigma\Delta$ modulator for speech coding with 80dB dynamic range," *IEEE J. Solid-State Circuits*, vol. SC-31, pp. 1873–1880, December 1996.
13. Y. Aiba, K. Tomioka, Y. Nakashima, K. Hamashita, and B. S. Song, "A fifth-order G_m-C continuous-time $\Delta\Sigma$ modulator with process-insensitive input linear range," *IEEE J. Solid-State Circuits*, vol. SC-44, pp. 2382–2391, September 2009.

14. P. Fontaine, A. N. Mohieldin, and A. Bellaouar, "A low-noise low-voltage CT $\Delta\Sigma$ modulator with digital compensation of excess loop delay," *Dig. Tech. Papers, IEEE Int. Solid-State Circuits Conf.*, pp. 498–499, February 2005.

15. S. Yan and E. Sanchez-Sinencio, "A continuous-time $\Sigma\Delta$ modulator with 88-dB dynamic range and 1.1-MHz signal bandwidth," *IEEE J. Solid-State Circuits*, vol. SC-39, pp. 75–86, January 2004.

16. Y. S. Shu, B. S. Song, and K. Bacrania, "A 65-nm CMOS CT $\Delta\Sigma$ modulator with 81dB DR and 8MHz BW auto-tuned by pulse injection," *Dig. Tech. Papers, IEEE Int. Solid-State Circuits Conf.*, pp. 500–501, February 2008.

7

High-Resolution Data Converters

Aggressive device scaling down to the nanometer range offers integrated circuit (IC) designers both opportunities and challenges. Digital designers benefit from the system flexibility and affordability, but analog designers struggle with flawed devices. Because scaled devices are faster and smaller, the incentive to use such strengths advantageously has prompted many efforts to overcome analog imperfection by digital means. Designers are introducing more digital signal processing (DSP) functionality to enhance the performance of analog systems, and more intelligence is being built into analog designs. Such pervasive design techniques with digital assisting will prevail in the analog-to-digital converter (ADC) design of both pipelined ADCs and continuous-time (CT) cascaded $\Delta\Sigma$ modulators. The design for the former is well established, and even the operational amplifier (opamp) nonlinearity is being calibrated. The latter exhibits many desirable features in broadband applications and gains momentum as it requires no anti-aliasing, and the signal-to-noise ratio (SNR) is improved not by the calibration accuracy but by the feedback. All calibration can be made more robust by applying a zero-forcing least mean square (LMS) servo feedback concept.

7.1 Nonlinearity of the Analog-to-Digital Converter

Either the reference voltage or current can be divided using passive or active circuit elements. Therefore, ADC resolution or linearity is ultimately limited by the matching accuracy of such components in the digital-to-analog converter (DAC). In complementary metal-oxide semiconductors (CMOSs), three components commonly used for DACs are resistors, capacitors, and transistors as shown in Figure 7.1.

The resistor DAC divides V_{ref} by the ratio of resistor values. The capacitor DAC does the same, but it should be operated dynamically. The current DAC also divides or replicates equal currents by the ratio of the device sizes. However, due to the inevitable mismatch in R_1-R_2, C_1-C_2, and I_1-I_2, the division by half is not accurate. The ratio mismatch is the fundamental limit for both ADCs and DACs unless trimmed, calibrated, or shuffled.

Figure 7.2 lists some passive components with process and matching information. Process data vary from foundries to foundries. All contact resistances are about 10% of the square resistance of the layer on which the contacts are made. The bottom-plate parasitic of the old poly-poly capacitor is about 10% of the capacitance, while that of the new metal-metal (MIM) capacitance is negligible as metal layers are raised high from the substrate.

Either poly-poly or MIM capacitors exhibit excellent matching and insensitivity to voltage and temperature changes. As a result, most high-resolution CMOS ADCs are based on capacitor-array DACs, and various performance-enhancing design techniques such as calibration and dynamic element matching techniques have evolved to overcome the limit that the matching accuracy imposes on the ADC resolution.

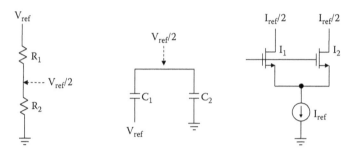

FIGURE 7.1
Three digital-to-analog converter (DAC) elements in complementary metal-oxide semiconductors (CMOSs) .

Type	Value (/μm²)	Mismatch (σ)	V.C. (ppm/V)	T.C. (ppm/°C)
Poly-Poly Cap	1.5fF	~0.04%	<100	<70
MIM Cap	1.2fF	~0.05%	<−50	<−50
N+ Poly Res	~80Ω	~0.5%	<50	<1000
P+ Poly Res	~300Ω	~0.5%	<−100	<1000
Hi R N- Poly Res	~800Ω	~0.5%	<100	<1000
N- Well Res	~1000Ω	~0.1%	<1000	~3000

FIGURE 7.2
Passive components in integrated circuits.

7.1.1 Inaccurate Residue in Pipelined Analog-to-Digital Converters

The resolution of all multistage ADCs such as pipeline and cascaded $\Delta\Sigma$ is fundamentally limited by the residue accuracy. The main cause of inaccuracy is the mismatch between the residue output range and the input range of the next stage. In the pipelined ADC, such inaccuracy results from the capacitor mismatch, opamp finite direct current (DC) gain, and opamp nonlinearity. On the other hand, in the cascaded $\Delta\Sigma$ modulator, it is mainly due to the time-constant mismatch of the analog filter and the digital noise cancellation filter.

Figure 7.3 explains the process of analog V_{ref} subtraction and the corresponding digital V_{ref} restoration at a comparator threshold point in the ADC transfer function. In the switched-capacitor residue amplifier, V_{ref} can be subtracted by flipping the bottom of one unit capacitor to V_{ref}. Consider only the linear error $(1 + \varepsilon)$, ignoring the opamp nonlinearity in the example. If the analog step of $(1 + \varepsilon)V_{ref}$ does not match the digital V_{ref}, the digital output can experience a small step discontinuity at major comparator threshold points as circled with the dashed line. The discontinuity leads to a differential nonlinearity (DNL) error occurring at that comparator threshold point. This reference mismatch is the main source of the DNL and integral nonlinearity (INL) errors in all multistep or subranging ADCs, including pipelines ADC. In particular, the inaccuracy of the first-stage residue gain contributes most to the nonlinearity, and large DNL errors may occur at major comparator threshold points. The accuracy requirement becomes less stringent in the later stages because the interstage gain of the residue is implemented.

Figure 7.4 explains the difference between digital correction and digital calibration when the digital V_{ref} is restored. The standard digital correction procedure is to restore the subtracted analog step $(1 + \varepsilon)V_{ref}$ by adding back the ideal digital V_{ref}, which corresponds to the ideal full-range most significant bit (MSB). This in effect raises the whole segment up by

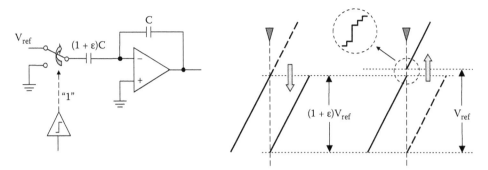

FIGURE 7.3
Analog V_{ref} subtraction and digital V_{ref} addition.

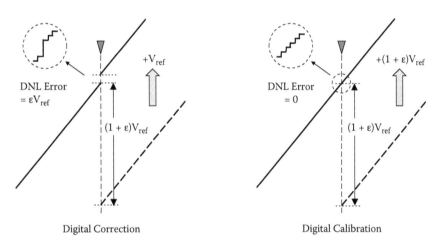

Digital Correction   Digital Calibration

FIGURE 7.4
Digital correction versus digital calibration.

V_{ref}. Then the residue range mismatch results in a discontinuity at that comparator threshold and creates the nonlinearity error. To eliminate the error, the residue output should be perfectly fitted into the input range of the next stage. Therefore, the digital calibration is to restore the subtracted analog step of $(1 + \varepsilon)V_{ref}$ with an exact digital step $(1 + \varepsilon)V_{ref}$, which is measured accurately.

7.1.2 Missing Codes and Nonmonotonicity

If the analog step is smaller than the digital step, all codes do not come out, as some codes are missing. In the standard ADC code-density test, such missing codes are rarely observed because noise works like dither and digital codes are spread over the neighboring ones. However, if larger, the opposite is true. It occurs when the residue gain is high or the input range of the next stage is small. The large residue creates a situation in which the residue goes out of the next ADC input range. Then the transfer function becomes nonmonotonic, which is usually measured by a large positive DNL. Both missing codes and nonmonotonicity are shown using output code densities in Figure 7.5.

Due to the overlapped duplicate range, the code density at the boundary is higher in the latter case, which can be interpreted as nonmonotonicity. But in the code density test, such

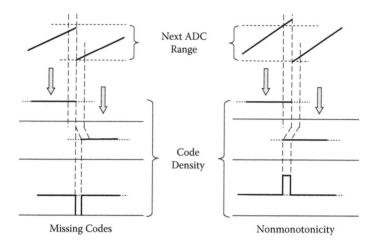

FIGURE 7.5
Missing codes and nonmonotonicity in the code density.

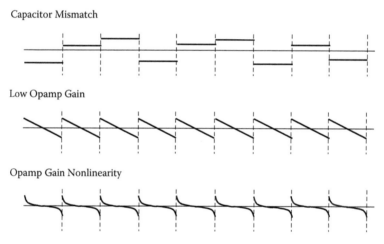

FIGURE 7.6
Integral nonlinearity (INL) resulting from three residue errors.

abnormally high code density indicates positive DNL. The DNL error at the comparator threshold point causes the subranging segment to be dislocated from the ideal straight-line ADC transfer function.

Figure 7.6 explains three cases of ADC nonlinearity in terms of INL. INL measures the amount of deviation from the straight-line transfer function. The capacitor mismatch makes the analog V_{ref} step inaccurate. So, each segment in the transfer function is shifted up or down as shown at the top, but errors within a segment stay uniform and constant if the later stages are assumed to be ideal. If there is a linear gain error, each segment is larger or smaller than the ideal one. Then the segment error will be constant, and the same error repeats in all segments as shown in the middle. Note that the error within each segment is still uniform. However, with the opamp nonlinearity considered, the error within each segment is no longer uniform as shown at the bottom. Calibration or oversampling is required to correct the gain nonlinearity.

7.2 Evolution of High-Resolution ADC Design

CMOS ADC design has evolved along with device scaling for three decades since the early 1980s. In its early days, the supply voltage was higher, the opamp had high gain while devices were slow, and the crude lithography limited the capacitor matching to 8 ~ 9b level. The standard Miller-compensated two-stage opamp was predominantly used at the low tens of kHz range for voice-band processing. The $\Delta\Sigma$ modulator was considered feasible, but digital filtering was very costly. This environment has changed since the 1990s as CMOS was aggressively scaled down toward the submicron range.

7.2.1 Device and Supply Voltage Scaling

In the middle period when the supply voltage was lowered from high 5 ~ 10 V to 1.8 ~ 3.3 V, devices were fast enough to digitize the video band and beyond. Two ADC architectures stood out: pipeline for high-speed communications and video, and $\Delta\Sigma$ for high-resolution audio. High-speed cascoded single-stage opamps emerged, and many ADC calibration techniques were developed to enhance the resolution of the pipelined ADC to above the 12b range. Now in the 2000s, CMOS is still being scaled down from the submicron to the nanometer range, and the supply voltage approaches sub 1 V. The real advantages of such scaled devices are raw digital logic speed, fine lithography, and almost free digital circuitry. The fine-line lithography also made the bare capacitor matching of over the 12b level feasible.

As scaled CMOS prevails, ADC engineers can start designs with faster and more accurate devices than earlier generations did, and most designs turn out to be already high speed and high resolution with low power. However, a couple of problems should be dealt with. With low supply voltages, the signal swing starts to limit the SNR, and the low gain defeats any design efforts to use the conventional analog design wisdom accumulated over decades. In addition, the gate leakage of the nanometer CMOS makes any accurate switched-capacitor design difficult. In fact, it appears that the analog design methodology starts all over again from the beginning. Two- or multistage opamps are back, but their gains are still very low and nonlinear. ADC designs such as algorithmic, successive approximation register (SAR), and time-interleaving are also being revisited. To avoid using low-gain nonlinear opamps, the new breeds of ADC architectures that use no opamps have started to emerge. Examples are ADCs based on time resolution. However, the main driver for all new ADC designs is the universal system-on-chip (SoC) trend that requires low power and high resolution in the data converter design.

7.2.2 Broadband High Spurious-Free Dynamic Range Applications

The industry has grown with the powerful broadband digital processing capability that enables large SoCs such as for cell phones, WiFi, and digital TV tuners. This new environment has created a demand for broadband ADCs such as intermediate-frequency (IF) quantizers that can quantize a weak desired channel and large blockers with very high spurious-free dynamic range (SFDR) for digital processing. Also for high-resolution graphic or imaging applications, high SNR over 80 dB and low-level linearity of over 15b at sampling rates over 50 MS/s are required to resolve even dark images in more detail. High-resolution ADCs at high sampling rates are only feasible with scaled technology with low supply voltages, and their performance is commonly characterized not by their effective number of bits (ENOB) but by their linearity measured by SFDR or total harmonic distortion (THD). Such ADCs with high linearity but poor SNR are allowed in most broadband systems that perform digital filtering.

It is still challenging to meet such high-linearity requirements with scaled low-voltage CMOS with both low-supply and broadband constraints. Low supply limits the upper headroom of the signal, while broad-banding raises the noise floor. This implies that the capacitance size should increase exponentially to reduce the noise floor given by the kT/C noise. Therefore, in all high-SFDR ADC designs, a trade-off of SNR for high linearity has inevitably been made.

7.2.3 High-Resolution ADC Techniques

High-resolution ADC sampling at 10 ~ 250 MS/s with 12 ~ 16b linearity has been implemented mostly with SAR, $\Delta\Sigma$, or pipeline architectures. SAR is very desirable for low-voltage and low-power applications because it repeatedly uses only one comparator, but its throughput rate is low due to the multiple decisions made per sampling. Pipeline offers a significant speed advantage due to the concurrent operation of the pipelined stages, and its throughput rate is high. The drawback is that due to its open-loop nature, the residue amplifier needs absolute accuracy in matching the amplified residue with the input range of the later stages. On the other hand, $\Delta\Sigma$ is more robust in achieving high resolution as its gain does not need to be designed high due to the feedback, but the oversampling requirement puts a cap on its throughput rate.

All evolutions related to high-resolution ADC techniques are shown in Figure 7.7. Historically, ADC resolution or linearity has been improved by trimming, and it will continue in production. Although measured errors were stored digitally, earlier self-calibration was done in the analog domain. Because errors are measured in separate cycles before the normal operation, it is called analog foreground calibration. The self-calibration concept quickly evolved into more elaborate digital background calibration. Analog or digital calibration implies which domain the measured errors are subtracted in, and foreground or background denotes when the errors are measured—before or during the normal operation. When the term *digital background* is used, errors are measured during the normal operation and subtracted in the digital domain.

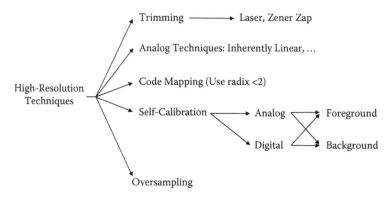

FIGURE 7.7
High-resolution ADC techniques.

Some inherently linear analog techniques were introduced to overcome analog imperfections, but worked only in specific ADC configurations. Self-calibration is more generic and applicable to most Nyquist-rate ADCs. However, self-calibration is still based on the open-loop principle. That is, the calibration accuracy is limited by the accuracy of the error measurement. The error measurement can be made more accurate using a zero-forcing LMS feedback algorithm. Suppressing the quantization noise by feedback leads to the oversampling ADC, which achieves high resolution in a more robust way. However, for broadband applications with low oversampling ratios, a compromise should be made using some of the calibration concepts, in particular in the CT $\Delta\Sigma$ modulator.

7.2.4 Inherently Linear Analog Techniques

A precise multiplying digital-to-analog converter (MDAC) residue gain of two can be obtained if two separate residue outputs spread in time but with opposite error polarities are averaged.

Figure 7.8 shows that the residue amplifier based on two-capacitor MDAC have two different gains. Assume that two identical capacitors are mismatched by the factor of α, but the opamp has a high DC gain and a wide unity-gain bandwidth for settling with sufficient accuracy. After sampling V_i, the MDAC amplifies V_i by 2, and subtracts bV_{ref} depending on the bit decision b. There are two possibilities of connecting two capacitors in feedback as shown on the right side.

$$V_o = (2+\alpha)V_i - (1+\alpha)bV_{ref} = 2V_i - bV_{ref} + \alpha(V_i - bV_{ref}).$$
$$V_o = (2-\alpha)V_i - (1-\alpha)bV_{ref} = 2V_i - bV_{ref} - \alpha(V_i - bV_{ref}).$$

(7.1)

Note that the error term containing α is reversed in polarity if two capacitors are swapped. If another amplifier takes the average of these two outputs, an ideal residue gain-of-2 can be obtained regardless of the capacitor mismatch [1].

$$V_o = 2V_i - bV_{ref}.$$

(7.2)

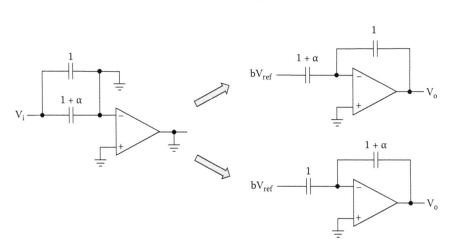

FIGURE 7.8
MDAC residue amplifier with a gain of two.

The capacitor-error averaging concept makes the residue inherently linear both in INL and DNL but requires one additional clock phase and one additional averaging amplifier. As a result, it would be 33% slower than the normal pipeline.

The capacitor-averaging technique is similar in concept to the dynamic element matching, which averages out mismatch errors over time in the oversampling ADC and DAC [2]. Any periodic or random shuffling warrants multibit DAC elements switched to the DAC output with equal probability. All oversampling $\Delta\Sigma$ modulators rely on dynamic element matching techniques such as shuffling, scrambling, or randomization applied to capacitor DAC and current-steering DAC, and the randomized DAC errors pushed out of band can be digitally filtered. However, in the Nyquist-rate ADC, dynamic element matching is not allowed because the randomized DAC error remains inside the signal band. However, if the randomized sequence is known, the DAC error can be subtracted by using the same correlation sequence as will be discussed later.

7.2.5 Self-Calibration of Successive Approximation Register ADC

The earliest effort to enhance the ADC resolution using errors stored in digital memory was an electronically programmable read-only memory (EPROM)-based code-mapping technique using a radix less than two, which warrants monotonicity and proper memory addressing as shown in Figure 7.9 [3].

Assume that an analog calibration signal equivalent to the exact digital input can be applied. The feedback through the up/down counter converges to the correct calibration DAC input to be stored in the memory. During normal operation, the DAC input reads the corresponding cal-DAC error for the SAR operation. However, the problem is that it is possible only at the factory, as the error programming process requires external precision test instruments. Factory trim may suffer from the effect of long-term process drift and environmental changes.

The first self-calibration for the SAR that does not use external test instruments measures capacitor mismatch errors, stores them digitally, and subtracts them during normal operation [4,5]. This self-calibrated SAR concept based on the charge-redistribution principle is shown in Figure 7.10.

After the top plate is initialized with 0 and V_{ref} on the bottom plates of two capacitors C_1 and C_2, respectively, the bottom-plate voltages are swapped, and the SAR finds the correct calibration voltage V_{cal} using the calibration DAC. The top-plate voltage V_x always converges to 0, and the comparator offset is ignored as it can be measured and canceled separately.

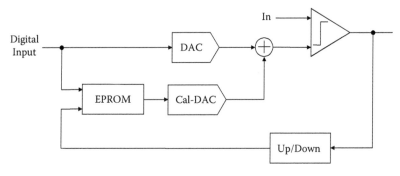

FIGURE 7.9
Code-mapping technique.

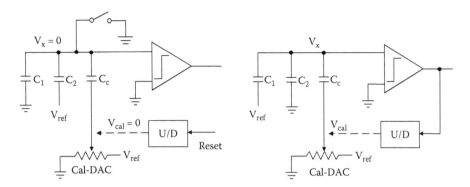

FIGURE 7.10
Self-calibration of capacitor mismatch error in the SAR ADC.

$$V_{cal} = \frac{(C_2 - C_1)}{C_c} V_{ref}.$$ (7.3)

The same measurement repeats until all capacitor mismatches of the binary-ratio capacitor array, and the measured V_{cal} are stored in digital memory. During the normal conversion, these V_{cal} values are used to correct capacitor array errors. The calibration can start from the MSB down or from the LSB up. The former top-down approach contains the cal-DAC range, but the latter bottom-up approach needs to cover a wider cal-DAC range.

Due to the digital truncation errors when calculating the total cal-DAC error, an extra 2 or 3b resolution is necessary, which requires long mismatch error measurement cycles. One flaw of high-resolution SAR ADCs is the comparator offset that varies slowly due to the stress inflicted upon the input differential pair of the comparator when decisions are made repeatedly. Comparators used in flash ADCs or pipelined ADCs are reset before use, but in the SAR-type ADC, preamplifiers experience large input voltage swings repeatedly without being reset. This poses a serious problem in high-resolution SARs.

When the differential pair is stressed with a large overdrive of several volts, its offset varies slowly with a long time constant of msec order, as shown in Figure 7.11. This slow-varying offset results mostly from the charge trapping in the oxide interface associated with the current [6]. A p-channel metal-oxide semiconductor (PMOS) is known to exhibit less threshold shift. To avoid this effect in high-resolution SARs, comparators can be either reset before every decision, or two coarse and fine comparators can be alternately used. Slow SAR ADCs are not suitable for critical high-resolution applications such as digital audio. Due to the feedback nature, oversampling $\Delta\Sigma$ modulators are more reliable and cover the voice or audio band with better resolution.

7.3 Digital Calibration of ADC

Nyquist-rate ADCs above the video band became a reality when the CMOS pipelined architecture was introduced, and the capacitor-array MDAC as a residue amplifier enabled the development of high-resolution ADCs. Calibration can be incorporated into the design

FIGURE 7.11
Comparator offset variation due to large overdrive stress.

naturally due to the subranging nature of pipelining. That is, later stages are used to digitize the residue errors from the earlier stages. An effort to perform the error subtraction in the digital domain has led to digital calibration [8,9], which avoids error subtractions using a separate calibration DAC.

7.3.1 Digital Calibration Concept

As in all subranging-type ADCs, each pipelined ADC stage resolves a certain number of bits and feeds the next stage with the amplified residue voltage that is defined as the unquantized portion of the signal. That is, the back-end ADC quantizes the smaller input range from the earlier stage repeatedly to cover the whole range. This implies that the back-end ADC transfer function repeats a number of times equivalent to the number of the quantization levels of the earlier stage. The back-end ADC transfer function can be defined as a segment. Then the whole ADC transfer function is made of many discontinuous segments. A simplified ADC transfer function on the left side and a typical real INL on the right side are shown in Figure 7.12.

Although only a few segments are shown in the transfer function, numerous segments are dislocated from the ideal transfer function depending on the ADC resolution due to the INL error as explained on the right side. All segments have the same nonlinearity because they are quantized by the same back-end ADC. If the back-end ADC is assumed to be ideal, the segment errors, which are defined as the amounts of segment dislocation from the straight line, can be measured, and the code errors can be calculated by accumulating segment errors. Subtracting code errors digitally is to move all dislocated segments back to the ideal locations so that the transfer function can be straightened [8].

7.3.2 Multiplying Digital-to-Analog Converter Capacitor Error Calibration

Figure 7.13 shows the concept of the MDAC capacitor error calibration. It is a 3b MDAC using the trilevel DAC shown in the single-ended form. The DAC is made of four identical capacitors, and during the residue amplification phase, one capacitor is in feedback, and three others take one of the three levels, V_{ref}, 0, and $-V_{ref}$, depending on the 3b digital input. In the example, the feedback cap is a unit C, and the other three are sized by $(1 + \alpha)$, $(1 + \beta)$, and $(1 + \gamma)$, respectively.

Figure 7.14 shows the mapping between the 3b digital input and the actual reference inputs to the three unit capacitors. The last two columns list the segment errors and code errors as sketched in Figure 7.13. Assuming $(\alpha + \beta + \gamma) = 0$ so that the gain error can be neglected, three segment errors become αV_{ref}, βV_{ref}, and γV_{ref}, respectively. Note that there are seven DAC levels, and the DAC errors are symmetric around the center code 011. The digital calibration is only to move the segments up and down by the amount of the code errors digitally. MDAC capacitor errors can be measured in the foreground by injecting V_{ref} into all capacitors one at a time as shown in Figure 7.15.

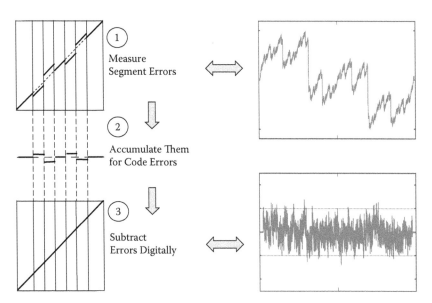

FIGURE 7.12
Digital calibration concept.

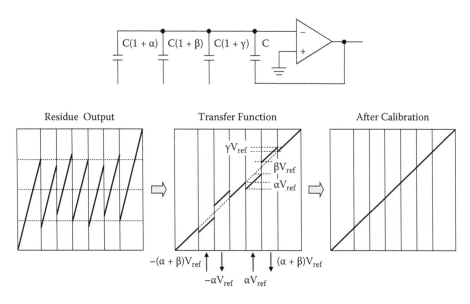

FIGURE 7.13
Calibration of MDAC capacitor errors.

In the sampling phase, V_{ref} and $-V_{ref}$ are first sampled on two capacitors $(1 + \alpha)C$ and C, respectively, while others remain switched to ground. In the amplification phase, the unit C is connected in feedback, and the $(1 + \alpha)C$ input is grounded. Then after the amplifier settles, the output becomes the segment error αV_{ref}. The errors of the other two capacitors can be measured similarly. Many measurement results are averaged as required to reduce the random quantization noise, which is reduced by the square root of the number of samples averaged. For accurate measurements, offset measurements should precede all error measurements. The measurement cycles stay the same for the differential implementation.

Coarse ADC Outputs	Inputs for Three DAC Caps			Segment Errors	Code Errors
	C1	C2	C3		
110	V_{ref}	V_{ref}	V_{ref}	γV_{ref}	$(\alpha + \beta + \gamma)V_{ref}$
101	V_{ref}	V_{ref}	0	βV_{ref}	$(\alpha + \beta)V_{ref}$
100	V_{ref}	0	0	αV_{ref}	αV_{ref}
011	0	0	0	0	0
010	$-V_{ref}$	0	0	$-\alpha V_{ref}$	$-\alpha V_{ref}$
001	$-V_{ref}$	$-V_{ref}$	0	$-\beta V_{ref}$	$-(\alpha + \beta)V_{ref}$
000	$-V_{ref}$	$-V_{ref}$	$-V_{ref}$	$-\gamma V_{ref}$	$-(\alpha + \beta + \gamma)V_{ref}$

FIGURE 7.14
Segment and code errors of a 3b digital-to-analog converter (DAC).

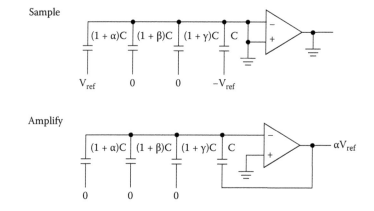

FIGURE 7.15
MDAC capacitor error measurement cycles.

7.3.3 Linear MDAC Gain Error Calibration

There are two ways to calibrate the linear residue gain error. One way is to match the analog V_{ref} step with the measured digital step [8], and the other is to find the exact gain [9]. The former uses only adders, but the latter requires multipliers. As a result, the ADC calibrated with the former may end up with a signal range slightly smaller or larger than the ideal range by the order of the gain error. However, both ways of calibrating gain errors perform the same basic function to eliminate the discontinuity in the ADC transfer function at the major comparator threshold points. The former is simple for the linear gain calibration, but the latter case is more suitable for complicated gain nonlinearity calibration. Extracting small gain nonlinearity terms involves far more sophisticated digital signal processing, and requires calculating multiple powers of the digital outputs.

Consider a simple case in which only the opamp has a finite DC gain a_o, but all capacitors are matched as shown in Figure 7.16. Then the segment will be uniformly smaller than the ideal back-end ADC range, and the same missing codes occur at the comparator threshold points. In this case, all segment errors are identical. Note that the opamp gain error can be measured together with the capacitor error. Once these analog step errors are measured, segments can be shifted up and down. The straightened transfer function has a slope slightly different from the ideal one as shown exaggerated. Therefore, as long as the finite MDAC gain error is linear,

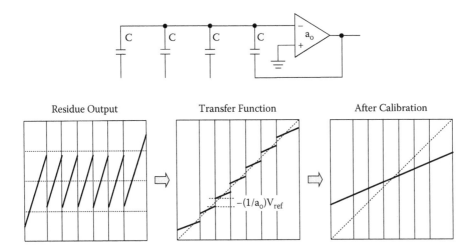

FIGURE 7.16
Calibration of MDAC gain error.

both MDAC capacitor and gain errors can be calibrated in one step using only digital adders. For the gain-error measurement, the MDAC output should produce a well-known voltage of V_{ref}, $V_{ref}/2$, or any of its fractions. However, due to the difficulty of generating accurate smaller references, the full magnitude V_{ref} is often used for convenience.

To measure the gain error together with capacitor errors, V_{ref} can be injected into each capacitor sequentially, and amplified so that the gain error of each DAC path can be measured as shown in Figure 7.17.

In this simplified case without capacitor errors, the gain error is just $-(1/a_o)V_{ref}$. The problem with this gain measurement is that the output may exceed the full back-end ADC range V_{ref} if capacitor errors are considered. There are a couple of ways to avoid it. The output signal range can be contained within the V_{ref} range if a radix smaller than two is used, or V_{ref} can be injected into two split capacitors to limit the output swing to $V_{ref}/2$. That is, two half V_{ref} errors can be summed to get the whole V_{ref} error. It is also possible to extend the output range to cover more than V_{ref}, or to initialize the MDAC output with any negative offset voltage V_{off} and measure $(V_{ref} - V_{off})$.

Note that both MDAC capacitor and gain errors are measured as segment errors and can be calibrated together using simple adders. It is to just straighten the ADC transfer function by piecing together all dislocated segments. However, in the digital signal processing approach, the calibration occurs in three steps. The MDAC capacitor errors are separately calibrated independently of the offset and gain errors. After that, the offset is calibrated before the gain error is corrected.

7.3.4 MDAC Gain Nonlinearity Calibration

The circuit complexity of the residue amplifier grows due to the high gain and wide bandwidth requirements. With low supply voltages, opamp nonlinearity is a dominant factor limiting the residue accuracy. The opamp nonlinearity effect appears in the residue output of the 3b trilevel MDAC example as shown in Figure 7.18.

This creates discontinuities in the transfer function, like missing codes. For example, the discontinuity is calibrated like other capacitor and gain errors if the analog steps at the comparator thresholds can be measured. However, the residual segment nonlinearity

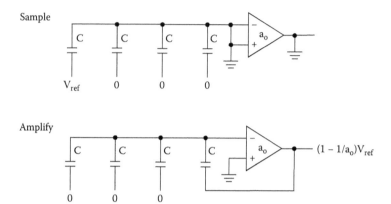

FIGURE 7.17
MDAC gain error measurement cycles.

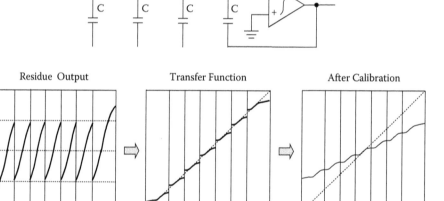

FIGURE 7.18
Calibration of MDAC nonlinearity.

remains, because the nonlinearity of the segment is affected, and the nonlinearity inside the segment repeatedly appears. The nonlinearity effect is less severe as more bits are resolved in the earlier stage, because the same opamp nonlinearity affects smaller segments. Nonlinearity calibration is far more involved than the linear gain calibration. Unlike the capacitor and gain errors, which require calibration only at major comparator thresholds, the opamp nonlinearity calibration is close to the code mapping for the entire transfer function that requires long training or measurement cycles.

7.4 Digital Background Calibration

Although early calibrations are done digitally, errors are still measured in separate foreground cycles. Numerous efforts have been made to measure errors in the background during normal operation. It is possible by duplicating either hardware or time. The former is to retire a redundant component for calibration [10,11], and the latter is to skip normal cycles randomly for error measurement and fill in missing data points by interpolation [12,13] or steal cycles for calibration by operating sample and hold (S/H) faster than the normal speed [14].

The latest background error measurement technique has evolved into a very sophisticated one, called pseudo-random (PN) dithering. The PN sequence is a pseudo-random binary pulse sequence with an equal probability of 1 or –1 over a long sample period, and its spectrum is sinx/x (SINC) shaped. It was used for pulse modulation for radar jamming, and also for military security communications known as direct sequence Spread Spectrum and Global Positioning System (GPS) [15], which are also used in commercial systems such as Code Division Multiple Access (CDMA) and GPS.

The spectrum of the PN sequence is spread uniformly like noise as shown in Figure 7.19. In the foreground calibration, offset measurements precede all error measurements. However, due to the pseudo-random nature of the PN sequence, DC offsets are chopped. Therefore, in the background calibration, no separate offset measurement cycles are necessary. If PN modulated, the error spectrum in the signal band decreases by the oversampling ratio. In oversampling systems, out-of-band errors can be filtered out, but in Nyquist-rate systems, they should be correlated out using the same PN sequence and subtracted digitally.

The first example of using the PN sequence to enhance the ADC resolution was to dither the ADC to smooth out DNL for low-level linearity, and the injected dither was subtracted digitally as shown in Figure 7.20 [16].

The PN-modulated dither is digitally correlated using the same dither, and low-pass filtered to recover the digital equivalence of the injected dither so that it can be subtracted

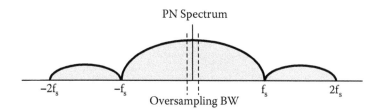

FIGURE 7.19
Spread spectrum of the pseudo-random sequence.

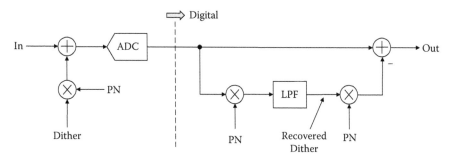

FIGURE 7.20
DNL improvement by pseudo-random (PN) dithering.

from the digital output. Dithering is transparent to the ADC operation, but the noise-like PN dither added to the signal effectively averages out low-level DNL errors. The end result is that INL is not affected by dithering while DNL is improved. Although the signal range is somewhat reduced to make room for the dither, dithering can be applied to measure the gain error of the residue amplifier. Because the PN sequence is not correlated to the signal, the PN sequence can be recovered.

As shown in Figure 7.21, the well-defined calibration voltage V_{cal} added to the input after being PN modulated experiences the same gain as the signal. Therefore, by measuring the digitized V_{cal} (DV_{cal}), the residue amplifier gain can be obtained. Shifting the input by the random dither amount results in signal loss.

Figure 7.22 shows that the interstage gain error of the MDAC can be measured in the background by injecting the same PN-modulated dither [17]. An additional slow but precise $\Delta\Sigma$ ADC is used to measure the dither magnitude. That is, V_{ref} can be adjusted to set the accurate residue gain of two by comparing the quantized dither with the accurately measured dither. This first calibration example using PN dithering is the analog background calibration.

7.4.1 Background Capacitor Error Measurement

One obvious advantage of the background calibration is its ability to track any long-term variations over process, voltage, and temperature (PVT) variations. Furthermore, the

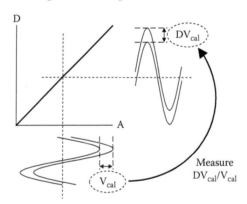

FIGURE 7.21
Pseudo-random (PN) dithering concept to measure the gain error.

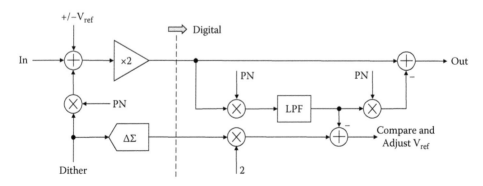

FIGURE 7.22
Residue gain calibration by PN dithering.

digital background calibration by PN dithering is preferred to others because the digital power and area overhead diminish as CMOS is scaled down. One drawback of the PN dithering is that the signal range shrinks to make room for dithers. It is inevitable because the signal plus dither should be contained within the linear signal range. Large dithers facilitate prompt convergence in averaging the PN-correlated output, and they shorten the calibration time. On the other hand, small dithers do not reduce the signal range, but they exponentially lengthen the averaging time of the correlated output. These two magnitude and time constraints in PN dithering should be taken into account in the design.

There are many ways to measure the MDAC mismatch and gain errors in the background. In particular, if only capacitor mismatch errors should be measured, they can be modulated and correlated using PN sequences in the background based on the shuffling or scrambling concept without injecting dithers.

Figure 7.23 shows a dynamic shuffling example of capacitor mismatch errors with individual PN sequences [18] applied to multibit MDACs. The concept is the same as the dynamic element matching for the multibit feedback DAC in the $\Delta\Sigma$ modulator. The shuffled noise is pushed out of the signal band in the oversampling modulator. However, in the Nyquist-rate ADCs, it still stays within the signal band and should be eliminated. The capacitor mismatch error of each capacitor can be correlated out with its own PN sequence and subtracted digitally. This way, only capacitor mismatch errors are shuffled, but PN dithering is required to measure the gain error. To shuffle each C with its own PN, two-level differential DAC elements are switched into the positive and negative summing nodes, respectively.

7.4.2 Gain Error Measurement by Pseudo-Random Dithering

A 1.5b pipeline stage PN dithered using V_{cal} is shown in Figure 7.24 with three error sources: MDAC capacitor error α, opamp gain error β, and back-end ADC error γ. Note that the MDAC has different gains for the signal and the reference. The calibrated condition is to ensure that the gain error $(\alpha + \beta)$ of the MDAC matches the input range error γ of the later stage. To measure the gain error of the full-scale V_{ref}, a well-defined PN-modulated calibration signal, V_{cal}, which is usually a fraction of V_{ref}, can be added as a dither into the

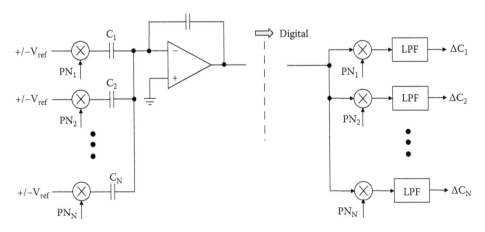

FIGURE 7.23
PN dithering for a multibit MDAC.

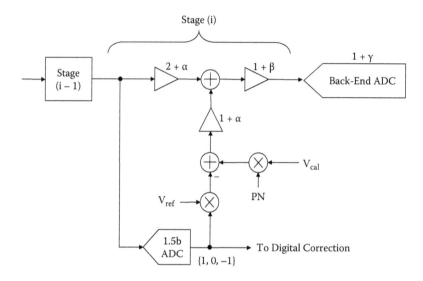

FIGURE 7.24
Gain errors of a 1.5b pipelined stage.

input of the stage to calibrate. In the multibit MDACs, the calibration should be repeated to all individual DAC capacitors.

A dither can be injected in two different ways before and after the ADC. The former is to dither the signal, and the latter is to dither the DAC output. However, the latter only applies to the multibit sub-DAC or MDAC. In the single-bit case, there are no adjacent DAC levels to dither to, and the input capacitor should be split to inject a dither smaller than V_{ref}. Although it is dithered digitally after the multibit ADC, it still relies on the assumption that the DAC is ideal, either calibrated or shuffled [19]. The magnitude of the dither should be well defined and set typically to be smaller than half the redundant digital correction range, which in effect reduces the over-range for digital correction by half.

Figure 7.25 explains the gain measurement concept by PN dithering, where the total gain errors of $(1 + \alpha + \beta + \gamma)$ are combined as one error $(1 + \varepsilon)$. Here all later stages are assumed to be ideal or previously calibrated. Because PN is a zero-mean sequence of 1 and −1, after being correlated by the same PN sequence, the digital output becomes

$$V_o = PN(1+\varepsilon)V_i + PN^2(1+\varepsilon)V_{cal} = PN(1+\varepsilon)V_i + (1+\varepsilon)V_{cal}. \tag{7.4}$$

Note that the input signal V_i is modulated by PN and translated into a noise, but the PN-modulated calibration signal is correlated by the same PN sequence. Therefore, a DC value of $(1 + \varepsilon)V_{cal}$ is obtained after being averaged in the low-pass filter because $PN^2 = 1$. If scaled by the ratio of V_{ref}/V_{cal}, the digitized gain of the full-scale V_{ref} can be found to be $(1 + \varepsilon)V_{ref}$.

$$V_o \times \left(\frac{V_{ref}}{V_{cal}}\right) = \left(\frac{V_i}{V_{cal}}\right)PN(1+\varepsilon)V_{ref} + (1+\varepsilon)V_{ref}. \tag{7.5}$$

However, note that as the bandwidth of the low-pass filter is limited, the noise-like PN-modulated V_i term remains a residual measurement error even after being low-pass

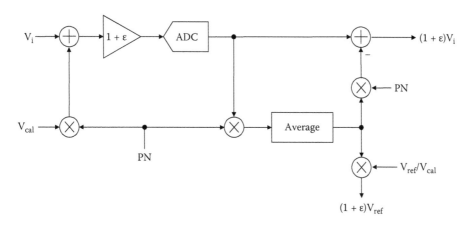

FIGURE 7.25
Principle of background gain error measurement by PN dithering.

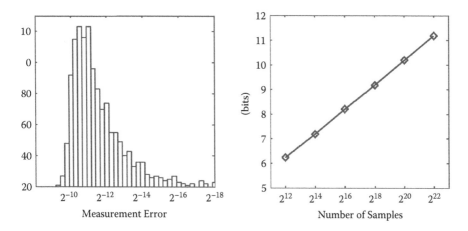

FIGURE 7.26
Simulations for correlation accuracy.

filtered. The simplest digital low-pass filter is an accumulator that averages a large number of samples. Ideally, the measurement error approaches zero if infinitely many samples are averaged. However, because the number of samples is limited in practice, V_i/V_{cal}, the signal-to-dither ratio, should be kept as small as possible in order to minimize this residual measurement error. Recovering small errors while decorrelating large signals is a process requiring a long accumulation time. The situation gets worse when calibrating for high resolution or using small dithers.

Figure 7.26 shows the error given by Equation (7.5) after averaging 2^{20} samples, where both $V_i/V_{cal} = 1$ and V_{ref} is normalized to 1. Simulations are repeated 1000 times. In the output histogram on the left side, 99% of the measurement errors are smaller than 2^{-10}, which is roughly of 10b accuracy. In the relation between the measurement accuracy and the numbers of samples averaged shown on the right side, note that four times more samples should be averaged to get one more bit of measurement accuracy. This is true if the PN-modulated V_i/V_{cal} in Equation (7.5) is treated as a white noise because the standard deviation of the white noise is reduced by the square root of 2 as the number of averaging samples is doubled. However, the result implies that 2^{30} samples should be averaged for

15-b accuracy. That is, it would take almost 1 minute to complete one measurement if the ADC sampling rate is 20 MS/s.

7.4.3 Constraints in PN Dithering

There are two constraints in PN dithering. One is measurement time, and the other is dither magnitude. PN-dithering an MDAC stage that resolves only one or two bits is more troublesome as it loses signal range to accommodate large dithers [20]. The measurement time constraint originates from the trade-off between the measurement accuracy and the averaging time.

In reality, the situation is worse, and it would take far longer to finish correlation with high accuracy because dithers should be far smaller than the signal ($V_i/V_{cal} \gg 1$). As a result, the dither magnitude constraint originates from the trade-off between the dither magnitude and the signal range. Figure 7.27 conceptually explains the dither magnitude constraint with large and small PN-modulated dither in normalized scale. The solid line is the signal, and two dashed lines indicate the boundary of the PN-dithered signal. The signal range is reduced accordingly to contain the total signal plus dither within the full-scale range. This leads to the reduction of the effective number of bit (ENOB).

The signal range reduction is not desirable in a system where the signal-to-noise ratio (SNR) is dominated by the thermal noise. For example, switched-capacitor circuits will need capacitors of twice the size to suppress the kT/C noise by 3 dB, thus resulting in a significant area and power penalty. Although a smaller dither makes the signal range larger, it takes much longer to achieve the same measurement accuracy because the ratio of V_i/V_{cal} in Equation (7.5) is large. This relation between the signal range and the measurement time makes it difficult to find a solution that resolves both constraints.

7.4.4 Signal-Dependent PN Dithering

A signal-dependent dithering relieves both the measurement time and dither magnitude constraints [21]. The dither magnitude can be adjusted depending on the signal level. For example, the normalized signal is divided into three subranges conceptually as shown in Figure 7.28.

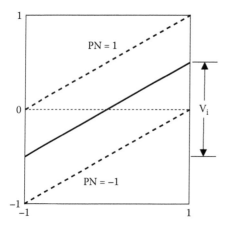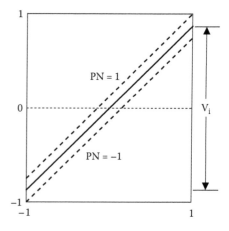

FIGURE 7.27
Normalized ADC transfer function with dithers.

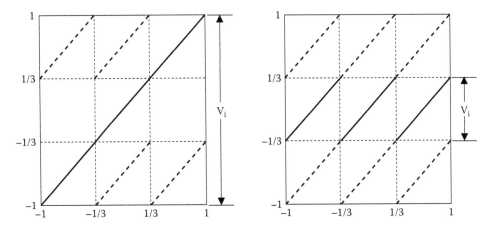

FIGURE 7.28
Signal-dependent dithering versus equivalent fixed magnitude dithering.

When the signal is higher than 1/3, no dither is injected if PN = 1, but a dither of −4/3 can be injected if PN = −1. Similarly, when it is lower than −1/3, dithers of 4/3 and 0 are injected if PN = 1 and −1, respectively. However, in the middle range, a normal dither of 2/3 can be injected depending on PN. As a result, the signal plus dither is equivalent to a smaller signal, which is between ±1/3, with a large fixed-magnitude dither of 2/3 as shown on the right side. This implies that because the ratio of V_i/V_{cal} can be kept low, the signal-dependent dithering shortens the measurement time while achieving the same level of measurement accuracy. That is, it allows a large dither to be injected with a full-scale signal. The signal-to-dither ratio V_i/V_{cal} can be further reduced if the signal range is divided into more subranges. Note that the difference of the MDAC outputs for two cases of PN = 1 and −1 should be generated using the same hardware in order to emulate the fixed-magnitude PN dithering.

Compare the signal-dependent dithering to the fixed-magnitude dithering with a dither of 1/4. First, because the dither of 1/4 reduces the signal range to 3/4, the ENOB improves by about 0.5b better. Second, the ratio of V_i/V_{cal} also decreases from 3 to 1/2. This improves the measurement accuracy further by 2.5b. So the signal-dependent dithering gives three extra bits of accuracy and can shorten the measurement time by 1/64. This implies that one measurement cycle that would otherwise take 1 minute to complete can be finished in 1 second.

Signal-dependent dithering achieves two goals simultaneously. One is that large dithers do not sacrifice the signal range, and the other is that the signal decorrelation time is greatly shortened due to the low signal-to-dither ratio. Examples shown in Figures 7.27 and 7.28 ignored the redundant signal range for digital correction. The half of the redundant range for digital correction can be used to make room for dithers, and comparators should resolve one more bit. Therefore, considering the digital correction range, the residue plot of a trilevel MDAC can be modified for signal-dependent dithering as shown in Figure 7.29.

The comparator thresholds in the sub-ADC are shifted from $\pm(1/4)V_{ref}$ to $\pm(3/8)V_{ref}$, and four more comparators are added with thresholds at $\pm(1/8)V_{ref}$ and $\pm(5/8)V_{ref}$ to divide the residue plot into seven subranges. Furthermore, this fine subdivision of the signal range makes it difficult to implement. If the signal is larger than $(5/8)V_{ref}$ and PN = −1, the residue should be $(2V_i − 2V_{ref})$, which requires an additional capacitor to subtract a reference voltage of $2V_{ref}$. However, dithering on large signals is not as effective because V_i/V_{cal} is high. Therefore, the penalty in the measurement time is insignificant though it is not dithered when the signal is large. That is, the comparators at $\pm(5/8)V_{ref}$ can be safely removed, and the

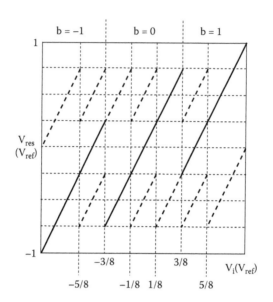

FIGURE 7.29
Modified residue with signal-dependent dithering.

V_i (V_{ref})	PN = −1	PN = +1
−1 ~ −3/8	0	0
−3/8 ~ −1/8	0	$+V_{ref}$
−1/8 ~ 1/8	$-(1/2)V_{ref}$	$+(1/2)V_{ref}$
1/8 ~ 3/8	$-V_{ref}$	0
3/8 ~ 1	0	0

FIGURE 7.30
Signal-dependent dithers.

signal-dependent dithering can still offer a substantial savings in the measurement time with low circuit complexity unless the input signal constantly stays higher than $\pm(3/8)V_{ref}$.

A dither of $-V_{ref}$, $-(1/2)V_{ref}$, 0, $(1/2)V_{ref}$, or V_{ref} can be injected depending on the PN values and the signal level as shown in Figure 7.30. The signal plus dither within $\pm(3/8)V_{ref}$ is in effect a large fixed-magnitude dither of $(1/2)V_{ref}$ with a small signal within the range of $\pm(1/4)V_{ref}$. The signal-to-dither ratio of V_i/V_{cal} is reduced to 1/2, and in simulations, 99% of the errors are smaller than 2^{-14} when only 2^{26} samples are averaged. If referred to the input after dividing by 2, it corresponds to 15b accuracy.

The standard trilevel MDAC can be modified for the signal-dependent dithering by adding two more comparators and splitting one of the capacitors into two as shown in Figure 7.31. Both C_1 and C_2 are switched between $-V_{ref}$ and 0 for the signal range from $-(3/8)V_{ref}$ to $-(1/8)V_{ref}$, and between 0 and V_{ref} for the signal range from $(1/8)V_{ref}$ to $(3/8)V_{ref}$ if PN is 1 and −1, respectively. When the signal lies in the middle range, C_1 and C_2 are alternately switched to $-V_{ref}$ if PN = 1 and V_{ref} if PN = −1 to inject a dither of $(1/2)V_{ref}$ equally through two capacitors. The mismatch between the two split capacitors contributes to noise after randomized and spread over the Nyquist band, and also needs to be subtracted digitally, though not significant.

An example of digital background calibration by PN dithering is shown in Figure 7.32, where each stage is assumed to resolve 1.5b. In this example, the signal-dependent dither

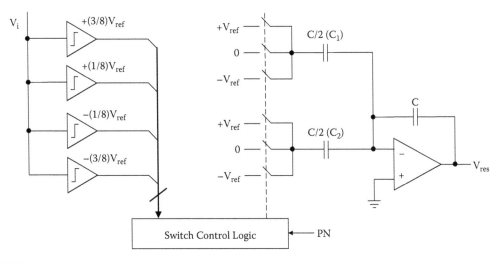

FIGURE 7.31
MDAC and comparators for signal-dependent dithering.

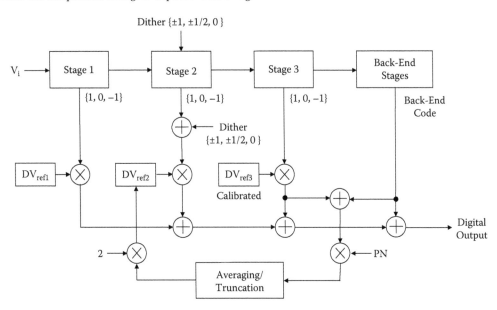

FIGURE 7.32
Digital background calibration by PN dithering.

is injected into the second stage. After the dither is digitized, it is PN correlated and averaged digitally so that the actual digital step of the second-stage DV_{ref2} can be measured and updated. Because V_{cal} is equivalent to $(1/2)V_{ref}$, it is multiplied by 2 to update it. Note that in the normal signal path, the injected dither is subtracted digitally in order to keep the digital output within the signal range. The calibration starts from the third stage and proceeds to the first stage. In this example, the third stage is fully calibrated assuming the back-end ADC is ideal, and the second stage is being calibrated. After the first stage is calibrated, the calibration cycle can start again from the third stage in order to track the DV_{ref} variations over PVT.

7.5 Digital Processing for Gain Nonlinearity

The gain nonlinearity calibration is based on the fact that the nonlinear residue transfer function can be approximated with a high-order polynomial. Canceling high-order distortions is far more involved and would require a prohibitively long time for error convergence. It all depends on how the high-order nonlinearity errors are modeled and how accurately they are measured. In the foreground measurement, several input levels can be tried to map the nonlinear transfer function, but in the background measurement, large dithers are not available. Unlike the linear gain error calibration that requires only adders, the gain nonlinearity calibration requires far more complicated numerical processing with multipliers.

7.5.1 Weakly Nonlinear Gain Error

Some compromises can be made to facilitate the background nonlinearity error measurement. The first assumption to be made is that the residue amplifier is weakly nonlinear so that the third-order distortion can be modeled as a dominant term. Heavily nonlinear cases require more substantial higher-order nonlinear curve fitting and longer calibration cycles for convergence. Assume that the gain nonlinearity of a residue amplifier is measured with an ideal back-end ADC as shown in Figure 7.33.

This intuitive way is to distribute the gain nonlinearity error over the full range. The nonlinear function $f(V_i)$ can be simplified only with a third-order term as follows:

$$f(V_i) = a_3 V_i^3 + a_5 V_i^5 + a_7 V_i^7 + \cdots \approx a_3 V_i^3. \tag{7.6}$$

Because typical pipeline stages are fully differential, even-order nonlinear terms in the amplifier transfer function are mostly canceled, and the dominant source of nonlinearity comes from odd-order nonlinearity. That is, the properly modeled gain nonlinearity error can be corrected if it is subtracted using the same nonlinearity error given by the same function $f(V_i)$.

However, the accurate value of the input V_i is not known, and the closest to V_i is the digitized output V_o. The digitized output includes all residue errors: offset, linear gain error,

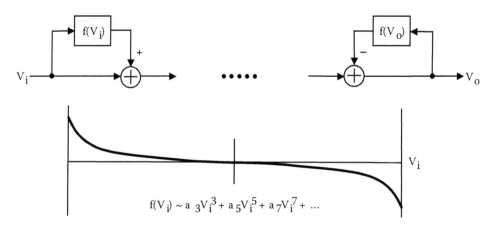

FIGURE 7.33
Gain nonlinearity modeling and correction.

gain nonlinearity error, and quantization noise. Assume that the offset is calibrated, and the random quantization noise contributes little error in the calibration. Then the output is approximately

$$V_o = (1 + a_1)V_i + a_3 V_i^3 + Offset + Q \approx (1 + a_1)V_i + a_3 V_i^3, \tag{7.7}$$

where a_1 is the linear gain error of the stage to calibrate. If Equation (7.7) is used for V_i in Equation (7.6), and high-order terms of a_3 are ignored, the estimated gain nonlinearity error becomes

$$f(V_o) = a_3 V_o^3 = a_3 \left\{ (1 + a_1)^3 V_i^3 + 3(1 + a_1)^2 a_3 V_i^5 + 3(1 + a_1) a_3^2 V_i^7 + a_3^3 V_i^9 \right\} \tag{7.8}$$

$$\approx a_3 (1 + a_1)^3 V_i^3 \approx a_3 V_i^3.$$

This implies that the digital output can be used for the gain nonlinearity error estimation with no significant penalty in the calibration accuracy.

7.5.2 Gain Nonlinearity Measurement by PN Dithering

If the input V_i is dithered with an analog calibration voltage V_{cal} using a PN sequence as is done for the linear gain error measurement, the digital output becomes

$$V_o = (1 + a_1)(V_i + PN \times V_{cal}) + a_3 (V_i + PN \times V_{cal})^3 \tag{7.9}$$

$$= (1 + a_1)V_i + a_3 V_i^3 + 3a_3 V_i V_{cal}^2 + PN \times \left\{ (1 + a_1)V_{cal} + a_3 V_{cal}^3 + 3a_3 V_i^2 V_{cal} \right\}.$$

If this output is correlated using the same PN sequence and is accumulated, the following constant error is obtained after being low-pass filtered:

$$V_e = LPF(PN \times V_o) = (1 + a_1)V_{cal} + a_3 V_{cal}^3 + 3a_3 V_i^2 V_{cal}. \tag{7.10}$$

The decorrelated random signal power decreases, while the correlated error term increases proportionally to the number of accumulated samples. Note that the gain nonlinearity error depends on the input power, and the linear gain error is constant. Therefore, the gain nonlinearity can be detected independently of the gain error, but it can be measured only with two input conditions at least. However, the gain nonlinearity error of $a_3 V_{cal}^3$ is also added to the linear gain error, and correcting the linear gain error with this gain nonlinearity present is more complicated than correcting the gain error only. That is why the former requires very sophisticated digital signal processing with multipliers, and the latter can be done with just adders.

The gain nonlinearity error can be measured by taking the difference of the values obtained at two measurement points, one where V_i is large and another where V_i is small. The two-point measurement requires random PN dithering of the residue.

Figure 7.34 shows the residue plot PN-dithered with random comparator threshold jumps. If the residue amplifier is linear, two straight-line residues would be generated with a constant vertical gap between them regardless of the gain [22]. The constant gap is defined as the segment error, and again, in this case, the linear gain error can be calibrated using only digital adders without using digital multipliers. However, with the gain non-linearity, the gap would not be uniform.

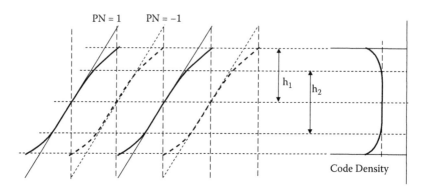

FIGURE 7.34
Nonlinearity measurement from code density.

The nonlinearity estimate is obtained by comparing the difference in the cumulative distribution functions of the output code densities. The gain nonlinearity makes $h_1 > h_2$. The difference is measured using the code density, and the third-order nonlinearity can be estimated for calibration. Then the measured errors can be distributed using a look-up table over the full range. However, the convergence and tracking speed of this measurement using code densities depend heavily on specific output codes occurring to estimate the gain error, and the measurement accuracy is also sensitive to the DC offset. Furthermore, this approach needs to assume that the busy random signal should constantly cover the input range to obtain sufficient code densities for all codes.

7.5.3 Measurement by Signal Correlation

It is more desirable to extract the gain nonlinearity information that does not rely on specific output codes. Because the PN-dither magnitude is modulated by the gain nonlinearity in Equation (7.10), it can be correlated again by the input power V_i^2. That is, the gain nonlinearity error can be derived from the following covariance between the error and the input [23].

$$Cov(V_e, V_i^2) = E(V_e \times V_i^2) - E(V_e)E(V_i^2)$$
$$= 3a_3 V_{cal}\{E(V_i^4) - E(V_i^2)E(V_i^2)\} = 3a_3 V_{cal} \times Var(V_i^2). \tag{7.11}$$

Because the variance of V_i^2 is always positive, the covariance has the same sign polarity as that of the error a_3. However, as shown in Equation (7.10), the correlation to get the linear gain a_1 is not straightforward in the presence of high-order gain nonlinearity. After high-order nonlinearity terms are measured, the linear term should be adjusted accordingly, taking high-order terms into account.

Although the exact value of the input signal V_i is not available during normal operation, the digital output of the later stage can be used as a substitute for that as in Equation (7.8). However, care must be taken as the digital output contains both the gain and nonlinearity components. The coefficient can be updated based on the LMS algorithm using the polarity information from Equation (7.11). This estimation approach has no relation to the input, but the DC input does not allow convergence because the DC signal has no variance. Therefore, for this case, it is assumed that the input signal still needs to be busy. Although the digital dither injected after the sub-ADC is effective, it requires the busy signal condition, and dithering the signal independently of the ADC would be more effective for the gain nonlinearity measurement.

7.5.4 Multilevel PN Dithering

Multilevel PN dithers can be used to extract the higher harmonic distortion terms selectively [24]. For example, m two-level, PN dithers with m digital sequences of PN_1, PN_2, ... , PN_m are needed for the mth-order harmonic distortion measurement. It is to separate each major harmonic term with its unique dither sequence. The mth-order distortion can be correlated out by the product of m dithers, $PN_1 PN_2 ... PN_m$ with magnitudes of $\pm V_{cal}^m$. In the third-order example, the PN-dithered residue after the dither is subtracted becomes

$$V_o = a_1 \left\{ V_i + (PN_1 + PN_2 + PN_3)V_{cal} \right\} + a_3 \left\{ V_i + (PN_1 + PN_2 + PN_3)V_{cal} \right\}^3. \tag{7.12}$$

If this is correlated by the sequence of $V_{cal}{}^3 PN_1 PN_2 PN_3$ and is low-pass filtered, the following gain nonlinearity term can be obtained, which is only proportional to the third-order nonlinearity error:

$$LPF\left(V_o \times V_{cal}^3 PN_1 PN_2 PN_3\right) = a_3 \times \left(3! V_{cal}^6\right). \tag{7.13}$$

Similarly, the same correlation would give the same result for the mth-order nonlinearity coefficient.

$$LPF\left(V_o \times V_{cal}^m PN_1 PN_2 ... PN_m\right) = a_m \times \left(m! V_{cal}^{2m}\right). \tag{7.14}$$

Because no accurate value of V_i is available, the digital output can also be used as in Equation (7.8). Note that $m = 1$ is the case of the linear gain error, which can be correlated with the random sequence PN_1 only. However, correlating out the linear gain error in the presence of high-order gain nonlinearity would not converge to a_1 directly as given by Equation (7.10).

Multilevel dithers for high-order harmonics should be made smaller than the redundant digital correction range as the sum of the calibration sequences is amplified along with the quantization error from the flash ADC. Otherwise, most of the overrange will be taken up by the calibration sequences, which leaves little or no overrange margin for other nonideal errors.

7.5.5 Accuracy Considerations for Background Error Measurement

The linear residue gain error can be measured with PN dithers in the background. Even multiple stages can be calibrated simultaneously with different PN dithers. However, applying it to the gain nonlinearity measurement is not straightforward. When small errors are recovered by correlation, two constraints of the measurement time and dither magnitude limit the measurement accuracy. Furthermore, the high-order nonlinearity that is proportional to the product of many small-signal terms becomes too small to detect. Due to the long convergence time required, detecting only the third-order nonlinearity with fine precision would be challenging.

In the standard digital correction, the extra range equivalent to the one-bit range (Δ) is used for redundancy. Considering two upper and lower ranges, a redundant overrange is about the half-bit range ($\Delta/2$). The offset difference of the residue amplifier and the comparator plus the calibration dither should be contained within the half-bit range.

$$\left| V_{os,amp} + V_{ft,amp} - V_{dac} - V_{os,comp} - V_{ft,comp} \right| + \left| \sum V_{cal} \right| < \frac{\Delta}{2}, \tag{7.15}$$

where all offsets from amplifiers, sampling circuits, and DAC are counted. For this reason, the total dither magnitude is typically set to be smaller than half of the overrange ($\Delta/4$). For example, to correlate this dither out while decorrelating signals with N-bit resolution, it requires accumulating more than 2^{2N} samples because the signal decorrelation decreases as a square-root function of the number of samples.

It is worse for gain nonlinearity errors, because they are small and proportional to the high power of the signal. Therefore, if the signal is small, the gain nonlinearity error would be very small. In such cases, PN dithering after ADC is not effective because it uses only the small portion of the residue transfer function where the gain nonlinearity is small, which would result in far longer correlation time. Therefore, the PN dithering at the input before the ADC would help to speed up the convergence. The nonlinearity calibration has yet to achieve such high linearity as is feasible with just the capacitor and gain error calibration. It has been proved to exhibit 12 ~ 13b resolution, which is sufficient to show the proper ADC operation using nonlinear opamps. However, modern scaled CMOS still demands challenging designs without sacrificing performance, and designers need to go this extra distance to further ensure that ADCs they design work in the future environment with very low voltage and power.

Four digital background calibration methods are discussed. One measures errors in the foreground, and the other three measure errors in the background. They differ drastically in the required number of measurement cycles. In the straightforward foreground error measurements, for example, errors should be accumulated so that the measurement noise can be averaged out and be smaller than errors. However, in the background measurement, PN-modulated errors are embedded in the signal, and the measurement by signal decorrelation requires far more samples to average out. For example, if a capacitor error V_e is embedded in the signal V_i, then the digitized output V_o becomes (V_i + PN*V_e). Exact digital V_i is unknown, and only the digitized residue V_o can be used. It is PN-correlated, and averaged for many samples to extract V_e. If n is the required number of samples to average, the variance of $\Sigma(PN*V_i)$ grows with $n^{1/2}$ while accumulated V_e grows with n. For the error measurement to be meaningful, the accumulated error nV_e should be greater than $n^{1/2}V_i$. That is, for $nV_e > n^{1/2}V_i$ to hold true, the following condition should be met:

$$n > \left(\frac{V_i}{V_e} \right)^2 = 2^{2N} \left(\frac{V_i}{V_{ref}} \right)^2. \tag{7.16}$$

Similarly, in the gain error measurements, the output of PN*$(1 + a_1)\{V_i + PN*V_{cal}\}$ needs to be averaged.

$$\sum PN*(1 + a_1)(V_i + PN*V_{cal}) = \sum PN*(1 + a_1)V_i + \sum V_{cal} + \sum a_1 V_{cal}. \tag{7.17}$$

Therefore, the required number of samples amounts to

$$n > 2^{2N} \left(\frac{V_i}{V_{cal}} \right)^2. \tag{7.18}$$

This implies that the background measurement by PN dithering takes far more samples than the foreground measurement because the calibration signal V_{cal} for dithering is far smaller than V_{ref} due to the dither magnitude constraints. Using a dither of $V_{ref}/8$ requires 64 times longer measurement cycles. Note that V_i/V_{cal}, which is the signal-to-dither ratio, should be minimized to reduce the number of samples to accumulate.

The situation is even worse in the gain nonlinearity measurement, because the nonlinearity error term is far smaller because the output is $PN*f(V_i + PN*V_{cal})$.

$$\sum PN * f\{V_i + PN * V_{cal}\} = \sum PN * \{(1+a_1)V_i + a_3 V_i^3 + 3a_3 V_i V_{cal}^2\}$$
$$+ \sum \{(1+a_1)V_{cal} + a_3 V_{cal}^3\} + \sum 3a_3 V_i^2 V_{cal}.$$

(7.19)

Then the required number of samples amounts to a prohibitively large number.

$$n > 2^{2N} \left(\frac{V_i}{V_{cal}}\right)^2 \left(\frac{V_{ref}}{V_i}\right)^4.$$

(7.20)

In the nonlinearity measurement, if the signal is small, it takes forever because there is no nonlinearity error for the small-signal input. So, the input signal should be made large to measure nonlinearity, and the signal-to-dither ratio becomes large. This implies that the nonlinearity measurement requires the large signal condition and requires far more samples to accumulate than the gain error measurement.

Dithers can be injected into the residue amplifier in either form of analog V_{cal} or digital DV_{cal} as shown in Figure 7.35.

The residue output with both analog and digital dithers can be written as

$$V_{res} = 2^N \left\{ V_i + PN * V_{cal} - \sum_{i=1}^{N} \left(b_i \frac{2^{i-1}}{2^N} V_{ref} + PN * DV_{cal} \right) \right\}.$$

(7.21)

Because their effects are the same in the residue output, the digital dither injection after the ADC is simpler and more commonly used. However, as mentioned in the gain non-linearity measurement, digital dithering is not effective if the input is small, and analog dithering would be more effective and necessary.

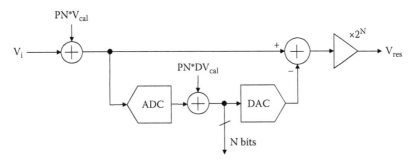

FIGURE 7.35
Analog versus digital dither injections.

7.6 Calibration by Zero-Forcing Least-Mean-Square Feedback

This last stage of evolution of ADC calibration is to update all errors based on an adaptive least-mean-square (LMS) algorithm that forces any error power to converge to zero. For example, the DAC gain calibration by PN dithering is to match analog V_{ref} step with an accurate digital V_{ref} step. The digital V_{ref} is updatable in two different ways. One way is just to measure DV_{ref} and subtract it. The other alternative is to measure only the polarity of the residual error and incrementally update it using small steps. The former way is an open-loop straightforward cancellation, and the latter is based on the zero-forcing feedback principle. The advantage of the closed-loop system over the open-loop one is obvious. The closed-loop implementation offers the robustness of all analog designs in any conditions over PVT variations, although the calibration process can be slower. The latter needs only the error polarity, while the former requires absolute accuracy for calibration.

7.6.1 LMS Feedback Concept

This servo-feedback algorithm does not require any specific signal condition. It behaves very similar to the classic LMS algorithm [25].

Applying the zero-forcing LMS feedback algorithm to enhance the ADC resolution requires the following three steps as shown in Figure 7.36. First, the gain or DAC error δ should be separated and embedded into the signal after being PN modulated. Second, after the same PN-modulated error estimate δ' is subtracted, the residual error $(\delta - \delta')$ needs to be correlated using the same PN sequence. The residual error is either quantized or just its sign is determined. Last, the residual error is updated based either on the quantized error or on the polarity of the residual error. This zero-forcing feedback is called servo-feedback due to the long time constant introduced by the LPF or averaging filter. The feedback forces the condition of $\delta = \delta'$, and it is stable as it has only a single very low-frequency pole given by the LPF in the loop, which is usually implemented with just a digital accumulator.

7.6.2 Self-Trimming

When applied to the digital background calibration, the LMS algorithm is equivalent to electrical trimming. Like analog designers adjusting trim potentiometers while watching instruments, electrical trimming can be carried out so that the observed error can be minimized, as can digital calibration. The sign polarity of PN-modulated errors can be correlated using the same PN sequence, detected, and used to close the LMS feedback loop.

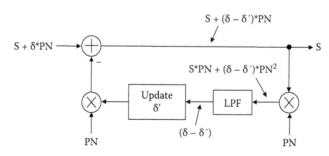

FIGURE 7.36
Least-mean-square (LMS)-based calibration concept with zero-forcing feedback.

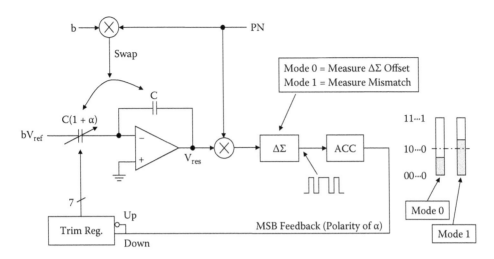

FIGURE 7.37
Capacitor trimmed using zero-forcing feedback.

One example of the zero-forcing feedback is shown in Figure 7.37 [26]. In the 1.5b pipelined ADC, the reference voltage added/subtracted during the normal operation is used as a dither to PN-modulate the capacitor mismatch error so that it can be embedded into the residue and be recovered later by correlating with the same PN sequence. The residue amplifier using the two-capacitor MDAC has an output of

$$V_o = 2V_i - b \times V_{ref},$$

(7.22)

which depends on the trilevel coarse bit decision b, which is 1, 0, or –1. The capacitor mismatch error appears in the residue output with the polarity inverted every time the two capacitors are swapped [1]. If the ratio of the two capacitors is $(1 + \alpha)$, the residue output is modified as

$$V_o = (2+\alpha)V_i - b \times (1+\alpha)V_{ref}.$$

(7.23)

Note that the signal gain $(2 + \alpha)$ is different from the reference gain $(1 + \alpha)$, and the error term polarity resulting from the capacitor mismatch is inverted when the two MDAC capacitors are swapped as

$$V_o = (2-\alpha)V_i - b \times (1-\alpha)V_{ref}.$$

(7.24)

Once α is trimmed out, the residue output contains no capacitor mismatch. If the two capacitors are swapped depending on the PN sequence, the error appears as

$$V_{error} = PN \times \alpha(bV_i - V_{ref}).$$

(7.25)

If this error is correlated with the same PN sequence and low-pass filtered, the output is

$$LPF(PN \times V_{error}) = \alpha(bV_i - V_{ref}).$$

(7.26)

This PN-correlated output can be low-pass filtered to estimate the αV_{ref} term. However, it is difficult to assume that the signal-dependent $\alpha b V_{in}$ term is averaged out to be zero. The polarity detection is far easier than digitizing the αV_{ref} term. The polarity of detected α is accurate and independent of the $\alpha b V_{in}$ term because $b V_{in} < V_{ref}$ no matter what the $b V_{in}$ value is.

This zero-forcing feedback is limited by the polarity detector accuracy and the incremental capacitor step size. The lower bound of the minimum detectable ratio error is set by the decorrelated signal average. This is true if the residue output has a zero-mean uniform distribution, but in reality, the accumulated PN-correlated residue ΣPN^*V_{res} depends heavily on the signal condition. To speed up the correlation process, the signal component can be subtracted before the PN correlation. The first-order $\Delta\Sigma$ modulator can be used for polarity detection. For example, it can easily achieve a high oversampling ratio of 2^{23}, which is good for over 15b resolution.

This adaptive zero-forcing feedback concept using an oversampling $\Delta\Sigma$ calibrator is useful for background self-trimming or self-calibration and can be applied to trim the current matching [27], resistor ratio [28], and offset [29] as sketched in Figure 7.38.

7.6.3 LMS Adaptation for Digital Background Calibration

The LMS feedback concept can be easily applied to any digital calibration, but the LMS-based adaptive calibration has a slightly different implication from that of the standard digital calibration. The normal digital background calibration by PN dithering operates in an open loop as shown in Figure 7.39.

Errors should be measured with absolute accuracy and subtracted once. After this feed-forward correction, residual errors can be left over. On the other hand, in the LMS feedback approach, errors are updated in a closed loop based on the residual errors that remain at the output after the correction as shown in Figure 7.40.

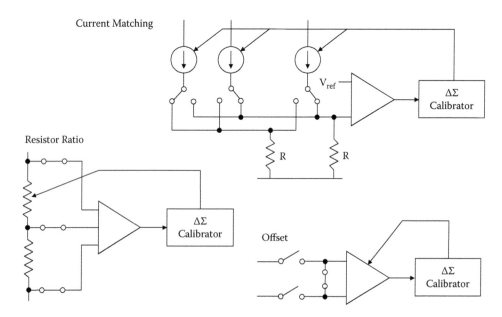

FIGURE 7.38
Self-trimming example using LMS feedback.

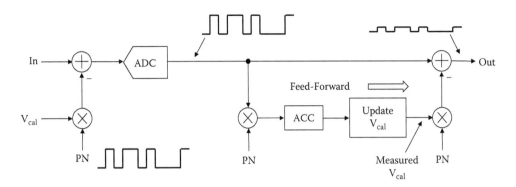

FIGURE 7.39
Calibration scheme by PN dithering.

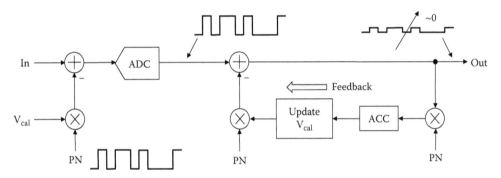

FIGURE 7.40
Calibration by LMS feedback.

Although the open-loop calibration can be performed repeatedly, residual calibration errors remain. On the contrary, in the closed-loop calibration, the residual error converges to zero by feedback. That is, the difference is that the open-loop calibration requires absolute accuracy in measuring the quantized error, while the closed-loop calibration needs to accurately detect only the error polarity. When applied to calibrate the pipelined ADC, the dithered DAC and gain error appear in the residue of the stage. Correlating only the small PN-modulated error out relies on an assumption that the large residue output averages out to be smaller than the error by 2^{-N} for N-bit resolution, for example. The correlated error term increases linearly as more samples are accumulated, but the decorrelated signal term randomly fluctuates. After the error polarity detection, the error subtraction can be done either in the analog or digital domain.

The sign-sign update algorithm is to detect the error polarity and update the error incrementally with the sign information only. It simplifies the digital implementation of the algorithm. The LPF can be implemented with a digital integrate-and-dump $\sin x/x$ (SINC) function with an extremely high oversampling ratio. Because the error is updated slowly with a negligible step change at a time, the loop gain is so narrowband, and the stability no longer becomes an issue. The same adaptive LMS algorithm has been applied to improve the analog system performance in many applications.

Blind LMS-based adaptation for multiple coefficients can also be applied as shown in Figure 7.41 [30]. Simultaneous adaptation to all nonlinearity coefficients including offset can be performed based on the reference $\Delta\Sigma$ ADC, which can be slow but accurate.

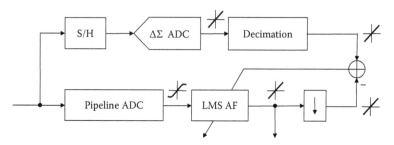

FIGURE 7.41
Blind equalization for gain nonlinearity.

Adaptive filters operating at a decimated rate minimize the error power of the pipelined ADC relative to the reference ADC. For this adaptation to work, the S/H should be linear, and the clock skew in sampling between the two paths is critical. The clock skew error should have a mean zero, and the input should stay busy.

7.7 Calibration of Time-Interleaving ADC

Time interleaving is the simplest way to multiply the sampling rate without using a new process technology but by duplicating hardware. However, the performance of time-interleaved ADCs is sensitive to various mismatches such as offset, gain, sample time, bandwidth, and nonlinearity in the time-interleaved channels. They are modulated up to the channel sampling frequency, and the spectrum folding and aliasing further complicate the situation. Any deviations from the ideal sampling instants can be represented as a sequence of sample-time errors that introduce errors in the input samples like the clock jitter. The input samples are phase modulated by the sequence of the sample-time errors in the ADC channels. In the frequency domain, this error produces copies of the input signal spectrum at the same frequencies as the spurious components resulting from the gain mismatches. The bandwidth mismatches also originate from the timing-constant mismatches of the RC-sampling networks, and they significantly degrade the performance for broadband ADCs. But all channel bandwidths commonly assumed to be wide enough to ignore the bandwidth mismatches. They cause the noise floor of the ADC to be elevated, thus lowering the system SNR. To improve the resolution of the time-interleaving ADCs, the remaining three issues related to the offset, gain, and sample-time mismatches should be addressed.

7.7.1 Offset Mismatch

The offset mismatches among the ADC channels contribute to a periodic additive pattern in the ADC output. In the frequency domain, this pattern appears as tones at integer multiples of the channel sampling rate f_s/N. The offsets in different channels are commonly subtracted by their values obtained by averaging many digital output samples. It is the high-pass filtering, and such offset calibration introduces notches in the ADC output spectrum at integer multiples of f_s/N. To eliminate the notches in the ADC output spectrum, a random-chopper-based offset calibration scheme can be used to cancel offset mismatches. Because the offset is PN modulated by alternating the signal polarity randomly, the signal is randomized during the offset correlation process.

7.7.2 Gain Mismatch

Gain mismatches among the parallel channels cause amplitude modulation of the input samples by the sequence of channel gains. In the frequency domain, this error causes copies of the input signal spectrum to appear centered around integer multiples of f_s/N. The gain mismatch of each channel can be calibrated using any of the ADC gain-error calibration schemes. Most notably by PN dithering, the gains of all ADC channels can be equated by comparing the correlated digital dither outputs, and then multiplied by the appropriate gain to correct for the gain mismatch.

Only for a two-path system [31], the gain mismatch and sample-time errors can be obtained through digital signal processing. Consider a single-tone example for a two-path system. Because the two channels are multiplexed at the rate of $f_s/2$, the normalized output is assumed to have the image as a function of the gain error α.

$$V_o = \cos\omega t + \frac{\alpha}{2}\cos\left(\frac{\omega_s}{2} - \omega\right)t. \tag{7.27}$$

If this is multiplied by $\cos(\omega_s/2)t$ to shift the image back to the same frequency of the signal, Equation (7.27) becomes

$$V_o \times \cos\left(\frac{\omega_s}{2}\right)t = \frac{\cos\left(\frac{\omega_s}{2} + \omega\right)t + \cos\left(\frac{\omega_s}{2} - \omega\right)t + \frac{\alpha}{2}\cos(\omega_s - \omega)t + \frac{\alpha}{2}\cos\omega t}{2}. \tag{7.28}$$

The chopper or switching gives the output $4/\pi$ times greater than this single tone case, and there are higher third, fifth, or seventh-order mixings present, which are difficult to trace. In this example, only the fundamental tone is considered like a sine wave. If Equations (7.27) and (7.28) are multiplied for squaring and low-pass filtered, the following DC term will give the gain mismatch error:

$$LPF\left\{V_o^2 \times \cos\left(\frac{\omega_s}{2}\right)t\right\} = \frac{\alpha}{4}. \tag{7.29}$$

However, the problem with the frequency shift by $f_s/2$ is that if there already exists any spectrum at $f_s/4$, Equation (7.28) will produce more correlated terms that will produce DC outputs. Therefore, the spectrum of Equation (7.28) should be notched at $f_s/4$ before mixing and squaring to get the magnitude as shown in Figure 7.42.

The gain mismatch error can be nulled with the sign of the gain mismatch in the zero-forcing LMS feedback. This is to calibrate only the average channel gain. However, if both DNL and INL of each channel are calibrated together with the gain error, the average gain calibration is not necessary.

7.7.3 Sample-Time Error

Now assume that two path gains are matched, but the sample-time errors are $\Delta t/2$ and $-\Delta t/2$, respectively. Similarly from Equation (7.27), the output becomes

$$V_o = \cos\omega t + \frac{\Delta t}{2}\sin\left(\frac{\omega_s}{2} - \omega\right)t. \tag{7.30}$$

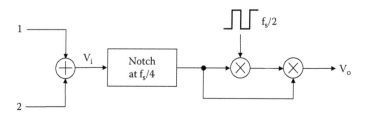

FIGURE 7.42
Gain mismatch correction algorithm.

The phase error is identical to the gain error except for the $-\pi/2$ phase shift, and is orthogonal to the signal, because the group delay is obtained by differentiation. If it is multiplied by $\cos(\omega_s/2)t$ to shift the image back to the same frequency of the signal, Equation (7.30) becomes

$$V_o \times \cos\left(\frac{\omega_s}{2}\right)t = \frac{\cos\left(\frac{\omega_s}{2}+\omega\right)t + \cos\left(\frac{\omega_s}{2}-\omega\right)t + \frac{\Delta t}{2}\sin(\omega_s-\omega)t + \frac{\Delta t}{2}\sin\omega t}{2}. \tag{7.31}$$

To detect the phase (time) error, Equation (7.31) should be phase shifted by $-\pi/2$ to make a sine function, which is the Hilbert filter. The sample-time error correction also requires all-pass time-delay filter to correct time error and a phase detector as shown in Figure 7.43.

The time delay can be simply implemented by adjusting the sampling time. However, the variable group delay can be digitally implemented using a finite impulse response (FIR) filter with symmetric coefficients of the sinx/x (SINC) function. In the two-path example, two channels are first up-sampled by inserting zero to produce two channel signals at the ADC sample rate of f_s, and the tap coefficients of the FIR filter are set to

$$c[n] = \frac{\sin\left(\pi\frac{\Delta t}{T}\right)}{\pi\left(\frac{\Delta t}{T}-n\right)}, \tag{7.32}$$

where Δt is the time error. In one path, a FIR with a fixed phase shift is added, but in the other path, the same FIR with programmable coefficients for variable group delay should be inserted. The FIR filter with fine time resolution steps requires an almost infinite number of filter taps, and some compromises in performance should be made when truncating the number of filter taps. The filter delay can be updated with the sample time error used in the zero-forcing LMS feedback. One channel is delayed by one clock period z^{-1} to be synchronized with the other. After two channels are summed, a notch filter of $(1 + z^{-2})$ adds a null at $\omega_s/4$. The frequency shift by $\omega_s/2$ is achieved by chopping the sum by multiplying it by a signal that alternates at $f_s/2$.

An ideal low-pass Hilbert filter that can cover the Nyquist band requires an infinite number of filter taps. However, it can be simplified in this case because the whole Nyquist band signal is not required for the phase detection. The Hilbert filter is approximated by a simple delay z^{-1}. It gives a phase shift ranging from 0 to $-\pi$ over the Nyquist band. It gives the desired phase shift of $-\pi/2$ at $f_s/4$, and most of the detection of the timing error

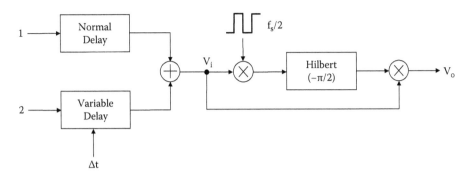

FIGURE 7.43
Sample-time error correction algorithm.

is determined by inputs near but far enough away from $f_s/4$ where the attenuation introduced by the notch filter is not significant.

Calibrating gain mismatch and sample-time errors in the two-path system is already a challenging task that requires precise group delay filtering with very fine time steps. However, it would be even more demanding to do it for general N-path systems. Furthermore, even after the gain and sample-time error calibration, there remain the problems related to the bandwidth and INL mismatches. Therefore, although it is desirable to distribute the S/H function into each channel, it is still common to avoid the sample-time errors with a very fast up-front S/H that operates at the sampling rate. S/H limits the overall speed and also the number of channels that can be interleaved in practice. In fact, the broadband high-accuracy S/H has been the most challenging circuit block in high-resolution ADC designs.

7.8 Calibrated Continuous-Time (CT) ΔΣ Modulators

It is difficult to design broadband Nyquist-rate ADCs for high resolution, but they are still suitable for high SFDR applications because the linearity can be improved by calibration. In general, they consume high power, require high voltage for large signal swings, and suffer from poor SNR as capacitor sizes are minimized for high-frequency switched-capacitor operation. However, it is advantageous to operate CT ΔΣ modulators with low oversampling ratios as they require no opamps settling with high accuracy, and offer a good alternative to reach the same goal. Although single-loop designs have been predominantly used for high-order modulators, cascading low-order stages becomes a viable alternative to implement high-order modulators because they are easier to stabilize.

7.8.1 Pipeline versus Continuous-Time ΔΣ Modulator

In broadband CT ΔΣ modulator designs, it is inevitable to reduce the oversampling ratio (OSR) to maximize the signal bandwidth. Single-loop high-order modulators are difficult to stabilize due to the excessive loop delay contributed by extra nondominant poles. Cascading low-order stages helps to alleviate the stability problem while effectively implementing high-order noise shaping, and achieves wide bandwidth with low OSR. What

CT Integrator in ΔΣ Modulator   MDAC in Pipeline

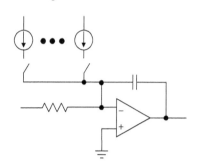 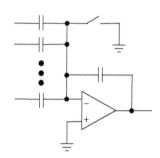

FIGURE 7.44
CT integrator versus MDAC.

single-loop modulator is to flash ADC is what cascaded modulator is to pipelined ADC. Therefore, the stability is a concern in the high-order single-loop design, but the residue accuracy becomes an issue in the cascaded design. As in the pipelined ADC, cascading modulators requires more accurate gain matching between stages. The single-loop design needs multibit DACs that require shuffling, but in the cascaded design it is also possible to use linear two- or trilevel DACs.

CT modulators give numerous advantages such as low voltage operation and no anti-aliasing requirement. CT integrators for ΔΣ modulators are quite different from MDAC residue amplifiers for pipelined ADCs, as shown in Figure 7.44.

The MDAC residue amplifier should settle with an absolute accuracy. However, the CT integrator needs no accurate settling because it operates inside the feedback loop, and its gain and nonlinearity errors are reduced by the loop gain. This opamp gain requirement is a great advantage over the open-loop system, but it also becomes a very limiting factor if interstage gain matching is required.

The speed advantage of the pipelined ADC over the ΔΣ modulator has always been by a factor of 2 ~ 4, but the gap is narrowed as technology is scaled down. However, closed-loop systems will most likely prevail. A good example of this trend is that the self-calibrated SAR was replaced with the ΔΣ modulator as an audio coder. The same competition between the pipelined ADC and the ΔΣ modulator would continue. While the pipelined ADC has been calibrated, the CT ΔΣ modulator has been updated with scaled digital technologies. For broadband applications, even the CT ΔΣ modulator may need some calibration as the OSR is lowered to 6 ~ 8 approaching the Nyquist rate.

7.8.2 Noise Leakage in CT Cascaded ΔΣ Modulator

In cascaded ΔΣ modulators, the digital noise cancellation filter (NCF) removes the quantization noises of the earlier stages and also shapes that of the last stage. If the NCF does not match with the analog filter, the quantization noise leaks to the output. The overall system noise performance will be degraded as a result, which defeats the whole purpose of cascading.

Figure 7.45 shows the simulated output spectrum of a 2-1-1 cascaded modulator with and without the quantization noise leakage from the first, second-order stage. Note that due to the leakage, the noise shaping is only second order. The noise leakage in the discrete-time (DT) cascaded modulator results mainly from the capacitor mismatch and finite opamp gain, but in the CT cascaded modulator there are more factors to consider. Two critical

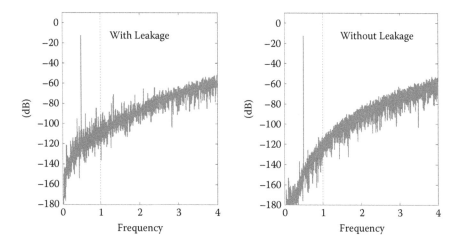

FIGURE 7.45
Output spectrum of 2-1-1 cascaded modulator with and without leakage.

ones are the inaccuracy of the digital NCF due to the approximation of the z-transform and the variation of the RC or C/G_m time constant over PVT variations. For high resolution, it is necessary to get an exact digital NCF from a CT-to-DT transform, and the analog filter time constant should be calibrated to match the digital NCF.

7.8.3 Continuous-Time to Discrete-Time Transform

Unlike the DT modulator, the loop filter function of the CT modulator is not well defined in the DT domain. Because the quantized feedback is in the DT sampled-data form, an effective way to analyze CT modulators is to use the CT-to-DT transform and to obtain their equivalent DT models. With the transform, the excess loop delay in the single-loop modulator can be adjusted using the feedback coefficients [32], and in the cascaded case, the quantization noise can be more accurately canceled by matching the digital NCF with the CT loop filter. However, the transform method has not been widely used because the transform actually depends on the shape of the feedback DAC pulse and is difficult to derive analytically.

Figure 7.46 shows the concept of the CT-to-DT transform. The CT modulator has a loop filter $H(s)$ and samples the signal before the quantizer. To model it as a DT modulator with a loop filter $H(z)$, the signals sampled by the quantizer $V(n)$ in both CT and DT cases should be identical when the same DAC input pulse $D(n)$ is applied. Because the DAC output in the CT modulator depends on the shape of the DAC pulse waveform, $V(n)$ sampled by the quantizer is different from the ideal DT DAC output, and so is the equivalent $H(z)$. The z-domain equivalence of the s-domain filter function can be obtained using the Laplace and z-transform of the CT DAC pulse and filter. If the DAC pulse shape is not well defined, any analytic solutions are prohibitively complicated to obtain.

The impulse invariant transform is based on the simple fact that the DAC pulses after filtered by the CT and DT filters should be identical in both cases when the quantizers sample them. Many research efforts have been made to analytically define the accurate transform, and they have been focused on the transform that is limited to the well-defined shapes of the DAC pulse. Some provide look-up tables for return-to-zero and non-return-to-zero (NRZ) DACs, and also for switched-capacitor (SC) DACs. The waveform affected by

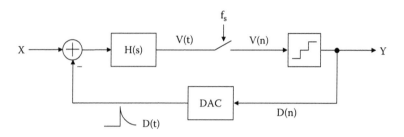

FIGURE 7.46
CT-to-DT transform concept.

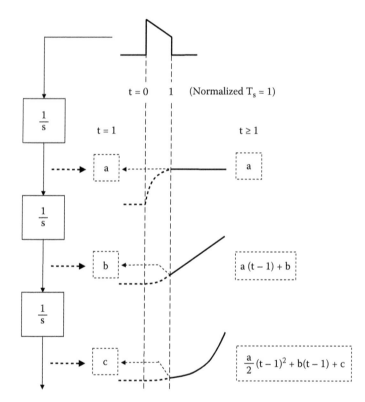

FIGURE 7.47
Parameters from integrator time responses.

nonideal factors is approximated with a pulse delay [33], and the transform is generalized for arbitrary DACs [34], but it is difficult to model arbitrary waveforms analytically. An intuitive way to avoid the mathematical complexity and to find the transform for arbitrary DAC pulse waveforms is to take a standard z-transform multiplied by the parameters set by the DAC pulse waveform [35].

7.8.4 CT-to-DT Transform of Integrators

Consider an arbitrary DAC pulse $t = 0 \sim T_s$ that passes through a series of integrators during the period T_s, where T_s is normalized to 1 as shown in Figure 7.47. The integrator outputs when $t \geq 1$ are simple polynomials of $(t - 1)$ as drawn with the thick solid lines. The high-order polynomials are obtained from the integrations of the previous ones. They are represented with the coefficients of a, b, and c, which are the outputs of the first, second, and third integrators when sampled at $t = 1$, respectively.

To find the DT equivalence, the time-domain polynomials are z-transformed. Because the z-transform of a unit digital pulse at $t = 0$ is 1, those DT equivalences are the DT functions having the same pulse responses as the CT integrators. Note that different DAC pulse shapes can lead to a different set of parameters of a, b, and c, but the polynomials and the DT equivalences remain the same. Therefore, the variation due to the DAC pulse shape can be simplified with a new set of parameters. These few key parameters at $t = 1$ ($t = T_s$) for arbitrary waveforms can be easily obtained using simulations with all DAC nonideal effects included. Based on the parameters, a look-up table of the CT-to-DT transform up to fifth-order integrator can be derived as shown in Figure 7.48.

Any loop filter lower than fifth order with all poles at DC can be expressed as a combination of these transforms. Simulation results with a half-clock delayed switched-capacitor DAC are shown in Figure 7.49.

The CT pulse responses of $1/s$, $1/s^2$, $1/s^3$, and $1/s^4$ are shown with the solid lines, and the DT counterparts are marked with the dashed lines. They are matched at every sampling point. That is, for the same $D(n)$ sequence in Figure 7.46, $V(n)$ sampled by DT and CT modulators are identical. Note also that although T_s is normalized to 1 in Figure 7.47, the z-domain equations stay the same for any T_s.

7.8.5 CT-to-DT Transform of Resonators

The other common circuit component used in the loop filter is a resonator that can distribute the in-band quantization noise evenly across the signal band. However, in cascaded architectures the resonators are rarely used because only simple first- or second-order stages are cascaded for stability. The same transform procedure can be generalized to include resonators. Figure 7.50 shows that the DAC pulse is fed into a resonator of $s/(s^2 + \omega^2)$, and then passes through a series of integrators. According to the Laplace transform, the pulse

$$\frac{1}{s} \leftrightarrow \frac{a}{z-1}$$

$$\frac{1}{s^2} \leftrightarrow \frac{a}{(z-1)^2} + \frac{b}{z-1}$$

$$\frac{1}{s^3} \leftrightarrow \frac{a(z+1)}{2(z-1)^3} + \frac{b}{(z-1)^2} + \frac{c}{z-1}$$

$$\frac{1}{s^4} \leftrightarrow \frac{a(z^2+4z+z)}{6(z-1)^4} + \frac{b(z+1)}{2(z-1)^3} + \frac{c}{(z-1)^2} + \frac{d}{z-1}$$

$$\frac{1}{s^5} \leftrightarrow \frac{a(z^3+11z^2+11z+1)}{24(z-1)^5} + \frac{b(z^2+4z+1)}{6(z-1)^4} + \frac{c(z+1)}{2(z-1)^3} + \frac{d}{(z-1)^2} + \frac{e}{z-1}$$

FIGURE 7.48
CT-to-DT transforms of integrators.

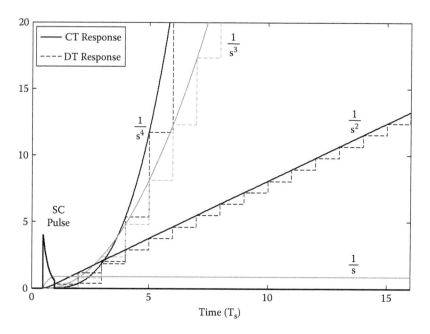

FIGURE 7.49
Integrator simulation with half-clock delayed switched-capacitor digital-to-analog converter (SC-DAC).

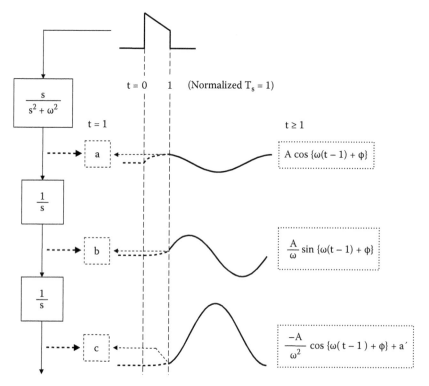

FIGURE 7.50
Parameters from resonator time responses.

responses of $s/(s^2 + \omega^2)$ and $1/(s^2 + \omega^2)$ are cosine and sine functions, respectively. In this example, the latter part integrates the resonator output.

Assume that the pulse response of $s/(s^2 + \omega^2)$ when $t \geq 1$ is $A\cos\{\omega(t-1) + \phi\}$, the pulse response of $1/(s^2 + \omega^2)$ when $t \geq 1$ is $(A/\omega)\sin\{\omega(t-1) + \phi\}$. The parameters of a and b when $t = 1$ are equivalent to $A\cos(\phi)$ and $(A/\omega)\sin(\phi)$, respectively. Thereby, the coefficients A and ϕ can be derived from the extracted parameters of a and b. Different DAC pulse shapes result in different sets of parameters a and b and also different values of A and ϕ.

For the function of $1/\{s^n(s^2 + \omega^2)\}$, where n is an integer, its pulse response when $t \geq 1$ is the integration of $1/\{s^{n-1}(s^2 + \omega^2)\}$. The pulse response of $1/\{s(s^2 + \omega^2)\}$ is also shown in Figure 7.50. It is a cosine function plus a constant of a', which is equal to $c + (A/\omega^2)\cos(\phi)$. Consequently, the CT-to-DT transform of $1/\{s^n(s^2 + \omega^2)\}$ can be made of two parts. One is a cosine/sine function, and the other is the integrator function. After z-transform, the CT-to-DT transform of resonators up to $n = 3$ are listed in Figure 7.51.

The parameters of c, d, and e are the output values of $1/\{s(s^2 + \omega^2)\}$, $1/\{s^2(s^2 + \omega^2)\}$, and $1/\{s^3(s^2 + \omega^2)\}$ at $t = 1$, respectively. When two resonators are cascaded, it can be considered as a sum of two individual resonators, for example,

$$\frac{s^2}{\left(s^2 + \omega_1^2\right)\left(s^2 + \omega_2^2\right)} = \frac{\omega_1^2}{\omega_1^2 - \omega_2^2}\frac{1}{s^2 + \omega_1^2} - \frac{\omega_2^2}{\omega_1^2 - \omega_2^2}\frac{1}{s^2 + \omega_2^2}. \tag{7.33}$$

However, if the resonant frequencies ω_1 and ω_2 are identical, an additional table of CT-to-DT transform is necessary, but the NTF zeros are usually placed at different frequencies in most $\Delta\Sigma$ modulator designs. Similar tables for integrators are also given in [32] and [36], and the table for resonators in [37] is with RZ and NRZ DAC pulses.

7.8.6 Half-Cycle Delay Effect

One common feedback DAC used in CT $\Delta\Sigma$ modulators is an NRZ pulse that is half-cycle delayed to avoid the comparator metastability problem. Such a pulse waveform that is not

$$\frac{s}{s^2 + \omega^2} \leftrightarrow \frac{az - a\cos\omega - b\omega\sin\omega}{z^2 - 2z\cos\omega + 1}$$

$$\frac{1}{s^2 + \omega^2} \leftrightarrow \frac{bz + \dfrac{a\sin\omega}{\omega} - b\cos\omega}{z^2 - 2z\cos\omega + 1}$$

$$\frac{1}{s(s^2 + \omega^2)} \leftrightarrow \frac{az - a\cos\omega - b\omega\sin\omega}{-\omega^2(z^2 - 2z\cos\omega + 1)} + \frac{c + \dfrac{a}{\omega^2}}{z - 1}$$

$$\frac{1}{s^2(s^2 + \omega^2)} \leftrightarrow \frac{bz + \dfrac{a\sin\omega}{\omega} - b\cos\omega}{-\omega^2(z^2 - 2z\cos\omega + 1)} + \frac{c + \dfrac{a}{\omega^2}}{(z-1)^2} + \frac{d + \dfrac{b}{\omega^2}}{z - 1}$$

$$\frac{1}{s^3(s^2 + \omega^2)} \leftrightarrow \frac{az - a\cos\omega - b\omega\sin\omega}{\omega^4(z^2 - 2z\cos\omega + 1)} + \frac{\left(c + \dfrac{a}{\omega^2}\right)(z+1)}{2(z-1)^3} + \frac{d + \dfrac{b}{\omega^2}}{(z-1)^2} + \frac{e - \dfrac{a}{\omega^4}}{z - 1}$$

FIGURE 7.51
CT-to-DT transforms of resonators.

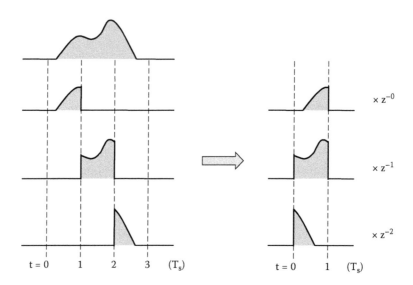

FIGURE 7.52
Arbitrary DAC pulse waveform.

bounded between $0 \sim T_s$ can be divided and spread over the consecutive sampling periods. An arbitrary DAC pulse spanning over 0 to $3T_s$ is shown in Figure 7.52.

It is divided into three pulses in the periods of $0 \sim T_s$, $T_s \sim 2T_s$, and $2T_s \sim 3T_s$, respectively. That is, the transform of the DAC pulse is a linear combination of them. The pulse between $T_s \sim 2T_s$ can be considered as a pulse between $0 \sim T_s$ delayed by one cycle. Therefore, it should be multiplied by z^{-1} after the CT-to-DT transform. Similarly, for the pulse between $2T_s \sim 3T_s$, the DT equivalence should be multiplied by z^{-2}.

Figure 7.53 shows simulation results of the resonators with a half-cycle delayed pulse between $T_s/2 \sim 3T_s/2$. The CT and DT pulse responses match well as expected at the sampling points as in the integrator case of Figure 7.49.

7.8.7 Noise Transfer Function (NTF) for Single-Loop CT ΔΣ Modulator

To find an exact NCF of the cascaded CT modulator, the NTF of each single-loop stage should be derived first.

Figure 7.54 shows a third-order single-loop CT modulator example. Its loop filter $H(s)$ refers to the filtering function from the DAC output to the quantizer input with an integrator and a resonator. Using Figures 7.48 and 7.51, the DT equivalence of $H(s)$ and the NTF of $1/\{1 + H(z)\}$ can be obtained as follows:

$$H(s) = \frac{a_1}{s} + \frac{a_2}{s^2 + \omega^2} + \frac{a_3}{s(s^2 + \omega^2)} + a_4(delay).$$

$$NTF(z) = \frac{1}{1 + CTD\{H(s)\}}.$$

(7.34)

where *CTD* refers to the CT-to-DT transform. When the DAC pulse is not bounded between $0 \sim T_s$, the portion after T_s results in extra delays to the loop filter $H(z)$. In such a case, an

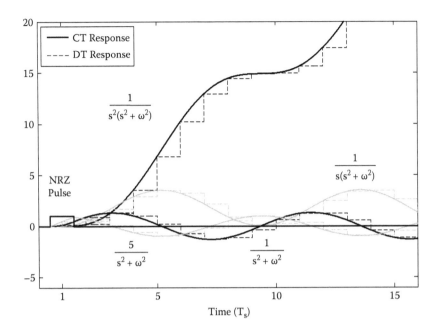

FIGURE 7.53
Resonator simulation with half-clock delayed NRZ DAC.

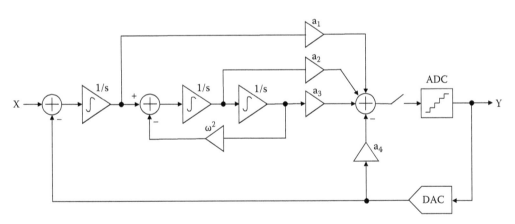

FIGURE 7.54
Single-loop modulator model to derive NTF.

extra feedback path of a_4 can be added to introduce an $a_4 z^{-1}$ term in $H(z)$ to compensate for the extra loop delay.

Note that the aliasing error in sampling is also suppressed by the NTF because the signal transfer function (STF) can be expressed as $H'(s)\text{NTF}(z)$, where $H'(s)$ can be defined as the filter function from the input to the sampling point at the quantizer input. $H'(s)$ does not need to be the same as the loop filter $H(s)$ in the feedback structure. Therefore, $H'(s)$ can be set differently from $H(s)$ so that the STF peaking in the feed-forward structure can be eliminated. However, the exact DT equivalence of $H'(s)$ cannot be defined precisely in the z domain because the signal is not a well-defined waveform like the DAC pulse.

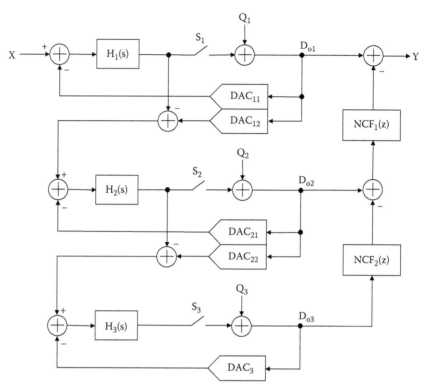

FIGURE 7.55
Cascaded modulator model to derive NCF.

7.8.8 NCF for CT Cascaded ΔΣ Modulator

Shown in Figure 7.55 is an example of a three-stage cascaded CT ΔΣ modulator. The NCF of the cascaded CT modulator is not as straightforward to derive as that of the DT modulator. It depends on the total filtering function of all the possible paths from the DAC output to the quantizer sampling point.

The output of each stage consists of the signal sampled by the quantizer and the quantization noise, which are filtered by its NTF. Assume that the input is 0 for simplicity. The output of the first stage D_{o1} contains only the quantization noise Q_1 and becomes

$$D_{o1} = Q_1 \times NTF_1(z). \tag{7.35}$$

The second-stage output D_{o2} contains both the quantization noise Q_2 and the signal sampled by the switch S_2. The signal sampled by S_2 has two components. One is the output of DAC_{11}, which is filtered by $H_1(s)$ and $H_2(s)$. The other is the output of DAC_{12}, which is filtered by $H_2(s)$. As a result, the D_{o2} can be expressed as

$$D_{o2} = \left[D_{o1} \times CTD\{H_1(s)H_2(s) + H_2(s)\} + Q_2 \right] \times NTF_2(z). \tag{7.36}$$

Similarly, D_{o3} includes the output of DAC_{11}, DAC_{12}, DAC_{21}, DAC_{22}, and the quantization noise Q_3. It is given by

$$D_{o3} = \begin{bmatrix} -D_{o1} \times CTD\{H_1(s)H_2(s)H_3(s) + H_2(s)H_3(s)\} \\ + D_{o2} \times CTD\{H_2(s)H_3(s) + H_3(s)\} + Q_3 \end{bmatrix} \times NTF_3(z). \tag{7.37}$$

To cancel Q_2 in D_{o2}, D_{o3} is multiplied by NCF_2 before summed with D_{o2}, and NCF_2 should be set to

$$NCF_2(z) = \frac{1}{CTD\{H_2(s)H_3(s) + H_3(s)\} \times NTF_3(z)} \approx \frac{\dfrac{1}{1+H_2}}{\dfrac{H_3}{1+H_3}} \approx \frac{NTF_2}{STF_3}. \tag{7.38}$$

Although NCF_2 looks complicated, it can be understood intuitively by ignoring the difference between s-domain and z-domain. Similar to the DT case, it is equivalent to NTF_2 divided by STF_3. To cancel Q_1 in D_{o1}, NCF_1 should be set to

$$NCF_1(z) = \frac{1}{CTD\{H_1(s)H_2(s)H_3(s) + H_2(s)H_3(s)\} \times NTF_3(z) \times NCF_2(z)} \tag{7.39}$$

$$\approx \frac{NTF_1}{STF_2}.$$

If properly canceled, the signal at the modulator output is

$$D_o = Q_3 \times NTF_3(z) \times NCF_2(z) \times NCF_1(z), \tag{7.40}$$

which implies that Q_3 is high-order noise shaped.

7.8.9 STF and Built-In Anti-Aliasing in Cascaded CT-DSM

The signal transfer function (STF) and the anti-aliasing effect of the CT cascaded modulator can be similarly analyzed. There are three samplings by switches S_1, S_2, and S_3. However, what is sampled by S_1 and S_2 is canceled digitally along with Q_1 and Q_2, and only the input component sampled by S_3 remains. Therefore, the input is filtered by $H_1(s)H_2(s)H_3(s)$ before being sampled by S_3. Consequently, the STF of the CT cascaded modulator can be represented by

$$STF = H_1(s)H_2(s)H_3(s) \times NTF_3(z) \times NCF_2(z) \times NCF_1(z) \approx STF_1. \tag{7.41}$$

Neglecting the difference between s-domain and z-domain, it is close to the first-stage STF and approaches unity at low frequencies. STF is a mixed function in both the s- and z-domains.

The built-in anti-aliasing function can also be analyzed with Equation (7.41). The aliased components at S_1 and S_2 are canceled by NCF_1 and NCF_2, respectively. The high-frequency input to be aliased is suppressed by $H_1(s)H_2(s)H_3(s)$ when sampled by S_3, and again by $NTF_3(z)NCF_2(z)NCF_1(z)$ after aliased into the low frequencies. Therefore, the built-in anti-aliasing effect stays the same. However, unless the digital NCF is accurately set, the aliased components at S_1 and S_2 leak to the output. As a result, the noise leakage calibration of the

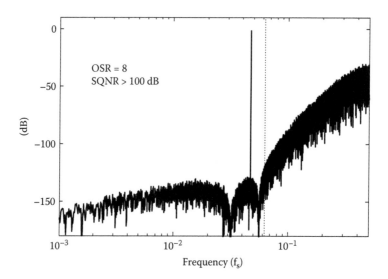

FIGURE 7.56
Simulated fast Fourier transform (FFT) of a 2-2-1 CT cascaded modulator example.

CT modulator not only helps to suppress the quantization noise but also to improve anti-aliasing function.

Based on the CT-to-DT transform and the derivations of NTF and NCF, a 2-2-1 CT cascaded modulator made of two resonators and one integrator is simulated as shown in Figure 7.56. Each resonator should be stabilized with two feed-forward zeros, and the comparator half-delay should also be canceled. In the example, the leaked quantization noise is canceled completely; therefore, the shapes of the in-band notches are clearly visible. This 2-2-1 fifth-order cascaded modulator example with a 4b DAC can achieve a signal-to-quantization-noise ratio (SQNR) of well over 100 dB with a −3 dB input at a low OSR of 8.

7.8.10 Noise Leakage Cancellation in CT Cascaded Modulators

The multistage cascaded design is more sensitive to the time-constant variation than the single-loop design. To minimize the in-band noise leakage, either the digital NCF or variable resistors can be calibrated with the modulator input grounded. PN dithering can also be used to calibrate the DT modulator by detecting its leakage in the background [38]. The PN-modulated random dither is spread over the wide band and is not effective for CT modulators because the frequency response of the CT loop filter is more influenced by the parasitic high-frequency poles and zeros. The filter time constant is effectively calibrated in the background using binary pulse injection [39], which focuses on the mismatch only around the frequency band of interest and is less sensitive to the broadband mismatch than the PN dither injection.

Figure 7.57 shows the simulation results of a 2-1-1 4b CT cascaded modulator with different time-constant errors when oversampling at ×10 and ×16. The noise leakage degrades the DR significantly, even with small time constant errors, but note that the stability is not affected even with large errors. The lower are the order and quantizer resolution of the first stage, the more sensitive is the DR to the time-constant error, because the leaked quantization noise to cancel is higher. Therefore, the filter time constants should be calibrated with a few percent tuning accuracy.

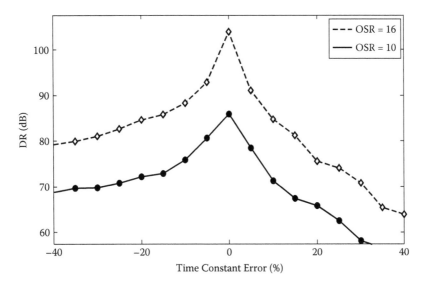

FIGURE 7.57
Dynamic range (DR) versus time constant error of a 2-1-1 CT cascaded modulator.

Ideally, the injected tones appearing in D_{o1} and the NCF output should have the same magnitude to cancel each other. However, the tone magnitude of D_{o1} is determined by the NTF_1, which approaches $1/H_1(s)$ at low frequencies and is sensitive to the filter time constant. On the other hand, the tone magnitude in the NCF output is fixed by digital filters and the STF_2, which is close to $H_2(s)/\{1 + H_2(s)\}$ and is about unity as long as $H_2(s)$ is high. This implies that after the noise cancellation, the polarity of the residual tone changes according to the error polarity of the mismatch between two tones. Therefore, the time constant can be tuned using the same zero-forcing adaptive LMS feedback similar to the zero detection in the single-loop modulator case discussed in Chapter 6.

That is, the noise leakage mechanism resulting from the mismatch between the analog transfer function and the digital NCF can be advantageously exploited to calibrate the very source of the mismatch based on the LMS algorithm. A single tone can be injected into the quantizer input to detect the incomplete noise cancellation as in the zero detection of the single-loop modulator. The injected tone is again considered as part of the quantization noise and should be canceled by the digital NCF. If the analog filter and the digital NCF are not matched, a residual tone remains in the digital output. Unlike the single-loop case, where the injected tone is rejected by the NTF zero, the quantization noise cancellation works at all frequencies, and the injected tone frequency can be chosen to be either inside or outside of the signal band. That is, if the filter time constant is calibrated by nulling the residual injected tone, the NCF also suppresses the noise leakage in the signal band.

7.8.11 Operational Amplifier Finite DC Gain and Bandwidth Effect

Although using low-order modulators greatly alleviates the stringent opamp gain and bandwidth requirement in the CT cascaded modulator, it is still an issue in achieving high DR over 90 dB.

Figure 7.58 shows system simulation results with both ideal integrators and nonideal integrators having a finite DC gain of 54 dB. Ideally, the modulator exhibits a clear 80 dB/dec

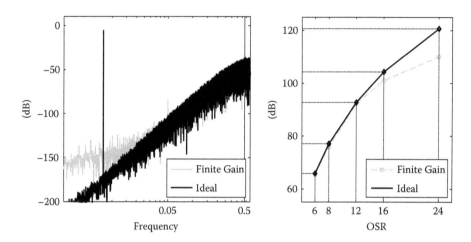

FIGURE 7.58
Simulated FFT and DR versus oversampling ratio (OSR) of a 2-1-1 cascaded modulator.

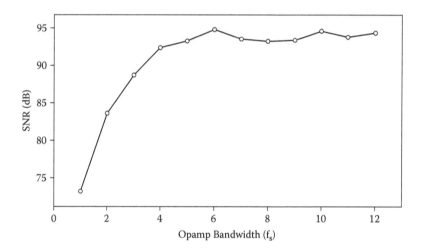

FIGURE 7.59
Operational amplifier (opamp) bandwidth effect.

noise shaping. The injected pulse at 1/20 sampling frequency and its harmonics are completely removed due to the exact NCF. With the finite DC gain, the leaked noise dominates the noise floor at low frequencies, and the residual tone appears. The influence of the finite DC gain is more prominent when OSR is high but is negligible when OSR is low. The modulator still achieves a simulated DR over 100 and 78 dB with an OSR of 16 and 8, respectively.

Figure 7.59 shows the SNR degradation of the same modulator with a −6 dB input due to the finite opamp bandwidth. The degradation on SNR mainly results from the inaccurate noise cancellation. The excess phase shift of a few degrees can contribute to the severe mismatch in the leakage cancellation. When the opamp bandwidth is about $4f_s$, the extra phase shift of the opamp at the unity loop-gain frequency of $f_s/2\pi$ will be

$$\tan^{-1}\left(\frac{f_s/2\pi}{4f_s}\right) \approx 2.3°. \tag{7.42}$$

That is, the opamp gain should be designed to be higher than 60 dB, and the opamp band-width should be wider than $4f_s$ to achieve over 90 dB DR with a 2-1-1 CT cascaded $\Delta\Sigma$ mod-ulator with a 4b feedback DAC. Furthermore, the time constant error should be trimmed within the accuracy range of 2 ~ 3%. These requirements are far less stringent than the residue matching requirements of any comparable pipelined ADCs.

7.9 Calibration of Current-Steering DAC

All stand-alone DACs suffer from both static and dynamic errors. In ADC, except for the S/H errors, the static nonlinearity error is dominant because the conversion is performed on sampled analog values. However, in DACs, because the DAC output is analog, no elabo-rate digital signal processing helps to improve the DAC performance.

7.9.1 Static DAC Nonlinearity Error

The static DAC error resulting from component mismatches can be improved by calibration or oversampling as is done in the ADC. Figure 7.60 shows the dynamic element-matching concept [2]. If two currents I_1 and I_2 are switched with a 50% duty and are low-pass filtered, two output currents can be matched because the static mismatch errors of two currents appear at the output with a 50% duty. This is to improve the accuracy of the binary ratio by multiplexing the ratio errors so that the complementary errors can be averaged out over time while the average values remain. It is the suppressed carrier balanced modulation of the error, and the modulated high-frequency energy can be filtered out. This technique relies on the accurate timing of the duty cycle. Any duty cycle error or timing jitter results in inaccurate matching, but the residual matching inaccuracy becomes a second-order error proportional to the product of the original mismatch and the timing error. In gen-eral, to match N independent elements, a switching network with N inputs and N outputs is required. The function of the network is to connect any one input to one output with an average duty cycle of $1/N$.

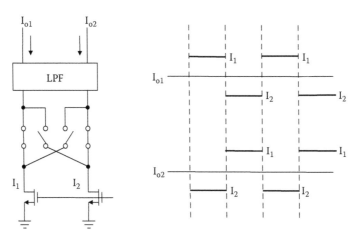

FIGURE 7.60
Dynamic current matching concept.

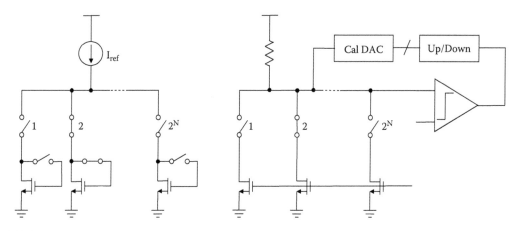

FIGURE 7.61
Current sampler and current digital-to-analog converter (DAC) calibration concept.

Figure 7.61 shows other improvement techniques for the static DAC error. The current sampler on the left is an electrical alternative to mechanical trimming [40]. The current sampler is often called a current copier. In this case, a voltage is sampled on the input capacitance of an MOS transistor to hold the current. The idea is to use the same current repeatedly to copy on other current sources so that the matching accuracy can be maintained as long as the sampling errors are kept constant. However, it is not practical to make a high-resolution DAC using the current copiers alone. Therefore, it is limited to generating MSB DACs for segmented DACs or subranging ADCs. The same is true to the current calibration concept on the right that measures current errors by quantizing and storing them as in the ADC calibration [27].

7.9.2 Dynamic DAC Nonlinearity Error

The dynamic linearity heavily depends on how the DAC output settles. Only configuring the DAC topology carefully helps to reduce such error sources. In fact, the high-linearity DAC design is more challenging than the ADC design. Generating wideband spectrum with high static and dynamic linearity is therefore not a trivial task. Furthermore, in scaled CMOS, devices are more nonlinear, and low supply voltages degrade linearity further. In the current-steering DAC, each current source is switched to the output at different times due to the clock skew, and adds parasitic capacitance to the output node, which results in code-dependent output settling. The clock skew and the code-dependent settling are the most prominent among many factors limiting dynamic DAC linearity. Isolating the output node from the current sources and synchronizing the code transitions are the simple solutions for these, but the clock skew is still a limiting factor when dealing with high precision. In theory, an output deglitcher can improve the dynamic linearity, but the deglitcher is functionally a track-and-hold circuit and is also difficult to implement at high sampling rates with high precision.

The deglitcher problem can be solved using the return-to-zero (RZ) scheme as shown in Figure 7.62. While the track-and-hold static error stays for the half clock period, the error in the RZ DAC exponentially decays. The logic behind this RZ scheme is rather than holding analog voltage accurately, it is easier to let the output keep changing exponentially because exponential settling is a linear process. This effect as RZ can be achieved if the output is

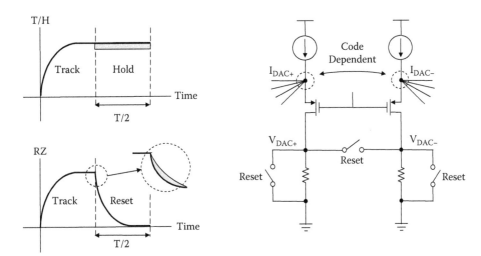

FIGURE 7.62
Track and reset return-to-zero (RZ) DAC.

also tracked and attenuated [27]. The time-domain errors as given by the shaded areas in Figure 7.62 in two cases are as follows:

$$Track \ and \ Hold \ Error = V_{error} \times \frac{T}{2}.$$

$$Track \ and \ Attenuate \ Error = V_{error} \times \tau \left(1 - e^{-\frac{T}{2\tau}}\right).$$

(7.43)

As shown, the differential DAC output current comes through the folded-cascode nodes, which are isolated from the output nodes. The output nodes track the signal during the half clock period but are shorted by three switches for attenuation during the remaining half clock period. The difference of the RZ DAC from the conventional one is that the output spectrum is nulled at multiples of twice the sampling frequency, which in turn provides less attenuation in the pass-band due to the sinx/x (SINC) function.

7.9.3 DAC in Feedback

Both Nyquist-rate ADC and DAC are operating in open-loop modes, and their performance is limited by most analog accuracies such as matching, absolute gain, and DAC nonlinearity. The output of the ADC is digital, and its performance can be enhanced by extensive digital postprocessing, such as calibration. However, the output of DAC is analog, and digital predistortion is not effective to enhance the DAC nonlinearity. Static DAC nonlinearity errors can be either calibrated or predistorted. In particular, it is difficult to predistort or calibrate errors resulting from dynamic nonlinearities. Only fast technology helps to reduce dynamic errors.

The performance of ADC can be improved if it can be placed inside a feedback loop as in the ΔΣ modulator. However, operating DAC in feedback requires a feedback ADC to complete the loop as compared in Figure 7.63. If oversampled, the ADC performance is improved by the amount of the loop gain, and the DAC error can be shuffled to be pushed out of band.

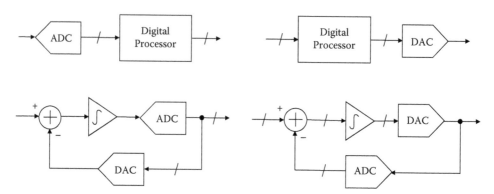

FIGURE 7.63
Open-loop and closed-loop operation of an ADC and a DAC.

However, in the closed-loop DAC operation, the ADC error is out of band, and its delay contributes adversely to the loop stability. Many efforts to correct analog nonlinearities have been tried using similar feedback approaches to linearize RF power amplifiers.

References

1. B. S. Song, M. F. Tompsett, and K. R. Lakshmikumar, "A 12-Bit 1-Msample/s capacitor error averaging pipelined A/D converter," *IEEE J. Solid-State Circuits*, vol. SC-23, pp. 1324–1333, December 1988.
2. R. van de Plassche, "Dynamic element matching for high-accuracy monolithic D/A converters," *IEEE J. Solid-State Circuits*, vol. SC-11, pp. 795–800, December 1976.
3. Z. G. Boyacigiller, B. Weir, and P. D. Bradshaw, "An error-correcting 14b/20 ps CMOS A/D converter," *Dig. Tech. Papers, IEEE Int. Solid-State Circuits Conf.*, pp. 62–63, February 1981.
4. J. C. Domogalla, "Combination of analog to digital converter and error correction circuit," U.S. Patent 4 451 821, May 1984.
5. H. S. Lee, D. A. Hodges, and P. R. Gray, "A self-calibrating 15 bit CMOS A/D converter," *IEEE J. Solid-State Circuits*, vol. SC-19, pp. 813–819, December 1984.
6. T. Tewksbury and H. S. Lee, "Characterization, modeling, and minimization of transient threshold voltage shifts in MOSFET's," *IEEE J. Solid-State Circuits*, vol. SC-29, pp. 239–252, March 1994.
7. B. S. Song, S. H. Lee, and M. F. Tompsett, "A 10b 15MHz CMOS recycling two-step A/D converter," *IEEE J. Solid-State Circuits*, vol. SC-25, pp. 1328–1338, December 1990.
8. S. H. Lee and B. S. Song, "Digital-domain calibration techniques for multi-step A/D converters," *IEEE J. Solid-State Circuits*, vol. SC-27, pp. 1679–1688, December 1992.
9. A. N. Karanicolas, H. S. Lee, and K. Bacrania, "A 15-b 1-Msample/s digitally self-calibrated pipeline ADC," *IEEE J. Solid-State Circuits*, vol. SC-28, pp. 1207–1215, December 1993.
10. D. W. J. Groeneveld, H. J. Schouwenaars, H. A. H. Termeer, and C. A. A. Bastiaansen, "A self-calibration technique for monolithic high-resolution D/A converters," *IEEE J. Solid-State Circuits*, vol. SC-24, pp. 1517–1522, December 1989.
11. J. M. Ingino and B. A. Wooley, "A continuously calibrated 12-b, 10-MS/s, 3.3-V A/D converter," *IEEE J. Solid-State Circuits*, vol. SC-33, pp. 1920–1931, December 1998.
12. U. K. Moon and B. S. Song, "Background digital calibration techniques for pipelined ADC's," *IEEE Trans. Circuits and Systems II*, vol. 44, pp. 102–109, February 1997.

13. S. U. Kwak, B. S. Song, and K. Bacrania, "A 15b 5Msample/s low-spurious CMOS ADC," *IEEE J. Solid-State Circuits*, vol. SC-32, pp. 1866–1875, December 1997.
14. E. B. Blecker, T. M. McDonald, O. E. Erdogan, P. J. Hurst, and S. H. Lewis, "Digital background calibration of an algorithmic analog-to-digital converter using a simplified queue," *IEEE J. Solid-State Circuits*, vol. SC-38, pp. 1059–1062, June 2003.
15. R. C. Dixon, *Spread Spectrum Systems*, 3rd ed., New York: John Wiley and Sons, 1994.
16. R. Jewett, K. Poulton, K. C. Hsieh, and J. Doernberg, "A 12b 128Msample/s ADC with 0.05LSB DNL," *Dig. Tech. Papers, IEEE Int. Solid-State Circuits Conf.*, pp. 138–139, February 1997.
17. J. Ming and S. H. Lewis, "An 8-bit 80-Msample/s pipelined analog-to-digital converter with background calibration," *IEEE J. Solid-State Circuits*, vol. SC-36, pp. 1489–1497, October 2001.
18. I. Galton, "Digital cancellation of D/A converter noise in pipelined A/D converters," *IEEE Trans. Circuits and Systems II*, vol. 47, no. 3, pp. 185–196, March 2000.
19. E. Siragusa and I. Galton, "A digitally enhanced 1.8 V 15 b 40 MS/s CMOS pipelined ADC," *IEEE J. Solid-State Circuits*, vol. SC-39, pp. 2126–2138, December 2004.
20. H. C. Liu, Z. M. Lee, and J. T. Wu, "A 15-b 40-MS/s CMOS pipelined analog-to-digital converter with digital background calibration," *IEEE J. Solid-State Circuits*, vol. SC-40, pp. 1047–1056, May 2005.
21. Y. Shu and B. S. Song, "A 15-bit linear 20-MS/s pipelined ADC digitally calibrated with signal-dependent dithering," *IEEE J. Solid-State Circuits*, vol. SC-43, pp. 342–350, February 2008.
22. B. Murmann and B. Boser, "A 12 b 75 MS/s pipelined ADC using open-loop residue amplification," *IEEE J. Solid-State Circuits*, vol. SC-38, pp. 2040–2050, December 2003.
23. J. P. Keane, P. J. Hurst, and S. H. Lewis, "Background interstage gain calibration technique for pipelined ADCs," *IEEE Trans. Circuits and Systems I*, vol. 52, pp. 32–43, January 2005.
24. A. Panigada and I. Galton, "A 130 mW 100 MS/s pipelined ADC with 69dB SNDR enabled by digital harmonic distortion correction," *IEEE J. Solid-State Circuits*, vol. SC-44, pp. 3314–3328, December 2009.
25. B. Widrow, J. McCool, and M. Ball, "The complex LMS algorithm," *Proc. IEEE*, vol. 63, pp. 719–720, April 1975.
26. S. T. Ryu, S. Ray, B. S. Song, G. H. Cho, and K. Bacrania, "A 14b-linear capacitor self-trimming pipelined ADC," *IEEE J. Solid-State Circuits*, vol. SC-39, pp. 2046–2051, November 2004.
27. A. R. Bugeja and B. S. Song, "A self-trimming 14b 100MS/s CMOS DAC," *IEEE J. Solid-State Circuits*, vol. SC-35, pp. 1841–1852, December 2000.
28. T. H. Shu, B. S. Song, and K. Bacrania, "A 13-b 10-Msample/sec ADC digitally calibrated with real-time oversampling calibrator," *IEEE J. Solid-State Circuits*, vol. SC-30, pp. 443–452, April 1995.
29. M. J. Choe, B. S. Song, and K. Bacrania, "A 13b 40MS/s CMOS pipelined folding ADC with background offset trimming," *IEEE J. Solid-State Circuits*, vol. SC-35, pp. 1781–1790, December 2000.
30. C. Tsang, Y. Chiu, J. Vanderhaegen, S. Hoya, C. Chen, R. Brodersen, and B. Nikolic, "Background ADC calibration in digital domain," pp. 21–24, *Custom Integrated Circuits Conf.*, September 2008.
31. S. M. Jamal, D. Fy, N. C. Chang, P. Hurst, and S. H. Lewis, "A 10-b 120Msample/s time-interleaved analog-to-digital converter with digital background calibration," *IEEE J. Solid-State Circuits*, vol. SC-37, pp. 1618–1627, December 2002.
32. J. Cherry and W. Snelgrove, "Excess loop delay in continuous-time delta-sigma modulators," *IEEE Trans. Circuits and Systems II*, vol. 46, no. 4, pp. 376–389, April 1999.
33. S. Loeda, H. M. Reekie, and B. Mulgrew, "On the design of high-performance wide-band continuous-time sigma–delta converters using numerical optimization," *IEEE Trans. Circuits and Systems I*, vol. 53, no. 4, pp. 802–810, April 2006.
34. O. Oliaei, "Design of continuous-time sigma–delta modulators with arbitrary feedback waveform," *IEEE Trans. Circuits and Systems II*, vol. 50, no. 8, pp. 437–444, August 2003.
35. Y. Shu, J. Kamiishi, K. Tomioka, K. Hamashita, and B. S. Song, "LMS-based noise leakage calibration of cascaded continuous-time ΔΣ modulators," *IEEE J. Solid-State Circuits*, vol. SC-45, pp. 368–379, February 2010.

36. M. Ortmanns, F. Gerfers, and Y. Manoli, "A case study on a 2-1-1 cascaded continuous-time sigma-delta modulator," *IEEE Trans. Circuits and Sysems I*, vol. 52, no. 8, pp. 1515–1525, August 2005.

37. O. Shoaei and W. M. Snelgrove, "Design and implementation of a tunable 40MHz–70MHz Gm-C bandpass $\Delta\Sigma$ modulator," *IEEE Trans. Circuits and Systems II*, vol. 44, no. 7, pp. 521–530, July 1997.

38. P. Kiss, J. Silva, A. Wiesbauer, A. Wiesbauer, T. Sun, U. Moon, J. Stonick, and G. Temes, "Adaptive digital correction of analog errors in MASH ADC's—Part II: Correction using test-signal injection," *IEEE Trans. Circuits and Systems II*, vol. 47, no. 7, pp. 629–638, July 2000.

39. Y. S. Shu, B. S. Song, and K. Bacrania, "A 65nm CMOS CT $\Delta\Sigma$ modulator with 81dB DR and 8MHz BW auto-tuned by pulse injection," *Dig. Tech. Papers, IEEE Int. Solid-State Circuits Conf.*, pp. 500–501, February 2008.

40. D. W. J. Groeneveld, H. J. Schouwenaars, H. A. H. Termeer, and C. A. A. Bastiaansen, "A self-calibration technique for monolithic high-resolution D/A converters," *IEEE J. Solid-State Circuits*, vol. SC-24, pp. 1517–1522, December 1989.

8

Phase-Locked Loop Basics

Phase-locked loop (PLL) is one of the most common functional blocks in modern digital communications systems. PLL has been used for demodulating amplitude modulation (AM) and frequency modulation (FM) signals and also for generating local carriers in radio systems. Its applications in digital communications cover clock and data recovery (CDR), clock generation and multiplication, clock synchronization, and clock distribution. In the system-on-chip (SoC) environment, the trend is that most PLL functions are implemented predominantly in the digital domain. Even high-speed CDR functions in the disk drive and the gigabit optical networking are being digitized. One thing to note is that PLL is a nonlinear feedback system like the $\Delta\Sigma$ modulator, and it is not possible to analyze it in exact mathematical terms. However, the small-signal analysis based on the linear model is still useful to understand PLL operation and predict its behavior. Like any other electrical systems such as feedback operational amplifiers (opamps) and $\Delta\Sigma$ modulators, PLL is a useful negative feedback system with only two poles and one zero. What has been discussed for other feedback systems such as Bode plot and phase margin can be directly applied to PLL.

8.1 Phase Noise

Any electrical signals can be represented as a sum of many sinusoidal waveforms. A sinusoidal waveform $A\cos(\omega t + \phi)$ is defined by its amplitude, frequency, phase, and time. Noise is defined as power per unit bandwidth of their random values (ΔA, Δf, $\Delta\phi$, and Δt). Therefore, they are related to each other as explained in Figure 8.1.

Voltage and current amplitude noises are handled with their power spectral densities like V^2/Hz or A^2/Hz. Similarly, units of power spectral densities used for phase, frequency, and time noises are defined as rad^2/Hz, Hz^2/Hz, and sec^2/Hz, respectively. Uncertainties in phase, frequency, and time are related in the time and frequency domains by $\phi = \omega t$ and $\omega = s\phi$, respectively.

In traditional analog systems, the fidelity of analog waveforms has been most important, and the amplitude noise defines the minimum resolvable signal. However, in modern digital communications systems, electronics operate at far higher frequencies, and information is mainly encoded into either phase or frequency because it is very difficult to maintain good waveform fidelity at high radio frequencies.

Figure 8.2 shows that the analog waveform is degraded by amplitude noise as shown on the left side. Noises in phase, frequency, or time matter more in high-speed electronic systems because as shown on the right side, phase noise and jitter occur only during the brief transient periods while the amplitude noise is suppressed due to nonlinearity or clipping most of the time.

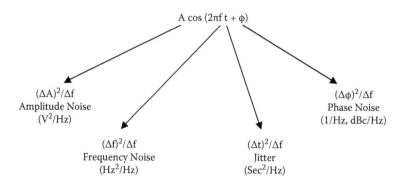

FIGURE 8.1
Noises in sine wave.

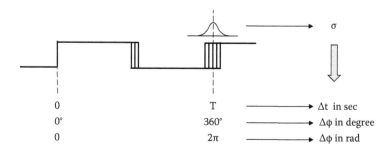

FIGURE 8.2
Amplitude versus phase, frequency, and jitter noises.

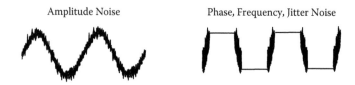

FIGURE 8.3
Cycle-to-cycle jitter and phase noise.

8.1.1 Jitter versus Integrated Root-Mean-Square Phase Noise

The waveform period T varies from cycle to cycle. Noise in time can be handled as a cycle-to-cycle jitter and defined as a root-mean-square (RMS) error of its period as shown in Figure 8.3.

Assume that the period error Δt of each cycle has a Gaussian distribution with a standard deviation σ. Then σ becomes an RMS jitter Δt_{rms}. This is equivalent to an integrated RMS phase noise in either degrees or radians when the period T is scaled by $360°$ or 2π, respectively.

For example, consider 1 MHz and 1 GHz carriers. If the RMS jitter is 100ps, their integrated RMS phase noises are different because the same jitter is normalized to the different periods of 1 μs and 1 ns, respectively.

$$\Delta\phi_{rms} = 2\pi f \times \Delta t_{rms} = 2\pi \times 1MHz \times 100ps \approx 0.63mrad \approx 0.036^{\circ}.$$

$$= 2\pi f \times \Delta t_{rms} = 2\pi \times 1GHz \times 100ps \approx 0.63rad \approx 36^{\circ}.$$

(8.1)

For the integrated RMS phase noise to stay the same, the jitter on the 1 GHz carrier should be 1000 times lower to be 0.1ps. This implies that it becomes harder to obtain the same level of integrated RMS phase noise at higher frequencies. Only faster technology can offer such an advantage as technology is scaled down. Therefore, for lower phase noise, it is advantageous to process signals at lower frequencies.

8.1.2 Amplitude-Modulation to Phase-Modulation Conversion

Although phase noise affects only zero crossings during the brief transient periods, both amplitude-modulation (AM) and phase-modulation (PM) signals have the same power spectral sidebands. Even additive amplitude noise appears as phase noise when the waveform is clipped or limited. It is called AM-to-PM conversion.

As shown in Figure 8.4, assume that a small side tone of ΔA is close but $\Delta\omega$ away from the carrier A. If $\Delta A \ll A$, the carrier waveform can be written as

$$v(t) = A\cos\omega_c t + (\Delta A)\cos(\omega_c - \Delta\omega)t$$

$$= \{A + (\Delta A)\cos(\Delta\omega)t\}\cos\omega_c t + (\Delta A)\sin(\Delta\omega)t\sin\omega_c t$$

$$= M\cos(\omega_c t - \theta).$$

(8.2)

$$\theta = \tan^{-1}\left\{\frac{(\Delta A)\sin(\Delta\omega)t}{1 + (\Delta A)\cos(\Delta\omega)t}\right\} \approx (\Delta A)\sin(\Delta\omega)t.$$

Note that the variation in the amplitude disappears after clipping or limiting, but the phase variation survives. The amplitude tone became a phase noise with the peak deviation and frequency of ΔA and $\Delta\omega$, respectively. AM carriers show two side tones of $(\Delta A)/2$, each appearing symmetrically at $\pm\Delta\omega$ away on both sides of the carrier.

8.1.3 Voltage-Controlled Oscillator Phase Noise

Another mechanism that converts voltage noise into phase noise is the voltage-to-phase, voltage-to-frequency, or voltage-to-time conversion. PLL is a negative feedback system in the phase domain, and a voltage-controlled oscillator (VCO) becomes a key element in

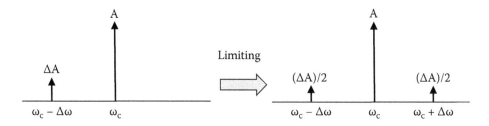

FIGURE 8.4
Amplitude-modulation to phase-modulation (AM-to-PM) conversion.

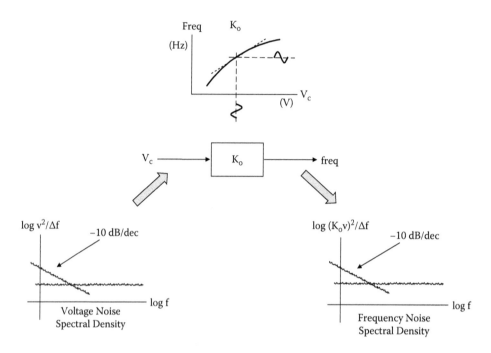

FIGURE 8.5
Voltage-to-frequency conversion by voltage-controlled oscillator (VCO).

the PLL system because it performs such an operation that links voltage and frequency as shown in Figure 8.5.

The output of the VCO is frequency, and the input is controlled by voltage. The gain K_o of the VCO is typically in the range of MHz/V ~ GHz/V. Assume that there is an equivalent thermal noise and $1/f$ noise at the input of the VCO as sketched. The VCO translates this voltage noise into the frequency domain after multiplying it by the VCO gain K_o. The power spectral density of the frequency noise is obtained from the input voltage noise as follows:

$$\frac{(\Delta f)^2}{\Delta f} = K_o^2 \times \frac{v^2}{\Delta f}. \tag{8.3}$$

As expected, the units of the input voltage noise and the output frequency noise are V^2/Hz and Hz^2/Hz, respectively.

From the frequency and phase relation of $\phi = \omega t = 2\pi ft$, the following phase noise can be derived after being normalized to $\phi = 1$ rad:

$$\Delta\phi = \frac{\Delta\phi}{\phi} = \frac{\Delta\omega}{\omega} = \frac{\Delta f}{f}. \tag{8.4}$$

That is, frequency and phase noises are related as frequency is integrated to be phase, or phase is differentiated to be frequency.

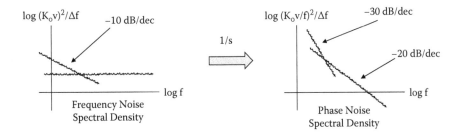

FIGURE 8.6
VCO phase noise.

If the phase noise is normalized to 1 rad, it becomes a unit-less ratio as in Equation (8.4). Therefore, it is common to use a relative scale of dBc/Hz for the phase noise spectral density, where dBc refers to dB relative to the carrier. Note that the integrated RMS phase noise is an absolute value with a unit of degrees or radians.

Figure 8.6 shows the VCO phase noise spectral density derived by integrating the frequency noise. From Equations (8.3) and (8.4), the phase noise spectral density of the VCO output is related to the VCO input noise as follows:

$$\frac{\left(\Delta\phi\right)^2}{\Delta f} = \frac{1}{f^2} \times \frac{\left(\Delta f\right)^2}{\Delta f} = \frac{K_o^2}{f^2} \times \frac{v^2}{\Delta f}. \tag{8.5}$$

Due to the integration, the slope of the phase noise is $1/f^2$, which is –20 dB/dec except at low frequencies, where the slope is –30 dB/dec due to the low-frequency flicker ($1/f$) noise. That is, the thermal noise at the input of the VCO produces the white frequency noise at the VCO output. The frequency noise is integrated to make the phase noise.

All noise spectral densities in the VCO are shown in Figure 8.7 with only a 1 kΩ thermal noise without the flicker noise at the VCO input and a VCO gain of 20 MHz/V. In this example, the 1 kΩ thermal noise is about 4 nV per unit bandwidth. Therefore, by multiplying it by the VCO gain, the VCO output phase noise spectral density of about $6.4 \times 10^{-11}/$ Hz is obtained, which amounts to –102 dBc/Hz at 10 kHz offset. Commonly, VCO phase noise spectral densities exhibit about –20 dB/dec slope.

8.1.4 Single-Sideband (SSB) and Double-Sideband (DSB) Phase Noises

Assume ω_c is the carrier frequency, and $\Delta\omega$ is the offset frequency away from ω_c. If the phase noise $|\phi_n(t)|$ is small, and has a magnitude of $\Delta\phi_m$ and a frequency of $\Delta\omega$,

$$|\phi_n(t)| = \Delta\phi_m \cos \Delta\omega t \ll 1 \ rad. \tag{8.6}$$

Then any carrier with a phase noise $\phi_n(t)$ can be rewritten using trigonometric relations as

$$v(t) = \cos\left\{\omega_c t + \phi_n(t)\right\} \approx \cos\omega_c t - \phi_n(t)\sin\omega_c t$$

$$\approx \cos\omega_c t - \frac{\Delta\phi_m}{2}\left\{\sin\left(\omega_c + \Delta\omega\right)t - \sin\left(\omega_c - \Delta\omega\right)t\right\}. \tag{8.7}$$

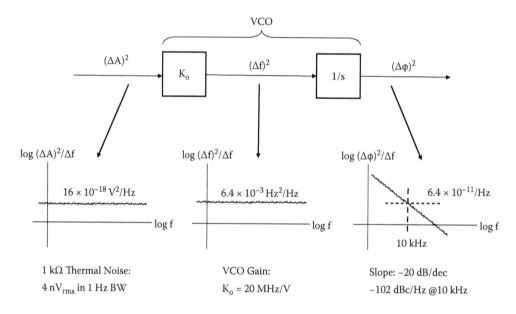

FIGURE 8.7
VCO noise spectral densities.

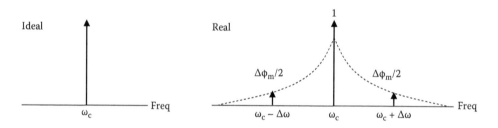

FIGURE 8.8
Ideal carrier versus real carrier with phase noise.

This implies that the carrier with a peak phase error of $\Delta\phi_m$ has symmetric side tones of $\Delta\phi_m/2$ at frequency offsets of $\pm\Delta\omega$ as shown in Figure 8.8.

The magnitude of this side tone decreases with a slope of -20 dB/dec as an inverse function of the offset frequency measured from the carrier. The single-sideband (SSB) phase noise spectral density at certain offset is therefore defined as phase noise power per unit bandwidth as follows:

$$SSB\ Phase\ Noise = \left(\frac{\Delta\phi_m}{2}\right)^2 (1/Hz)$$

$$= 10 \times \log\left(\frac{\Delta\phi_m}{2}\right)^2 (dBc/Hz).$$

(8.8)

Note that the unit of the phase noise spectral density is unit-less 1/Hz because it is already normalized to 1 rad, and dBc/Hz in log scale as explained in Equation (8.4).

In radio-frequency (RF) receivers, the SSB phase noise sets a lower limit on how much blocker channels are mixed down to the baseband. On the other hand, the double-sideband (DSB) phase noise is used to determine the total integrated RMS phase noise that affects signal-to-noise ratio (SNR) in the demodulation. The DSB phase noise spectral density is just the RMS noise power density adding two SSB powers on both sides:

$$DSB\ Phase\ Noise = 2 \times \left(\frac{\Delta\phi_m}{2}\right)^2 (1/Hz)$$

$$= 10 \times \log \frac{(\Delta\phi_m)^2}{2} (dBc/Hz).$$

(8.9)

Mathematically, the phase noise in Equation (8.4) is the same definition as the modulation index of the frequency-modulation (FM) system. If modulated with a peak voltage of ΔV_{peak}, VCO makes an FM signal with a modulation index of

$$m = \frac{\Delta f_{peak}}{f_m} = \frac{K_o \times \Delta V_{peak}}{f_m},$$

(8.10)

where Δf_{peak} is the peak frequency deviation, and f_m is the modulating frequency. The narrowband FM spectrum with $m \ll 1$, the Bessel polynomial gives the coefficients of

$$J_0(m) = 1, \qquad J_1(m) = \frac{m}{2}, \qquad J_2(m) = 0,.... .$$

(8.11)

Therefore, the SSB FM side tone spectrum can be defined as

$$SSB\ FM\ Side\ Tone = \left(\frac{m}{2}\right)^2 (1/Hz),$$

(8.12)

which is identical to Equation (8.8) if $m = \Delta\phi_m$. That is, the narrowband FM spectrum has two side tones of $m/2$ each at a frequency offset of $\pm f_m$. Note that the side tones in the spectrum are identical to those of the carrier modulated by AM and FM noises.

8.1.5 Effect of Frequency Division on Phase Noise

Phase noise is the modulation index of the FM, but it is scaled by the carrier frequency. Consider a carrier with its phase defined using the modulation index as in Equation (8.10).

$$\phi(t) = \omega_c t + \left(\frac{\Delta f_{peak}}{f_m}\right) \sin \omega_m t.$$

(8.13)

Its frequency can be obtained by differentiating the phase.

$$\omega(t) = \omega_c + \left(\frac{\Delta f_{peak}}{f_m}\right)\omega_m \cos \omega_m t. \tag{8.14}$$

Assume that a frequency divider divides this frequency by N without adding any of its own phase noise. Then the divided frequency becomes

$$\frac{\omega(t)}{N} = \frac{\omega_c}{N} + \left(\frac{\Delta f_{peak}}{N f_m}\right)\omega_m \cos \omega_m t. \tag{8.15}$$

This implies that the modulation index is divided, and the phase noise is improved by the divider ratio N. This is because the modulating frequency stays the same while the carrier frequency is divided. That is, the period of the carrier is longer after divided, and the phase noise gets smaller as a result.

8.2 Phase-Locked Loop Operation

Like other negative feedback systems such as $\Delta\Sigma$ modulators, PLL is a nonlinear system, but the small-signal analysis can help to understand its operation. It is made of three functional blocks: a phase detector (PD), a low-pass loop filter (LPF), and a voltage-controlled oscillator (VCO) as shown in Figure 8.9.

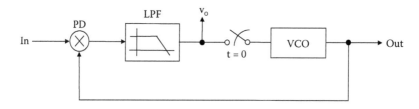

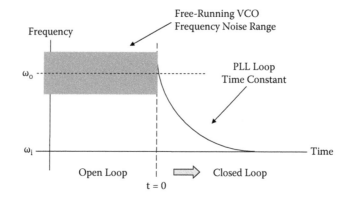

FIGURE 8.9
Operation of PLL.

The PLL input can be defined as either the input phase ϕ_i or frequency ω_i. In steady state, they are related as follows:

$$\phi = \frac{\omega}{s}. \tag{8.16}$$

For most applications, such as clock recovery and synthesis, both input and output are phases ϕ_i and ϕ_o, respectively. However, for FM demodulation the output can be taken from the VCO input v_o while the input is frequency ω_i.

PD detects the phase difference between input phase ϕ_i and VCO phase ϕ_o, and outputs a voltage proportional to the phase difference $(\phi_i - \phi_o)$. VCO is an oscillator, whose frequency is controlled by this voltage. Assume that the loop is initially opened. Then, there is no feedback, and the VCO free-runs at its own frequency ω_o, which can be different from the input frequency ω_i but within the frequency noise range. If the loop is closed at $t = 0$, the phase (frequency) difference is fed back as an error to correct the VCO frequency to be ω_i.

For example, the simplest PD is an analog multiplier. If the input frequency is $\cos(\omega_i t)$, and the VCO output frequency is $\cos(\omega_o t)$, the multiplied output will have two frequency components. One is $(\omega_i + \omega_o)$, and the other is $(\omega_i - \omega_o)$. The high-frequency component $(\omega_i + \omega_o)$ is attenuated (filtered out) by the loop LPF, and the phase (frequency) difference error $(\omega_i - \omega_o)$ remains at the input of the VCO. Initially, the LPF output is a beat frequency $(\omega_i - \omega_o)$ in the open-loop condition, but after the loop is closed, the beat frequency should approach zero as the VCO frequency follows the input frequency. Therefore, PLL can be considered as a phase/frequency follower like an opamp voltage follower in unity-gain feedback. This transient for frequency error correction and phase acquisition decays exponentially with a time constant of the PLL if the small-signal linear analysis holds.

8.2.1 Linear Model of PLL

A linear mathematical model of PLL can be drawn as shown in Figure 8.10, where $F(s)$ is the loop-filter gain, and the units of the PD gain k_d and the VCO gain k_o are V/rad and Hz/V, respectively.

Note that the VCO model is divided into two parts as explain in Figure 8.7. One is voltage-to-frequency conversion, and the other is frequency-to-phase conversion, which is just an integration 1/s as in Equation (8.16).

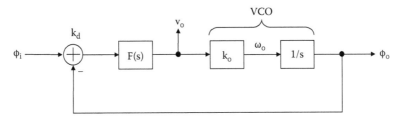

FIGURE 8.10
Linear model of PLL.

The following closed-loop transfer function $H(s)$ can be derived in a standard form:

$$H(s) = \frac{\phi_o(s)}{\phi_i(s)} = \frac{T(s)}{1 + T(s)} = \frac{k_d F(s)\dfrac{k_o}{s}}{1 + k_d F(s)\dfrac{k_o}{s}}, \tag{8.17}$$

where $T(s)$ is the loop gain defined as

$$T(s) = k_d F(s)\frac{k_o}{s}. \tag{8.18}$$

The loop gain $T(s)$ is very high at low frequencies because it includes at least one integrator gain. Therefore, if $T(s) \gg 1$ at low frequencies,

$$\frac{T(s)}{1 + T(s)} \approx 1. \tag{8.19}$$

That is, the PLL function is basically a low-pass filter for the low-frequency input.

For FM demodulators, the same transfer function can be rewritten from Equation (8.17) as

$$\frac{v_o(s)}{\omega_i(s)} = \frac{1}{k_o} \times H(s) = \frac{1}{k_o} \times \frac{k_d F(s)\dfrac{k_o}{s}}{1 + k_d F(s)\dfrac{k_o}{s}}. \tag{8.20}$$

Equations (8.17) and (8.20) are identical except for the VCO gain k_o term. Their AC and transient responses should be identical. The former is just a standard closed-loop transfer function with a unit-less gain, but the latter is the frequency-to-voltage transfer function, which is to invert the gain of the VCO. That is, the VCO output is FM modulated, and inverting the VCO function is to demodulate the FM signal.

8.2.2 Second-Order PLL

The PLL performance is greatly affected by the loop filter function $F(s)$. If $F(s) = 1$, PLL is a first-order feedback loop with a single pole contributed by the pole of the VCO function, and the closed-loop transfer function in Equation (8.17) becomes a one-pole low-pass response.

$$H(s) = \frac{k_d k_o}{s + k_d k_o} = \frac{k}{s + k}, \tag{8.21}$$

where k is the unity loop-gain frequency defined as $k = k_d k_o$.

The Bode gain plot of the closed-loop transfer function given by Equation (8.21) is shown in Figure 8.11. The open-loop gain of the first-order PLL is just an integrator with a unity loop-gain frequency of k.

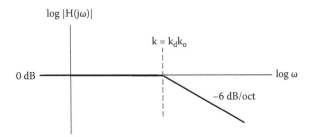

FIGURE 8.11
Closed-loop transfer function of the first-order PLL.

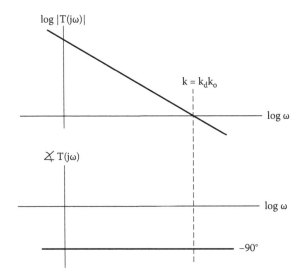

FIGURE 8.12
Open-loop gain and phase of the first-order PLL.

$$T(s) = \frac{kF(s)}{s} = \frac{k}{s}. \tag{8.22}$$

The unity loop-gain frequency k becomes the closed-loop bandwidth, and its Bode plots are shown in Figure 8.12.

It is convenient to handle the stability of PLL using the phase margin (PM) at a unity loop-gain frequency as in the feedback opamp analysis. The first-order PLL is absolutely stable with a PM close to 90°.

However, though stable, the first-order PLL is seldom used because it has a finite loop gain at low frequencies. If the loop filter has one pole at ω_p, the PLL becomes a second-order loop. There are two poles inside the loop. One is the loop-filter pole, and the other is the VCO integrator pole at DC. Therefore, the second-order closed-loop transfer function is obtained from Equation (8.17) as follows:

$$H(s) = \frac{k\omega_p}{s^2 + \omega_p s + k\omega_p}. \tag{8.23}$$

It is the same standard two-pole transfer function used in all electronic systems such as opamps and second-order $\Delta\Sigma$ modulators. Also, the second-order loop gain becomes

$$T(s) = \frac{k}{s\left(1 + \dfrac{s}{\omega_p}\right)}. \tag{8.24}$$

Figure 8.13 shows the closed-loop transfer function of the second-order PLL. The unity loop-gain frequency ω_k becomes the closed-loop bandwidth, but the stability depends on the PM at ω_k.

The Bode plots of the loop gain are sketched in Figure 8.14. Due to the pole ω_p of the LPF, the loop phase delay is almost 180° at ω_k, and PM approaches zero if the pole is placed close to the unity loop-gain frequency k of the first-order loop. If $\omega_p = k$, PM = 45°. Then, because the loop gain decreases at the same –20 dB/dec slope up to the unity loop-gain frequency k, there is no advantage of using the second-order loop. Therefore, the actual ω_p should be placed lower as shown, and the loop stability is at risk.

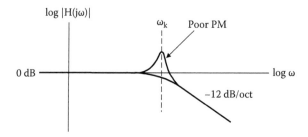

FIGURE 8.13
Closed-loop transfer function of the second-order PLL.

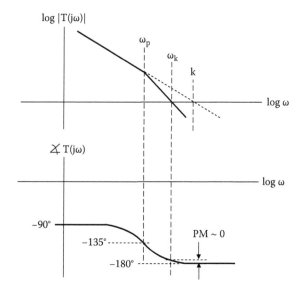

FIGURE 8.14
Open-loop gain and phase of the second-order PLL.

8.2.3 Stability of Second-Order PLL

The purpose of placing a pole at frequencies lower than the unity loop-gain frequency is to achieve a higher loop gain at low frequencies with a high slope of −40 dB/dec. All second-order PLLs need one zero ω_z placed below the unity loop-gain frequency to make them stable with a good PM as shown in Figure 8.15.

The second-order PLL can achieve more loop gain while keeping the same bandwidth as the first-order PLL. With a zero inserted, the unity loop-gain frequency ω_k is scaled to other parameters as follows:

$$\omega_k \approx k\frac{\omega_p}{\omega_z}. \tag{8.25}$$

With one pole and one zero in the loop filter, PM can be estimated as follows:

$$PM \approx 90^o - \tan^{-1}\frac{\omega_k}{\omega_p} + \tan^{-1}\frac{\omega_k}{\omega_z}. \tag{8.26}$$

The main purpose of the loop filter pole ω_p at low frequencies is to raise the loop gain with a steeper gain slope of extra −20 dB/dec or −12 dB/oct. This implies that the lower the pole is, the higher the low-frequency loop gain will get. Therefore, if the pole is moved to DC, the low-frequency loop gain can be maximized while keeping the same bandwidth as shown in Figure 8.16.

As discussed, the loop-filter gain is the most important parameter in the PLL design. A DC pole maximizes the low-frequency loop gain. Then from Equation (8.26), PM is solely determined by the zero frequency because the pole at DC contributes 90°.

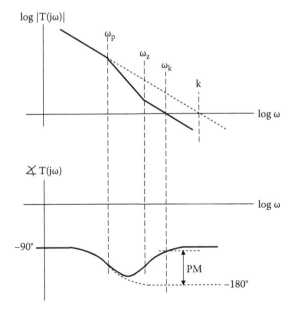

FIGURE 8.15
Second-order PLL gain and phase with a zero.

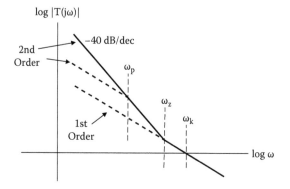

FIGURE 8.16
Second-order loop filter with a direct current (DC) pole.

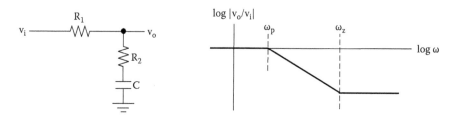

FIGURE 8.17
Passive loop filter with a pole and a zero.

$$PM \approx +\tan^{-1}\frac{\omega_k}{\omega_z}. \tag{8.27}$$

As the advantage is so obvious, second-order PLLs are implemented with a DC loop filter pole. Only an integrator can make a pole at DC. Depending on the phase detector used, either an active integrator or a charge pump can perform the integrator function.

8.2.4 Loop Filter with a Pole and a Zero

When using a multiplier as PD, either passive or active loop filter can be used. An example of a passive RC loop filter and its frequency response are shown in Figure 8.17.

It has one pole with a zero. Above the zero frequency, it works like a voltage divider because the capacitor is shorted at high frequencies. Its transfer function with a pole and a zero is defined as follows:

$$F(s) = \frac{1+\dfrac{s}{\omega_z}}{1+\dfrac{s}{\omega_p}}, \quad \omega_p = \frac{1}{(R_1+R_2)C}, \quad \omega_z = \frac{1}{R_2C}. \tag{8.28}$$

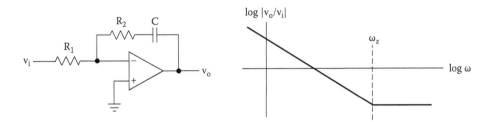

FIGURE 8.18
Active integrator with a zero.

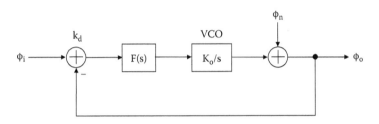

FIGURE 8.19
Linear PLL model with VCO phase noise.

However, passive RC filters implement only negative real poles. To place a pole at DC, an active RC integrator using an opamp is required.

Figure 8.18 shows an active RC loop filter. At low frequencies where C is open, it makes an integrator, and its zero is created at the same frequency where C starts to be shorted. Its transfer function with a pole and a zero using an ideal opamp is defined as follows:

$$F(s) = -\frac{1}{s}\left(1 + \frac{s}{\omega_z}\right), \qquad \omega_p = 0, \qquad \omega_z = \frac{1}{R_2 C}. \tag{8.29}$$

However, with an up/down state-machine PD, a simple charge-pumping circuit can implement the same integrator function with a zero using a capacitor as an integrating element.

8.3 Phase Noise Transfer Function

It is difficult to analyze the second-order PLL analytically, and it is common for PLL designers to analyze PLL through more graphical means. Because PLL is a feedback system in the phase domain, understanding the phase noise transfer function helps to understand the PLL operation properly. It is the same concept as the quantization noise transfer function in the $\Delta\Sigma$ modulator.

Figure 8.19 shows the linear PLL model with the VCO phase noise ϕ_n added in the loop. Then the output phase noise is related to other noises through

$$\phi_o = H(s)\phi_i + \{1 - H(s)\}\phi_n. \tag{8.30}$$

$H(s)$ is the standard closed-loop transfer function known to be the low-pass function. However, the VCO phase noise is shaped by $\{1 - H(s)\}$ to appear at the output. What is commonly called phase noise transfer function (PNTF) is a high-pass filter (HPF) function given by

$$1 - H(s) = \frac{s}{s + k_d k_o F(s)} = \frac{s}{s + k}.$$

(8.31)

It is the same as the quantization noise transfer function (NTF) of the $\Delta\Sigma$ modulator. Using Equation (8.5), the spectral density of the VCO output phase noise is shaped as follows:

$$S(f) = \{1 - H(j2\pi ft)\}^2 \times \frac{(\Delta\phi)^2}{\Delta f} = \{1 - H(j2\pi ft)\}^2 \times \frac{K_o^2}{f^2} \times \frac{v^2}{\Delta f}.$$

(8.32)

Note that the VCO input-referred thermal noise spectral density appears at the output after shaped by $1/f$ and $\{1 - H(s)\}$. Because PLL works as a HPF to the VCO phase noise or jitter, the low-frequency VCO phase noise or jitter is attenuated by the loop gain. On the other hand, the high-frequency VCO phase noise or jitter passes through without attenuation because the PLL is open. Therefore, PLL should be made wide band to suppress the in-band VCO phase noise or jitter. Two signal and noise transfer functions are explained in Figure 8.20.

With the open-loop gain shown at the top, the closed-loop gain $H(s)$ becomes the lower portion of the Bode plot below the line drawn at the unity loop-gain frequency ω_k, which becomes the −3 dB closed-loop bandwidth shown on the lower left side. The inverted upper portion of the open-loop gain is the PNTF $\{1 - H(s)\}$ as shown on the lower right side.

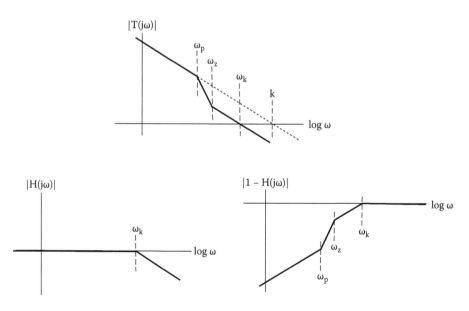

FIGURE 8.20
Open-loop gain, closed-loop gain, and phase noise transfer function (PNTF) of the second-order PLL.

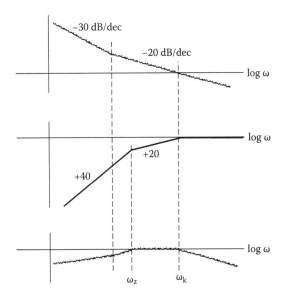

FIGURE 8.21
VCO phase noise, PNTF, and shaped VCO phase noise.

Figure 8.21 shows the open-loop VCO phase noise, PNTF, and closed-loop shaped VCO phase noise from the top. If the loop filter pole is at DC, PNTF shapes the VCO phase noise with 40 and 20 dB/dec slopes as shown. The open-loop VCO phase noise has a −20 dB/dec slope except for −30 dB/dec slope of the low-frequency $1/f$ noise. In reality, typical VCO phase noises exhibit even steeper slope at extremely low frequencies. Although the 40 dB/dec slope cuts more of the low-frequency VCO phase noise, the shaped noise will be relatively flat within the unity loop-gain frequency ω_k, which is also the PLL closed-loop bandwidth. Note that an integrator with a DC pole is more desirable than the low-pass filter, because the loop gain at low frequencies is maximized, thereby suppressing low-frequency phase noise further.

8.3.1 SSB Phase Noise Effect on Blocker

The SSB phase noise represents a side-tone power at an offset frequency per unit bandwidth Hz. For example, assume that a −2 dBm carrier has a side tone of −70 dBm at 1 MHz offset measured with 1 kHz resolution bandwidth. Then, normalizing the spectral density to 1 Hz bandwidth, the SSB phase noise at 1 MHz offset becomes

$$\frac{(\Delta\phi)^2}{\Delta f}@1MHz = 2 - 70 - 30 = -98\,(dBc/Hz), \tag{8.33}$$

where −30 dB is to scale the power bandwidth from 1 kHz to 1 Hz.

Existence of such side tones affects the RF receiver performance significantly. If a local carrier is generated using a PLL frequency synthesizer, the SSB phase noise at blocker

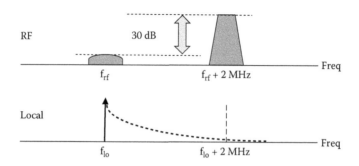

FIGURE 8.22
Single-sideband (SSB) blocker example in radio-frequency (RF) receivers.

frequencies down-converts blockers into the signal baseband. In particular, all RF systems require 30 ~ 60 dB stronger blockers on the adjacent and alternate channels.

Figure 8.22 explains one blocker situation of the Bluetooth standard. Assume that an RF receiver with 1 MHz bandwidth needs an SNR of 20 dB. If a 30 dB stronger blocker is at 2 MHz offset, the SSB phase noise requirement of the local carrier is

$$\frac{(\Delta\phi)^2}{\Delta f}\,@2MHz = -30 - 20 - 60 = -110 \ (dBc/Hz), \tag{8.34}$$

where −60 dB is to scale the power bandwidth from 1 MHz to 1 Hz, and −20 dB is to ensure the down-converted blocker is below the range of 20 dB SNR.

8.3.2 Integrated Root-Mean-Square Phase Noise Effect on Phase Modulation and Frequency Modulation

The phase noise of the local carrier directly affects angle-modulated baseband signals such as PM and FM. For example, the quadrature amplitude modulation (QAM) signal is a complex vector $(a + jb)$ carried by a complex carrier as follows:

$$\mathrm{Re}\left\{(a + jb)e^{j\omega_c t}\right\} = a \times \cos\omega_c t - b \times \sin\omega_c t. \tag{8.35}$$

The discrete levels of a and b are ±1, 3, 5, 7, ..., and the quadrature phase shift keying (QPSK) is a special case when $a = b = \pm 1$. Synchronous demodulation eliminates the carrier and recovers the two-dimensional original vector called constellation. The effect on FM will be the same because the modulation index is identical to the phase noise.

Figure 8.23 shows the received QPSK constellation and one quadrant of the ideal 64-QAM constellation. If the local carrier has phase noise, the received vector becomes

$$\mathrm{Re}\left\{(a + jb)e^{j\Delta\phi}\right\} = a \times \cos\Delta\phi - b \times \sin\Delta\phi. \tag{8.36}$$

Although amplitude noise appears in the radial direction, the phase noise rotates the received constellation. In complicated constellations such as 64- or 256-QAM, the

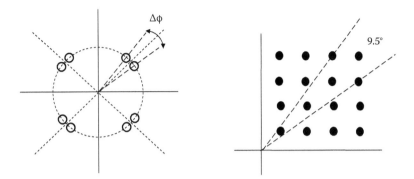

FIGURE 8.23
Rotation of constellation due to phase noise.

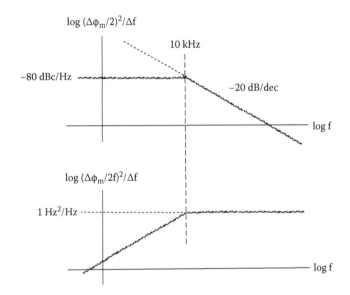

FIGURE 8.24
Spectral densities of VCO phase and frequency noises.

neighboring vectors are close, and the phase noise degrades SNR in demodulation. In 64-QAM, the closest constellations are only 9.5° apart. RF systems using such complicated constellations cancel the phase noise of the local carrier using low-speed bipolar PSK (BPSK) pilot channels.

The demodulator performance heavily depends on the integrated RMS phase noise. There are two ways to estimate it. One is to integrate the SSB noise over the band-pass bandwidth, or the DSB noise over the low-pass bandwidth. Because the phase noise spectrum is modified by PNTF, after phase locking, the phase noise and frequency noise spectral densities are modified as shown in Figure 8.24.

For simplicity, the phase noise density is assumed to be flat within the PLL bandwidth. In this example, the open-loop SSB VCO phase noise is about −80 dBc/Hz at 10 kHz offset.

If the PLL bandwidth is set to 10 kHz, the integrated RMS phase noise can be estimated as follows:

$$\Delta\phi_{rms} = \sqrt{2 \times \left(\frac{\Delta\phi_m}{2}\right)^2 \times BW}$$

$$= \sqrt{2 \times 10^{-8} \times 10^4 \times \frac{\pi}{2}} \approx 0.018 \; rad \sim 1^o, \tag{8.37}$$

where the factor $\pi/2$ results from integration of the phase noise above the bandwidth to infinity.

Similarly, the integrated RMS frequency noise is approximated after scaling the phase noise by f^2. This implies that the frequency noise is not band limited and stays constant at high frequencies. The high-frequency (hissing) noise is the common problem of the FM system. Most practical FM systems use very sharp channel selection filters to band-limit this high-frequency noise. Assume that a sharp channel filter with a cut-off frequency of 500 kHz is used. Then the in-band frequency noise can be estimated:

$$\Delta f_{rms} = \sqrt{2 \times \int \left(\frac{\Delta\phi_m f}{2}\right)^2 df}$$

$$\approx \sqrt{2 \times 500 \times 10^3} \approx 1 \; kHz. \tag{8.38}$$

For example, a Gaussian frequency-shift-keying (GFSK) system like Bluetooth uses a peak frequency deviation of 150 kHz with 1 MHz channel spacing. Then the maximum achievable SNR using this local carrier is estimated to be

$$SNR < \frac{150kHz/\sqrt{2}}{1kHz} \approx 41 \; dB. \tag{8.39}$$

8.3.3 PLL as FM Demodulator

Note that PLL is a phase/frequency follower in a unity-gain feedback. However, it also works as an FM demodulator as given by Equation (8.20). Its closed-loop transfer function from the input frequency to the VCO input is the same low-pass filter with a gain of $1/k_o$ that functionally inverts the operational gain of the VCO. For example, assume that the input is an FSK signal. It is a FM-modulated signal, but the modulating signal is digital 1 or 0 rather than the analog signal. The digital signal modulates the carrier frequency in a binary way: ω_1 if digital bit is 1, and ω_0 if 0, as shown in Figure 8.25.

Assume the following parameters for PLL and FSK input:

$$k_o = 100kHz/V, \quad \omega_k = 2\pi(1MHz),$$

$$\omega_1 = 2\pi(200kHz), \quad \omega_0 = 2\pi(100kHz). \tag{8.40}$$

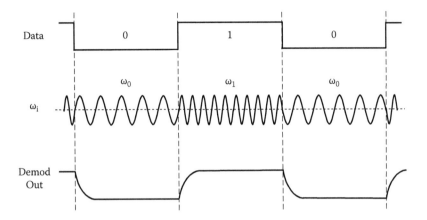

FIGURE 8.25
PLL response to frequency-shift-keying (FSK) signal.

Then the PLL settling time constant and output can be estimated as follows:

$$\frac{1}{\omega_k} = \frac{1}{2\pi(1MHz)} \approx 160ns.$$

$$\Delta V_o = \frac{\omega_1 - \omega_0}{2\pi k_o} = 1V. \tag{8.41}$$

PLL demodulates the FSK input and follows its frequency change with a time constant of $1/\omega_k$, like an opamp voltage follower in a unity-gain feedback. As in other feedback systems, good PM warrants exponential settling in transients, while poor PM will cause peaking and ringing.

8.4 Phase Detector

There are four types of phase detectors commonly used in PLL: analog multiplier, digital multiplier, J-K flip/flop, and state machine. Most PLLs are implemented digitally and use an up/down state machine, as a PD called phase-frequency detector (PFD).

8.4.1 Multiplier as Phase Detector

A common PD is an exclusive-OR (EX-OR) digital multiplier that takes two digital inputs of the input V_i and the VCO output V_o, as shown in Figure 8.26. Assume that the phase difference between the two is $\Delta\phi$. Digital EX-OR output is 1 only when two outputs are either 10 or 01. Therefore, by averaging (low-pass filtering) the EX-OR output, an average output proportional to the area difference $(A_1 - A_2)$ is obtained as follows if the digital logic has a magnitude of V_x:

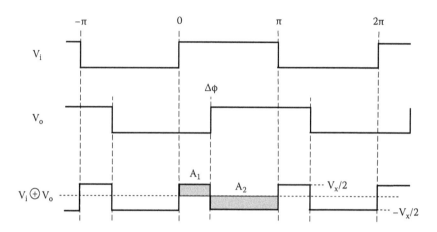

FIGURE 8.26
EX-ORing two digital inputs.

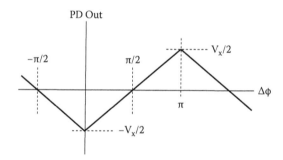

FIGURE 8.27
Transfer function of digital multiplier phase difference (PD).

$$V_o = \frac{A_1 - A_2}{\pi} = \frac{V_x}{\pi}\left(\Delta\phi - \frac{\pi}{2}\right). \tag{8.42}$$

Then, the digital multiplier-type PD has a transfer function as shown in Figure 8.27.

The transfer function has a null point with a $\pi/2$ phase offset. The slope is the PD gain defined by

$$k_d = \frac{V_x}{\pi}. \tag{8.43}$$

Analog multiplier also performs similarly as a PD. Multiplier-type PDs find the phase null point at an offset phase of $\pi/2$. The PD output has two frequency components: one at DC and the other at the doubled input frequency. After being low-pass filtered by the loop filter, the former becomes the phase error, but the latter stays as a reference tone. When locked, $\Delta\phi = \pi/2$ and $\omega_i = \omega_o$, and the EX-OR digital PD output has the remaining high-frequency harmonic tones.

$$v_i(t) \oplus v_o(t) = \frac{2V_x}{\pi}\left\{\cos(2\omega_i t) + \frac{\cos(6\omega_i t)}{3} + \frac{\cos(10\omega_i t)}{5} + \cdots\right\}. \tag{8.44}$$

These harmonics occurring at odd multiples of $2\omega_i$, in particular the largest fundamental tone of $2V_x/\pi$, should be suppressed with additional out-of-band loop filter poles. High harmonic tones modulate the VCO frequency and show up as spurious side tones.

Although multiplier-type PDs are simple, they are not suitable for PLLs that need to generate low-spurious RF local carriers, because it is difficult to meet the SSB blocker requirement with them. However, they are acceptable for other general clock recovery and generation purposes since the spurious tone is not a critical design factor.

8.4.2 Up/Down State Machine as Phase Detector

The up/down state machine is predominantly used as PD in most high-performance PLL implementations because it has lower spurious tones and can achieve high PNTF at low frequencies. It is used together with a charge-pump circuit, as is called charge-pumped PLL. It is to use an integrator as LPF for higher low-frequency loop gain by placing a pole at DC. Although it is still a second-order PLL, it has higher loop gain than other second-order PLLs with a real low-pass pole.

Figure 8.28 shows the up/down PD implemented digitally. It is made of two set-reset (SR) flip/flops. The SR flip/flop can be triggered either by the positive or negative edge of the logic transitions. Out of four states of the up/down outputs, PD eliminates one state when both Up = 1 and Down = 1. Both flip/flops are reset by the *and* output of the up and down states. If the input triggers the PD earlier than the VCO output, the up output of the PD changes from low to high, which will boost the VCO control to raise the VCO frequency and will advance the VCO clock edge. If the VCO output triggers the PD first, the opposite will happen. The down output changes from low to high. Then the VCO frequency will drop, and its clock edge will be delayed. Both up and down outputs are reset after the input and VCO output clock edges, and a new cycle resumes.

This operation is explained in Figure 8.29, where early, lock, and late cases are shown, and only three PD states of up, down, or none exist. When locked, this PD needs to generate a very short pulse with an almost zero width as shown on the left side, because both flip-flops are reset as soon as they reach the state when both up and down are high. Due to the electronics speed, this short pulse requirement limits the PD linearity for small phase errors. This can create a dead zone in the PD transfer function where the PD gain is low

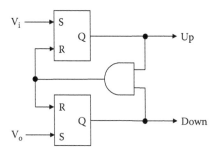

FIGURE 8.28
State-machine up/down PD.

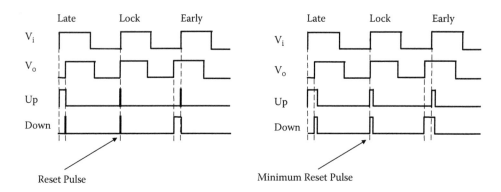

FIGURE 8.29
Minumum reset pulse.

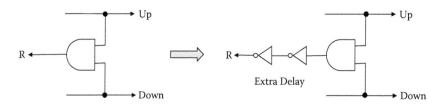

FIGURE 8.30
Extra delay in the reset loop to avoid the dead zone.

and nonlinear. To eliminate the dead zone, some delay in the reset path should be added so that a pulse with a minimum width can always be output even when the phase error is zero, as shown in Figure 8.30.

The reset delay is effective in operating the up/down PD in class-A mode without cross-over distortion, but turning on up/down pulses longer when PLL is in lock has one side effect. The current noise of the charge pump is integrated for a longer period and will degrade the VCO phase noise. The charge pump current noise is one of the critical parameters in the low phase noise design of frequency synthesizers.

The transfer function of the up/down PD with the dead zone is shown in Figure 8.31. It has a wide input range of $\pm 2\pi$ from its locked phase. If $\Delta\phi = 2\pi$, the up output stays high continuously, while if $\Delta\phi = -2\pi$, the down output is high continuously. Because the PD transfer function should repeat every 2π, the transfer function beyond $\pm 2\pi$ has no meaning and is not drawn here. Therefore, the PD gain is defined only within the phase error range from $-2\pi \sim 2\pi$. The PD gain is $V_x/2\pi$, but it can be redefined later, including the charge pump and loop filter gains.

8.4.3 Phase Frequency Detector

One advantage of using the up/down PD is that it detects the frequency error and requires no frequency acquisition. For example, if the VCO frequency is low compared to the input frequency, there are more up pulses while the down pulse stays continuously low as shown in Figure 8.32.

The same happens when the VCO frequency is high. Then the down pulse will be mostly high while the up pulse stays low. Because it is also sensitive to the frequency error, the

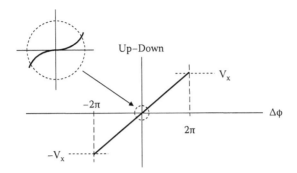

FIGURE 8.31
Transfer function of up/down PD with a dead zone.

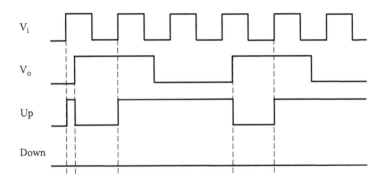

FIGURE 8.32
Up/down PD detects frequency.

up/down PD is called phase-frequency detector (PFD). Note that the multiplier-type PD forces the VCO phase to be locked to $3\pi/2$ and $-\pi/2$ if the phase error $\Delta\phi$ exceeds π and 0, respectively, as shown in Figure 8.27.

8.5 Charge-Pumped Phase-Locked Loop

The up/down PFD together with the charge-pumping circuit greatly simplifies the loop filter topology. PLL with an up/down PFD and a charge pump requires no frequency acquisition as the PFD detects the frequency error. Because the charge pump is turned on briefly, the low-pass-filtered reference spurs are far smaller than those of the multiplier-type PD. The loop filter pole is moved to DC so that extra loop gain can be attained to further suppress the VCO phase noise. As a result, the charge-pumped PLL has been predominantly used for low-jitter clock synthesizers.

Figure 8.33 shows the charge-pumped PLL system. Note that the input is commonly named a *fixed reference* (Ref) clock in clock synthesizers. The charge pump is two up/down current sources switched to the loop filter. It is driven by the up and down pulses from the PFD. Its function is to dump a packet of positive or negative charge that is proportional to the phase error onto the loop filter. Because the up/down current pulses from the charge

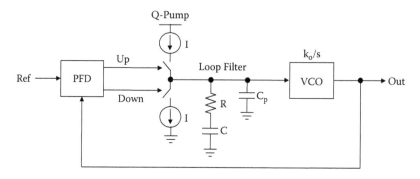

FIGURE 8.33
Charge-pumped PLL.

pump are pulse-width modulated (PWM) by the phase error, the loop filter can be made of a capacitor C, which effectively works as a current integrator. That is, the low-frequency integrator pole is now moved to DC. As a result, the loop gain at low frequencies has a slope of -40 dB/dec or -12 dB/oct, providing higher gain than in other second-order PLLs with a low-pass filter.

8.5.1 Stability of Charge-Pumped PLL

Due to two poles at DC, an in-band zero should be placed below the unity loop-gain frequency. It is done with a resistor R connected in series with the integrating capacitor C. An extra parallel capacitor C_p is also needed so that high-frequency components from the charge pump such as reference spurs can be suppressed. Therefore, the unity loop-gain frequency ω_k, zero ω_z, and extra out-of-band pole ω_p are defined as follows:

$$\omega_k = \frac{IR}{2\pi}k_o, \qquad \omega_z = \frac{1}{RC}, \qquad \omega_p = \frac{1}{RC_p}. \tag{8.45}$$

Then the phase margin at ω_k can be estimated as follows:

$$PM = \tan^{-1}\frac{\omega_k}{\omega_z} - \tan^{-1}\frac{\omega_k}{\omega_p}. \tag{8.46}$$

Note that the out-of-band pole degrades PM when placed too close to the unity loop-gain frequency. PM becomes about $62°$ if the zero and pole are placed symmetrically around the unity loop-gain frequency like $\omega_z = \omega_k/4$ and $\omega_p = 4\omega_k$. Additional high-frequency poles can be added to suppress the reference spur further, but they will degrade the PM. This PLL topology benefits greatly from its simple passive loop filter.

8.5.2 Loop Filter of Charge-Pumped PLL

Figure 8.34 illustrates the loop filter as a function of frequency. First, PLL is in an open-loop condition for frequencies higher than the unity loop-gain frequency ω_k. In the closed-loop condition at frequencies lower than ω_k, the VCO phase noise is modified by the PNTF,

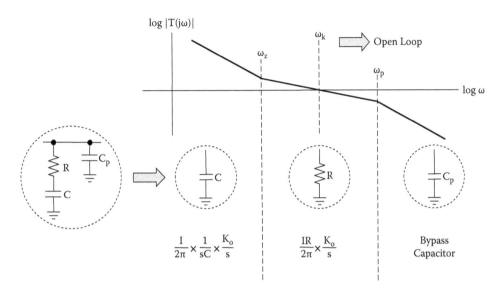

FIGURE 8.34
Loop filter function versus frequency.

which exhibits two slopes of s (20 dB/dec) and s^2 (40 dB/dec). In the open-loop condition, the reference tone is suppressed only by the out-of-band pole ω_p with the $1/s$ (−20 dB/dec) slope.

The loop filter made of R, C, and C_p looks different in three different frequency ranges. In the low-frequency range of $\omega < \omega_z$, the loop gain decreases by −40 dB/dec. Because the current output from the charge pump is integrated by the capacitor C, the loop gain is given by

$$T(s) = \frac{I}{2\pi} \times \frac{1}{sC} \times \frac{k_o}{s}. \tag{8.47}$$

In the middle frequency range of $\omega_z < \omega < \omega_p$, the loop gain decreases by −20 dB/dec. Because the loop filter is just a resistance R, the loop gain becomes

$$T(s) = \frac{IR}{2\pi} \times \frac{k_o}{s}. \tag{8.48}$$

In the high-frequency range of $\omega > \omega_p$, there is no loop gain, and the VCO free-runs. The VCO input is just bypassed by a capacitor C_p. Any high-frequency components including the reference spur can be filtered at the VCO input in the open-loop condition.

The loop gain becomes unity in the middle range. Therefore, from Equation (8.48), two most important parameters, the PFD gain k_d and unity loop-gain frequency ω_k, are defined as

$$k_d = \frac{IR}{2\pi}, \qquad \omega_k = k_d k_o = \frac{IR}{2\pi} \times k_o. \tag{8.49}$$

Note that ω_k becomes the closed-loop bandwidth of the PLL. The unit of the closed-loop bandwidth solely depends on the unit of k_o, because k_d is unit-less. It is Hz/V if k_o is in Hz/V, or rad/s if k_o is in rad/s/V.

8.5.3 Reference Spur

Setting the loop bandwidth is the most critical design issue in the PLL design. If PLL bandwidth is set to be narrow, the VCO phase noise is suppressed less while the reference spur is suppressed more. Furthermore, the closed-loop transient response settles slowly with a long time constant. The narrowband PLL is easier to stabilize with good phase margin, but because the bandwidth is too narrow, it is difficult for PLL to lock in. If PLL bandwidth is set wide, the opposite is true, and the reference spur cannot be suppressed much. The reference spur is also one of the most important parameters in the RF frequency synthesizer design, though it is not important in the clock recovery.

  This trade-off situation is explained in Figure 8.35. The in-band VCO phase noise is suppressed by PNTF, but the out-of-band reference spur can only be filtered with the extra pole at ω_p. Note that within the frequency range from the unity loop-gain frequency ω_k to the pole ω_p, PLL does nothing. These two frequency corners should be set carefully, as they affect both the VCO phase noise and the stability of PLL.

  Figure 8.36 shows the origin of the reference spur. Up/down pulses are mismatched mainly because of the current mismatch. For example, due to the inevitable mismatch between two currents I_1 and I_2, the turn-on times t_1 and t_2 of the up/down currents are also mismatched as sketched. When PLL is in lock, the pumped up/down charges should be equal, and the following condition should be met:

$$I_1 \times t_1 = I_2 \times t_2. \tag{8.50}$$

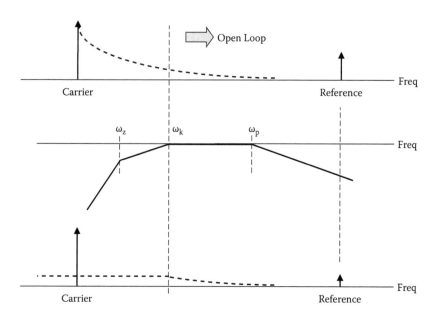

FIGURE 8.35
Filtering of phase noise and reference spur.

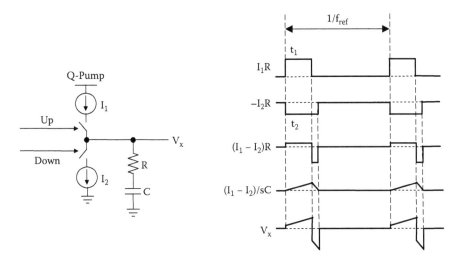

FIGURE 8.36
Charge pump and the reference spur at the VCO input.

These mismatched pulses make the VCO control voltage disturbed every reference period. The fundamental tone energy of this waveform survives to create the reference spur in the VCO output. Switching time or pulse width mismatches that result from the finite rise/fall times of the up/down logic result in a constant phase offset and do not contribute to the mismatch.

The bypass capacitance C_p at the input of the VCO can short this high-frequency waveform with the out-of-band pole ω_p.

$$\omega_p = \frac{1}{RC_p} = \frac{1}{\tau}. \tag{8.51}$$

For simplicity, assume that the loop filter is a simple low-pass filter with the above time constant at the reference frequency. Adding extra out-of-band poles helps to cut the reference spur more as shown in Figure 8.37.

It is difficult to analyze the reference spur analytically with real loop filter elements. For simple approximation, assume that the mismatch current is a rectangular periodic waveform with a magnitude of I with pulse duration of Δt. Then the reference current tone at ω_{ref} can be approximated by

$$I_{ref}(t) \approx \left(I \times \frac{\Delta t}{T} \right) \times \sin \omega_{ref} t, \tag{8.52}$$

where T is the reference period $1/f_{ref}$, and the sinx/x (SINC) function of the line spectra is ignored because $\Delta t \ll T$. Intuitively, the reference tone is very small because the energy within Δt is spread evenly over the longer period T. So the reference tone at the VCO input can be approximated as

$$\Delta V_{peak} \approx IR \times \frac{\Delta t}{T}. \tag{8.53}$$

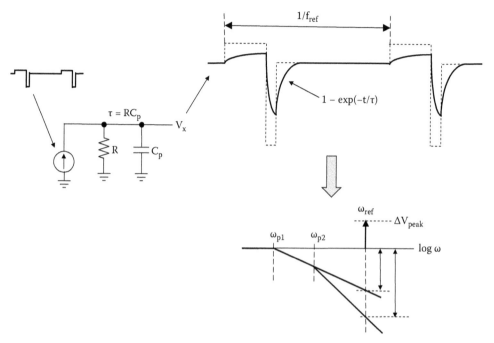

FIGURE 8.37
Reference spur filtering by out-of-band poles.

Two poles ω_{p1} and ω_{p2} can attenuate the reference spur by $(\omega_{p1}/\omega_{ref})$ and $(\omega_{p2}/\omega_{ref})$, respectively. Therefore, if the peak of the reference tone from the charge pump is ΔV_{peak}, the carrier to SSB reference spur as a phase noise is defined in the same way as is done for the FM modulation index as follows:

$$\frac{\Delta\phi_{ref}}{2} = \frac{\Delta V_{peak}k_o}{2f_{ref}} \times \frac{\omega_{p1}\omega_{p2}}{\omega_{ref}^2}.$$

(8.54)

Therefore, for low reference spur, the minimum pulse width should be made as narrow as possible, and the VCO gain k_o should be low so that the same-level reference tone may affect the VCO phase noise less. The low VCO gain requirement demands that the charge pump gain should be set higher. One way to improve up/down current matching is to use a negative feedback to adjust one current to match the other current.

An example of the charge-pump circuit using simple differential pairs is shown in Figure 8.38. Up/down current matching is similar to the common-mode feedback for differential amplifiers. The charge pump is only briefly turned on to charge the loop filter, and most of the time it is turned off like tristate logic. Therefore, if a slow opamp adjusts the common mode of the charge pump to match the loop filter voltage, the charge pump up/down currents can be matched. The feedback loop should be applied through a small portion of the split bias current to ensure the start-up of the charge-pump circuit.

8.5.4 Charge-Pump Circuits

Charge pump is a bidirectional current switching network like a 1b DAC, and the simplest charge pump is made of N-channel metal-oxide semiconductor (NMOS) and P-channel

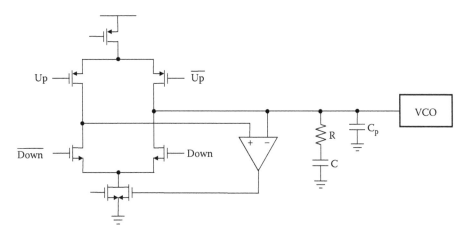

FIGURE 8.38
Current-matched charge pump.

metal-oxide semiconductor (PMOS) differential pairs switched as shown in Figure 8.38. Like the current-steering DAC, the charge pump performance is affected by many factors, such as digital rise/fall times, time skew, glitch, switch feed-through, and up/down current mismatch, which contribute to higher charge-pump nonlinearity and reference spur. There are many variations in charge-pump circuits, but the simple charge pump is widely used because the charge pump accuracy is mainly limited by the up/down current mismatch and its low-frequency performance.

The common design strategy of the charge pump is to enhance the clock edges and to switch the current on and off symmetrically. As in the standard current DAC design, the current mirror and source should be biased with large V_{GS}, using longer channel device for better matching. Because the current output is sensitive to the output voltage, cascoding of the current source is desirable. However, in low-voltage CMOS, the supply voltage is low, and the VCO input control range is reduced if a small device with large V_{GS} is used and cascoded.

Figure 8.39 shows one variation that switches current directly rather than steering current. Although the clock feed-through effect will be reduced, the rise and fall times will be longer due to the turn-on delay of the cascode transistor from the off state.

To improve the power supply rejection, a differential version of the charge pump can be used as shown in Figure 8.40. Although the VCO control voltage is differentially taken, it is rather a difference circuit using two charge pumps. For fully differential PLL operation, the VCO control voltage should also be made differential, and the common-mode feedback is required for proper operation. The complexity and power consumption are doubled, and the up/down current mismatch problem still exists.

8.6 PLL Bandwidth Constraints

Several factors should be considered to operate PLL with adequate bandwidth. They are RMS frequency and phase noises, settling requirement, reference spur, and nonlinearity. Their relations to the loop bandwidth need to be closely looked at.

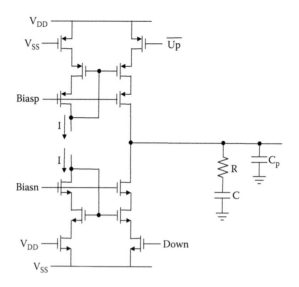

FIGURE 8.39
Cascoded charge pump.

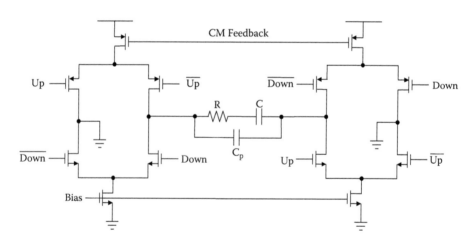

FIGURE 8.40
Differential charge pump.

8.6.1 Capture and Lock-In Ranges

Before PLL acquires locking, VCO should free-run at a frequency close to the input frequency. If VCO frequency is too far off, PLL cannot lock in. The capture range is defined as a frequency range of the input onto which the loop can acquire locking. Furthermore, even in locked condition, if VCO frequency suddenly jumps too far from the input frequency, PLL can lose locking. Therefore, the lock-in range is similarly defined as a frequency range of the input over which PLL can maintain locking.

In PLLs using multiplier-type PDs, both capture and lock-in ranges are defined as the maximum frequency ranges VCO can cover in capturing and lock-in situations. The maximum PD output multiplied by the VCO gain is defined in two situations as follows:

$$Capture\ Range = \frac{\pi}{2}k_d k_o \left| F(\omega_i - \omega_o) \right|.$$

(8.55)

$$Lock - In\ Range = \frac{\pi}{2}k_d k_o.$$

This is because the maximum output from PD is $(\pi/2)k_d$, and the capture range is always narrower than the lock-in range because the PD output of the low beat frequency $(\omega_i - \omega_o)$ is attenuated by the loop filter transfer function of $F(\omega_i - \omega_o)$. The VCO should free-run within the capture range. This requires either initial frequency acquisition or PFD for capturing. The lock-in condition sets the minimum PLL bandwidth. The RMS frequency noise $\Delta\omega_{rms}$ of the VCO should be smaller than the lock-in range for the PLL to sustain locking. Using the loop gain of Equation (8.25) for the second-order PLL,

$$\omega_k >> \frac{2}{\pi}\left(\frac{\omega_p}{\omega_z}\right)\Delta\omega_{rms}.$$

(8.56)

That is, if the multiplier-type PD is used, the loop bandwidth ω_k should be set to meet Equation (8.56) so that PLL can stay locked.

The PD type affects the free-run frequency range of the VCO and the PLL bandwidth. However, in charge-pumped PLLs, the up/down PD is a PFD detecting both phase and frequency. Therefore, the capture range cannot be defined because any frequency within the VCO oscillation range can be captured. Similarly, the lock-in range can be the VCO frequency range. However, if the VCO phase jumps suddenly by more than $\pm 2\pi$, the PLL becomes nonlinear. To warrant PLL to operate linearly, it is still logical to define the lock-in range for charge-pumped PLLs as a linear range of operation. From Equation (8.49), the lock-in range for the charge-pumped PLL can be defined as

$$Lock - In\ Range = 2\pi k_d k_o = IRk_o.$$

(8.57)

Note that the linear (lock-in) range of the charge-pumped PLL is about 2π times wider than the PLL bandwidth. For larger phase jumps, the PLL becomes nonlinear.

8.6.2 Settling Requirement

Another condition that sets the lower bound of the bandwidth is the settling requirement of the PLL. Systems such as frequency hopping require very agile frequency synthesis. Like feedback amplifiers, PLL responds to the transient input with a time constant of $1/\omega_k$. Therefore, the PLL bandwidth should be wide to meet

$$\omega_k > \frac{1}{\tau_{settle}}, \qquad (8.58)$$

where τ_{settle} is a minimum time constant to finish settling within a given accuracy. For the fine exponential settling within ±0.1%, it takes about $6.9\tau_{settle}$. If PM is smaller than 60°, the transient response starts to ring.

That is, the lower limit of the bandwidth is set by the noise and settling requirement, but the upper limit of the PLL bandwidth is set by factors such as stability, reference spur, and nonlinearity. As in the second-order $\Delta\Sigma$ modulator with two poles and one zero, PLL operation can be nonlinear, and its performance degrades if the bandwidth is too wide relative to the reference frequency.

8.6.3 PLL versus Second-Order $\Delta\Sigma$ Modulator

The second-order PLL is compared with the second-order $\Delta\Sigma$ modulator in Figure 8.41. In the $\Delta\Sigma$ modulator, the quantization noise is suppressed by the loop gain, which is defined as the quantization noise transfer function (QNTF). It can be redrawn in the middle with the quantization noise as an input. Then with the X input removed, it is identical to the charge-pumped PLL at the bottom.

Figure 8.42 compares how the quantization noise and phase noise are suppressed. In PLL, the VCO phase noise is suppressed by the loop gain, which is defined as PNTF. Both the QNTF in $\Delta\Sigma$ modulators and the PNTF in PLLs are defined as $\{1 - H(s)\}$, where $H(s)$ is the low-pass closed-loop transfer function. The out-of-band quantization noise of the $\Delta\Sigma$ modulator is further filtered in the digital domain, but in PLLs, out-of-band reference spurs are only filtered in the analog domain by extra out-of-band poles. The difference is

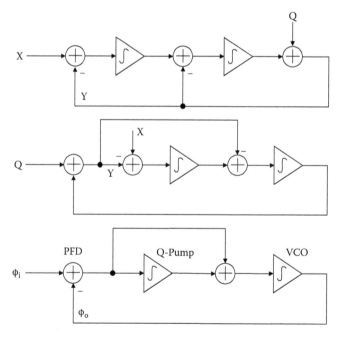

FIGURE 8.41
Similarity between PLL and $\Delta\Sigma$ modulator.

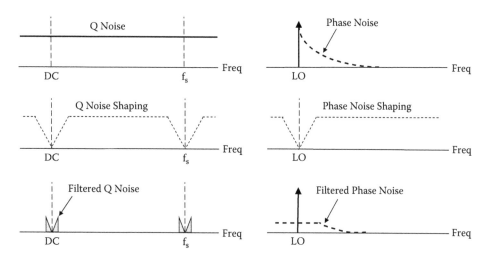

FIGURE 8.42
Quantization noise shaping versus phase noise shaping.

that the $\Delta\Sigma$ modulator is a sampled-data system, and the noise shaping happens at every multiple of the sampling frequency.

In $\Delta\Sigma$ modulators, it is common to set the oversampling ratio higher than π (>4). Similarly, in PLLs, the unity loop-gain frequency is typically set to be lower than 1/8 of the reference frequency.

$$\omega_k < \frac{\omega_{ref}}{8}.$$
(8.59)

In numerical simulations, PLL with even wider bandwidth may work, but in reality, the phase feedback error tends to get too large, and the loop may become nonlinear and even unstable.

8.7 High-Q LC VCO

VCO is a key component in the PLL system that affects its closed-loop phase noise and RMS jitter performance. It is also the most sensitive element in the system. It picks up noises from power and bias lines and converts them into phase noise and jitter. For clock synthesis, high-Q inductor-capacitor (LC) oscillators are used for low phase noise and jitter, but in clock recovery, low-Q ring or relaxation oscillators can be used. To vary the oscillation frequency, component values should be either voltage controlled or current controlled.

8.7.1 LC Components in CMOS

For high-Q oscillation, low-loss energy storing elements are needed. In CMOS, spiral inductors using top-layer thick metal or bonding wires are used as inductors, while MOS transistors or junction diodes are used as variable capacitors (varactors).

Bonding wire inductor has higher Q (30 ~ 50) and self-resonates at higher frequencies, but it is not easily reproducible (>±10%). Although well controlled within ±2% ~ 3%, spiral inductor has low Q (5 ~ 10) in the frequency range of 1 ~ 5 GHz. The low Q results from many losses due to series resistance, substrate, eddy current, and skin effect. There are many empirical models for spiral inductors, and process foundries also supply their inductor models. Circle inductor exhibits higher Q than other octagonal, hexagonal, and square inductors. Furthermore, symmetric inductor is known to have higher inductance values due to the mutual inductance. It also achieves a 50% higher Q within a 30% smaller area, and is used in most differential amplifier topology [1].

Figure 8.43 shows an example of a circular spiral inductor model. The inductor value is about 5nH obtained from four turns of the sixth-layer metal. Q is about 5.7 at 2.4 GHz. The parallel equivalent resistance of the inductor is about 422 Ω.

The losses in an inductor L and a capacitor C are modeled using a series resistance R_s and a parallel resistance R_p, respectively, as shown in Figure 8.44. The quality factor Q is defined as a ratio of the stored energy to the dissipated energy. Because the same current flows through the series circuit or the same voltage is applied to the parallel circuit, Q can be defined as one of either the impedance ratio or the admittance ratio as shown.

$$Q = \frac{\omega L}{R_s}, \quad Q = \frac{R_p}{\omega L}, \quad Q = \frac{1}{\omega R_s C}, \quad Q = \omega R_p C. \quad (8.60)$$

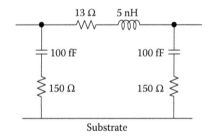

FIGURE 8.43
Example of a spiral inductor model.

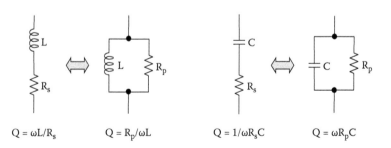

FIGURE 8.44
Definition of Q for L and C.

Then the following relation holds to relate the parallel resistance to the series resistance:

$$R_p = Q^2 R_s. \tag{8.61}$$

The series-to-parallel conversion of the series inductor example yields the following parallel circuit, and note that the above approximation is valid if Q is high. The same conclusion can be drawn for the capacitor case.

$$L_{eq} = \frac{R_s^2 + (\omega L)^2}{\omega^2 L} \approx L, \qquad R_{eq} = \frac{R_s^2 + (\omega L)^2}{R_s} \approx R_s. \tag{8.62}$$

Two varactors used in CMOS are shown in Figure 8.45. Their Qs are higher than the inductor Q. The diode junction capacitance varies abruptly when forward biased, and the reverse-biased capacitance varies little. On the other hand, the MOS capacitance varies symmetrically in the middle range. Because the maximum tuning range depends on the limited supply voltage, the MOS capacitance C-V characteristic is more desirable and offers a wider tuning range than the diode junction capacitance [2].

8.7.2 Oscillation Condition for LC VCO

As discussed in Chapter 1, even a negative feedback system becomes unstable if the loop gain is higher than unity at a frequency where the excess loop phase delay is 180°. Oscillator is a positive-feedback system that is intentionally made unstable. However, two conditions should be met for oscillation to start. One is the gain condition that the loop gain should be greater than 1, and the other is the phase condition that the total loop phase delay should

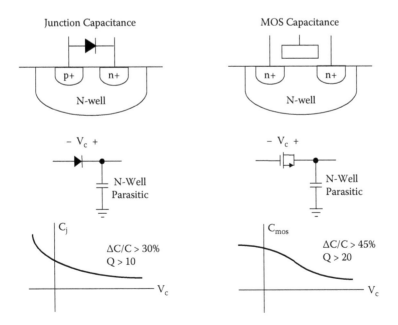

FIGURE 8.45
Two varactors in a complementary metal-oxide semiconductor (CMOS).

be 0 or -2π ($-360°$) at a resonant frequency. Therefore, oscillators are made of high-gain amplifiers connected in positive feedback.

The linear model of a common-source (CS) amplifier with an inductor load L is shown in Figure 8.46. The VCO design starts from the process parameters of the inductor. Typical spiral inductors of $L = 5$ nH exhibit $Q = 6 \sim 8$ at 2 GHz. This is directly translated to the series resistance value of $R_s = 5 \sim 10\ \Omega$. Then from Equation (8.61), R_p falls in the range of $300 \sim 400\ \Omega$, and the output resistance r_o is usually larger than R_p. This demands that $g_m = 1/100 \sim 1/200\ \Omega$ so that the small-signal gain of the amplifier at the resonant frequency ω_o of the tank circuit can be higher than unity.

$$g_m R \approx 2 \sim 4, \qquad \omega_o = \frac{1}{\sqrt{L(C + C_p)}}, \tag{8.63}$$

where C_p is the total output parasitic including the next stage loading capacitance.

The device size and bias current can be set considering a few constraints. The oscillation magnitude is either current or supply limited. High g_m can reduce noise, but it requires more power. Large W/L reduces V_{DSsat} to get larger output, but it would increase parasitic capacitance to limit the VCO tuning range.

Therefore, if two of this amplifier are connected back to back as shown on the left side of Figure 8.47, the loop phase condition is met at the resonant frequency f_o, because each stage contributes $180°$. Then the loop gain condition demands

$$(g_m R)^2 > 1. \tag{8.64}$$

If the same circuit is biased with a common tail current as shown on the right side, it becomes the most common high-Q LC VCO widely used [3].

The tail current is often added at the top using a PMOS current source because PMOS has lower 1/f noise than NMOS. The cross-coupled differential pair can also be considered as a current controlled negative resistance device with a differential driving-point small-signal output resistance of $-2/g_m$. The current-controlled negative resistance device makes an oscillator if loaded with a parallel LC tank circuit. More stages, for example, four stages can be cascaded to achieve higher Q, which will lower the VCO phase noise because there

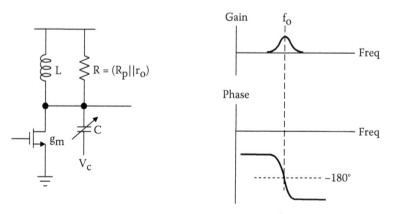

FIGURE 8.46
Inductor-loaded common-source (CS) amplifier and its Bode plots.

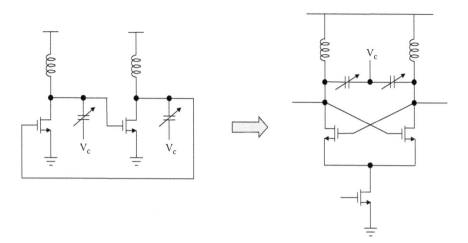

FIGURE 8.47
Positive feedback for oscillator.

are twice as many tuning stages. However, there are also twice as many noise sources, and power and chip areas are doubled.

By meeting the loop gain and phase conditions, the oscillation can grow from the small noise at the oscillation frequency, but the magnitude growth will stop either by limited current or supply voltage. In this magnitude-limited steady state, the oscillation sustains a constant level of magnitude. Although limiting is a nonlinear operation, it is still useful to assume that the small-signal linearity is good for the steady-state output. That is, if only the fundamental oscillation tone is considered, the loop gain is higher than unity as in Equation (8.64), and the oscillation grows, but the loop gain is reduced after being limited. Therefore, for oscillation to grow, $(g_m R)^2 > 1$. However, in the steady state after being limited, this relation holds to sustain the oscillation.

$$\left(g_m R\right)^2 = 1. \tag{8.65}$$

This steady-state unity loop-gain condition helps to derive the VCO phase noise based on the small-signal linear model of an oscillator.

8.7.3 Phase Noise of LC VCO

The VCO phase noise was intuitively analyzed earlier to have a −20 dB/dec slope relative to the input referred noise. However, it is not easy to refer all noise sources to the input side. So it is necessary to derive it from the VCO circuit directly.

Figure 8.48 shows the half circuit of the small-signal linear model of VCO with the total noise i_n, which is given by

$$i_n^2 = 4kT\left(\frac{2g_m}{3} + \frac{R_s}{R_p^2} + g_m^2 R_g\right)\Delta f + 1/f\ Noise, \tag{8.66}$$

where R_g is the gate resistance of the transistor, which can be reduced by laying out with transistors connected in parallel. The gate resistance is typically 2 ~ 3 Ω per square.

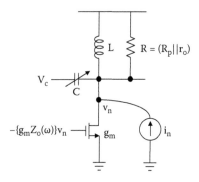

FIGURE 8.48
Half circuit of inductor-loaded differential pair.

Because the driving-point impedance seen by i_n is reduced by the feedback loop gain, i_n can be related to the output noise v_n. Assume that $Z_o(\omega)$ is the impedance of the parallel R, L, and C load. Then the transimpedance function is obtained as

$$\frac{v_n}{i_n}(\omega) = \frac{Z_o(\omega)}{1 - \{g_m Z_o(\omega)\}^2}. \tag{8.67}$$

Note that the numerator is the open-loop transimpedance, and the denominator is the standard closed-loop gain form of $\{1 - T(j\omega)\}$ for the positive feedback system. The LC tank circuit will resonate at

$$\omega_o = \frac{1}{\sqrt{LC}}. \tag{8.68}$$

The parallel load impedance can be approximated at offset frequencies $\omega = \omega_o + \Delta\omega$ close to the resonant frequency ω_o.

$$Z_o(\omega) = \frac{1}{\frac{1}{R} + \frac{1}{j\omega L} + j\omega C} = \frac{R}{1 + jQ\left(\frac{\omega}{\omega_o} - \frac{\omega_o}{\omega}\right)} \approx \frac{R}{1 + j2Q\left(\frac{\Delta\omega}{\omega_o}\right)} \approx R \times e^{-j\theta(\omega)}. \tag{8.69}$$

This implies that the tank load is purely R at ω_o with no phase shift, but with frequency offset $\Delta\omega$, the phase is delayed slightly by

$$\theta(\Delta\omega) = \tan^{-1}\left\{2Q\left(\frac{\Delta\omega}{\omega_o}\right)\right\} \approx 2Q\left(\frac{\Delta\omega}{\omega_o}\right). \tag{8.70}$$

Using Equation (8.69), Equation (8.67) can be rewritten assuming the phase shift is small.

$$\frac{v_n}{i_n}(\omega) = \frac{R\left\{1 + jQ\left(\dfrac{\omega}{\omega_o} - \dfrac{\omega_o}{\omega}\right)\right\}}{\left\{1 + jQ\left(\dfrac{\omega}{\omega_o} - \dfrac{\omega_o}{\omega}\right)\right\}^2 - (g_m R)^2} \approx \frac{R\left(1 + j2Q\dfrac{\Delta\omega}{\omega_o}\right)}{j4Q\left(\dfrac{\Delta\omega}{\omega_o}\right)}, \tag{8.71}$$

where the steady-state unity loop-gain condition $(g_m R)^2 = 1$ of Equation (8.65) is used. Also in the approximation, higher-order terms are ignored because $\Delta\omega \ll \omega_o$.

Using Equation (8.71), the phase noise can be defined as a power ratio of the noise v_n to the output v_o, which becomes a function of an offset frequency $\Delta\omega$.

$$\frac{v_n^2}{v_o^2}(\Delta\omega) = 2 \times \frac{i_n^2 R^2 \left\{1 + 4Q^2\left(\dfrac{\Delta\omega}{\omega_o}\right)^2\right\}}{16Q^2\left(\dfrac{\Delta\omega}{\omega_o}\right)^2} \times \frac{1}{v_o^2} \approx \frac{i_n^2 R^2}{8Q^2\left(\dfrac{\Delta\omega}{\omega_o}\right)^2} \times \frac{1}{v_o^2}. \tag{8.72}$$

The factor of two is to include the other transistor noise in the feedback, and the output v_o is now the differential output. The slight gain change due to the phase shift is also neglected. This is the standard equation for the VCO phase noise [4]. As expected, the phase noise slope is −20 dB/dec for offset frequency. For low phase noise, Q should be high, and v_o should be large.

As Q increases, the phase noise decreases, and active devices contribute less noise. This directly leads to lower power consumption in high-Q oscillators. To get the phase noise that Equation (8.72) predicts, consider typical numbers as follows: $Q = 8$, $f_o = 2.5$ GHz, $v_o = 1\ V_{rms}$, $R_s \sim$ few tens of Ω, $R_p \sim 300\ \Omega$, $g_m \sim 1/100\ \Omega$, and $i_n^2/\Delta f = 4\ kT(2g_m/3) \sim 10^{-22} A^2/$Hz. Then from Equation (8.72), the phase noise at 1 MHz offset will be

$$\frac{v_n^2}{v_o^2}(\omega) \approx \frac{10^{-22} \times 300^2}{8 \times 8^2 \left(\dfrac{1MHz}{2.5GHz}\right)^2} \sim -129.5 dBc/Hz\ @\ 1MHz. \tag{8.73}$$

The spectrum is sketched in Figure 8.49 for two cases when the center frequency is tuned to 2.5 and 5 GHz, respectively. The SSB phase noise is −132.5 dBc/Hz, which is 3 dB lower than the DSB phase noise. Because the phase noise is also a function of oscillation frequency, the same phase noise will be at twice the offset frequency of 2 MHz if the frequency is doubled.

FIGURE 8.49
VCO phase noise at different frequencies.

8.7.4 1/f Noise Up-Conversion

Although the VCO phase noise is dominated by the thermal noise amplified, the up-conversion of the 1/f noise of the tail current is problematic at low frequency offsets. When the oscillation starts to be current-limited, the differential pair switches the tail current of M_3 alternately into the M_1 and M_2 branches as shown in Figure 8.50.

Note that the frequency axis around f_o is not scaled properly in the log or linear scale. It is just to show the shift of the 1/f spectrum to f_o. It is, in effect, equivalent to chopping the bias current, and the low-frequency 1/f noise is up-converted to the oscillation frequency as in the chopping mixer. However, 1/f noises of M_1 and M_2 do not come out at the output. Also note that frequency is doubled at the common node.

Figure 8.51 shows the typical VCO phase noise spectrum. In the middle range, the characteristic –20 dB/dec slope as predicted by Leeson's is shown, and at the low end

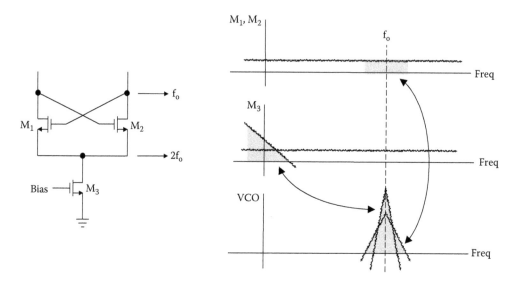

FIGURE 8.50
Up-conversion of the tail current 1/f noise.

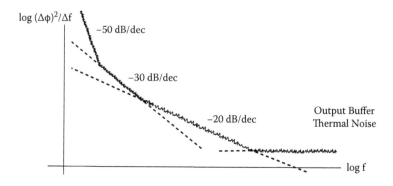

FIGURE 8.51
High-Q VCO phase noise.

of the spectrum, 1/f noise of the tail current with −30 dB/dec slope is prominent. Noise coupling through the bias line is also suspected for higher slope. In reality, the slope is steeper than −30 dB/dec at extremely low offset frequencies. Note also at the high offset frequencies, the spectrum can go flat due to the noise of low-impedance output buffer. In RF applications, VCO drives the switching mixer, which exhibits time-varying input impedance. To protect the output of the VCO from the mixer transient, a source follower buffer with low output impedance is added to the VCO output. The output buffer noise is typically below −140 dBc/Hz, and does not affect most applications. If the buffer is designed to have high output impedance, the slope will continue to be −20 dB/dec at high offset frequencies.

8.7.5 Low Phase Noise Design for LC VCO

As given by Equation (8.72) and shown in Figure 8.51, the VCO phase noise is limited by the process parameters, power supply voltage, and 1/f noise of the tail current. With the bias current set, the output swing should be maximized for low phase noise. The output swing of the differential VCO shown in Figure 8.47 is the largest because its DC bias sits on the positive power supply, and the output can swing higher than the supply rail. The output magnitude can be larger than the supply if the VCO is on a chip packaged separately. However, in integrated systems, the output swing should stay within the supply rails. Therefore, VCO design for low phase noise focuses on reducing 1/f noise of the tail current and increasing the output magnitude. In addition, to increase g_m for any given bias current, both NMOS and PMOS transistors can be used like a CMOS inverter.

There are several ways to remove or reduce the 1/f noise of the tail current as shown in Figure 8.52. The first option is to operate VCO without the tail current, but the bias condition depends too much on process, voltage, and temperature (PVT) variations. AC short using a capacitor can set the bias correct, but it still exhibits sensitivities to common mode and power supply. Inductor isolation uses chip area and limits output swing. Although the DC bias current can be filtered while keeping the low AC ground, the most common remedy is just to add a PMOS tail current at the positive supply, which exhibits lower 1/f noise.

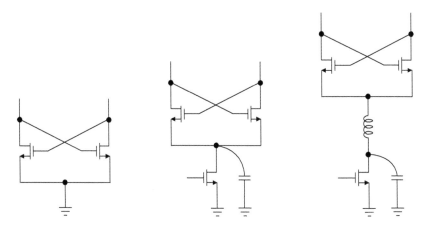

FIGURE 8.52
Reduction of the tail current 1/f noise.

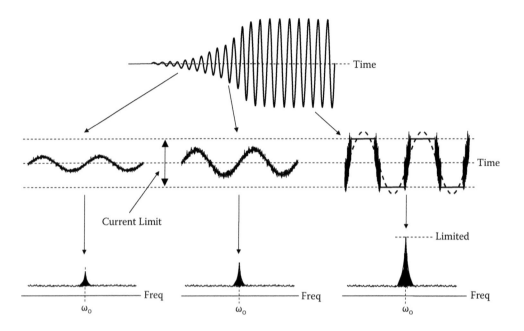

FIGURE 8.53
VCO oscillation growth both in time and frequency domain.

8.7.6 Current versus Voltage Limiting

The VCO phase noise is, in fact, the thermal noise around the resonant frequency of the tank circuit. As the oscillation grows, only the thermal noise around ω_o is selectively amplified. Note that there is no 1/f noise present at ω_o. During the linear amplification phase, both the current and voltage waveforms are sinusoidal. However, the oscillation stops growing when the current is limited by the bias current.

Figure 8.53 explains the limiting process of the oscillator. If the output starts to exceed the supply rail, the output is voltage limited or clipped. The current-limited output is sketched by the thick dotted line when it grows to be supply limited. Although the output stops growing, noise still grows but only during the brief transient periods. If the current to the load is limited, the current noise is low because active devices are switched either on or off. The phase noise performance is optimized at the boundary where the current-limited magnitude is close to the supply limit, and Equation (8.72) based on the linear model reasonably well predicts the phase noise of the supply-limited VCO output.

Both NMOS and CMOS differential VCO configurations with PMOS tail current are shown in Figure 8.54. The output magnitude is a direct function of the bias current when limited. Therefore, it is of paramount interest to set the bias current level properly. Assume that the positive feedback amplifier injects the tail current switched on and off hard into the tank load as shown at the top equivalent circuit. The square-wave current is filtered in the RLC tank circuit, yielding a sine wave with a slightly higher magnitude. The current switching paths are shown with solid lines.

In the VCO on the left side, the tail current flows through two transistors and the inductor L. However, in the VCO using differential CMOS inverters on the right side, one more device is stacked, and the output swing is reduced accordingly. For the oscillation to be contained within the supply rails, the current-limiting condition is about

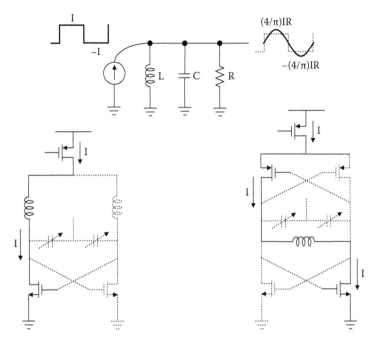

FIGURE 8.54
Current and voltage limiting.

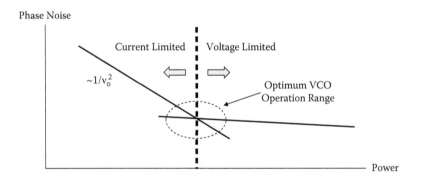

FIGURE 8.55
Phase noise versus power.

$$V_{DD} \approx \left(\frac{4}{\pi}\right) IR + n V_{DSsat},$$

$$(8.74)$$

where n is the number of transistors in the current path from V_{DD} to ground. As given by Equation (8.72), if the differential CMOS inverter is used, both NMOS and PMOS g_m are added together, and high g_m improves phase noise. However, the improvement over NMOS-only VCOs would be marginal as the output magnitude decreases.

Figure 8.55 shows that the phase noise dependence on the bias current. From Equation (8.72), the phase noise decreases as a square function of the output magnitude, and higher power offers lower phase noise. However, once the magnitude reaches the supply rail, the phase noise levels off, and the phase noise gain diminishes though the bias current is increased further. Therefore, with technology and supply voltage given, the VCO power is set.

8.7.7 Other Noise Sources in PLL

In most PLLs, the VCO phase noise is the dominant noise source. The noise from the charge pump is a burst noise that is injected into the loop filter only when the charge pump is on. In narrowband PLLs, it tends to be averaged by the loop filter. However, if PLL bandwidth is wider than 100s of kHz, other noise sources that begin to contribute are such noises as loop filter noise, coupling from bias and supply lines, and even jitters from digital logic for PFD, frequency divider, and prescaler, as shown in Figure 8.56.

Any noise coupled to the VCO input modulates the VCO and shows up as phase noise. For broad-banding PLL, the charge pump current is raised, and its current noise increases accordingly. If the minimum pulse width is widened to avoid the dead zone in the PFD, its contribution will increase further. Bypassing current mirrors also helps to cut the current source noise mirrored into the charge pump current. Any residual current after being filtered by the loop filter will modulate the VCO as noise.

Similarly, the noise from the loop-filter resistance R is injected into the VCO control voltage after being low-pass filtered by the following transfer function:

$$H(s) = \frac{1}{1+\dfrac{C_p}{C}+sRC_p}.$$
(8.75)

The noise bandwidth is about the out-of-band pole of PLL, and the gain is also about unity because $C \gg C_p$. Therefore, the loop-filter thermal noise directly affects PLL without being filtered.

Low charge pump current reduces noise, but it results in large loop filter resistance. Therefore, the values of I and R are chosen carefully after the PLL bandwidth is set. The product of IR is typically limited to about a Volt or two. Small R also leads to large C. That is why most loop filter capacitors are supplied externally. Integrating small C on a chip inevitably sacrifices the noise performance because either larger loop filter R or higher charge pump current I would contribute higher noise.

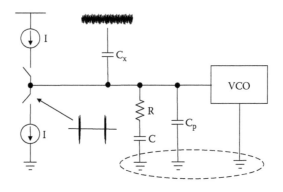

FIGURE 8.56
Noise sources in PLL.

In most system chip designs, the VCO gain is minimized to reduce the VCO sensitivity to digital coupling noises from bias and supply lines. It is important to ground the VCO and loop filter together. Their ground difference will modulate the VCO like noise. The VCO input should stay open with no capacitive coupling in the tristate mode. Noise coupling through parasitic capacitances is reduced if the VCO input node is well isolated by careful layout. Large loop filter C also reduces the coupling effect. Large C, in turn, requires either small resistance R or high charge pump current I and will increase the power consumption.

8.8 Low-Q Ring-Oscillator VCO

High-Q VCO can meet the stringent low phase noise requirement in frequency synthesis. However, in other applications such as clock recovery, regeneration, and distribution, low phase noise is not required, and low-Q ring or relaxation-type oscillators have been used due to their simplicity.

8.8.1 Oscillation Condition for Ring-Oscillator VCO

The oscillation conditions stay the same as the high-Q LC oscillator. To make a total loop phase delay of 360°, each inverter stage needs to make the extra phase shift of $-180°/N$, where N is the number of inverters. The lowest number for N is 3, and odd numbers of stages are commonly used for ring oscillators. The minimum N for differential ring oscillators is still three because the maximum phase shift of the inverter is only 90°.

Figure 8.57 shows an example of a three-stage ring oscillator. For oscillation to grow, each stage should contribute 60° of the phase delay, and the loop gain should be higher than unity at the resonant frequency ω_0. That is,

$$\{g_m Z_0(\omega_0)\}^3 > 1, \qquad \tan^{-1}(\omega_0 RC) = 60°, \tag{8.76}$$

where $Z_0(\omega)$ is the impedance of the parallel R and C load.

$$Z_0(\omega) = \frac{R}{1 + j\omega RC} = \frac{R}{\sqrt{1 + (\omega RC)^2}} e^{-j\tan^{-1}(\omega RC)}. \tag{8.77}$$

At low frequencies, the ring oscillator is stable due to the negative feedback. However, if there is a loop gain, it will oscillate at a frequency where the extra loop phase reaches $-180°$. The oscillation in the ring oscillator starts to grow initially in the small-signal linear mode as in the high-Q LC oscillator. Similarly, when limited by the supply rails, the oscillation stops growing, and it gets into the steady state. Assume that the linear small-signal model is still valid for the oscillation tone as in the LC oscillator. At the oscillation frequency,

$$Z_0(\omega_0) \approx \frac{R}{2} e^{-j60°}, \qquad \omega_0 = \frac{\sqrt{3}}{RC}. \tag{8.78}$$

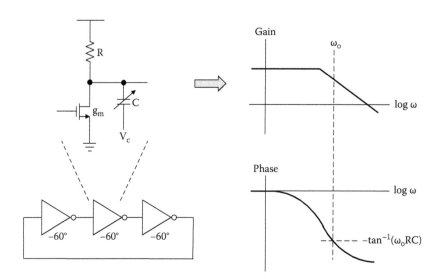

FIGURE 8.57
Inverter and its Bode plots.

The impedance is reduced to $R/2$, but the phase is delayed by 60° to meet the loop phase condition. Then the loop gain becomes

$$\left\{ g_m Z_o (\omega_o) \right\}^3 = \left(\frac{g_m R}{2} \right)^3 e^{-j180^o} = -\left(\frac{g_m R}{2} \right)^3. \tag{8.79}$$

This extra phase inversion by −180° makes the negative-feedback system unstable. As it becomes a positive-feedback system for oscillation to grow, the small-signal gain should meet the loop-gain requirement of $g_m R > 2$, but after limited in the steady state, the gain of the fundamental tone is reduced to meet the steady-state loop gain condition of

$$g_m R = 2. \tag{8.80}$$

Again this steady-state unity-loop gain condition is to ensure that the oscillation in the steady state can be sustained with a large constant magnitude.

8.8.2 Phase Noise of Ring-Oscillator VCO

The phase noise of ring-oscillator VCO can be derived similarly to that of LC VCO with the following noise injected into the output node. As in the LC oscillator, a small-signal linear model of one inverter in a three-stage ring oscillator can be drawn with a noise source as shown in Figure 8.58.

All transistor noise sources are summed as a current noise.

$$i_n^2 = 4kT \left(\frac{2g_m}{3} + g_m^2 R_g \right) \Delta f + 1/f \; Noise. \tag{8.81}$$

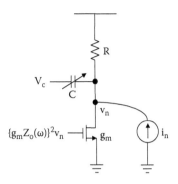

FIGURE 8.58
One inverter stage in a three-stage ring oscillator.

The load resistance R is an equivalent output resistance of the inverter. Because it is not a physical resistance, its noise is not included in Equation (8.81).

Due to the lack of the reactive inductor component, the inverter behaves like an amplifier with one pole at the output node. The transimpedance function that relates i_n to the output noise v_n is

$$\frac{v_n}{i_n}(\omega) = \frac{Z_o(\omega)}{1 + \{g_m Z_o(\omega)\}^3}.$$

(8.82)

Therefore, using Equations (8.77), (8.79), and (8.80), it can be written in the steady state as

$$\frac{v_n}{i_n}(\omega) = \frac{R(1 + j\omega RC)^2}{(1 + j\omega RC)^3 + (g_m R)^3} \approx \frac{R\{1 + j(\omega_o + \Delta\omega)RC\}^2}{3(1 + j\omega_o RC)^2 (j\Delta\omega RC)}.$$

(8.83)

Because the gain of the closed-loop transfer function should be infinite at the oscillation frequency, the following condition should be met from Equation (8.78):

$$(1 + j\omega_o RC)^3 + (g_m R)^3 = 0.$$

(8.84)

The numerator is modified by the gain change due to the phase shift and can be approximated as follows:

$$\{1 + j(\omega_o + \Delta\omega)RC\}^2 = 1 + (\omega_o + \Delta\omega)^2 R^2 C^2 \approx 4 + 2\sqrt{3}\left(\frac{\Delta\omega}{\omega_o}\right).$$

(8.85)

Therefore, the phase noise can be approximated as

$$\frac{v_n^2}{v_{o(\Delta\omega)}^2} = 3 \times \frac{i_n^2 R^2 \left\{4 + 2\sqrt{3}\left(\dfrac{\Delta\omega}{\omega_o}\right)\right\}^2}{(3 \times 4 \times \sqrt{3})^2 \left(\dfrac{\Delta\omega}{\omega_o}\right)^2} \times \frac{1}{v_o^2} \approx \frac{i_n^2 R^2}{9\left(\dfrac{\Delta\omega}{\omega_o}\right)^2} \times \frac{1}{v_o^2},$$

(8.86)

where the factor of 3 is to include the noises contributed by the three stages. Also note that the small gain change due to the phase shift is ignored. Due to the similarity to the LC VCO phase noise given by Equation (8.72), the same conclusions can be drawn.

8.8.3 Q Effect on VCO Phase Noise

If implemented differentially, the phase noise of ring-oscillator VCO will be further lowered by 3 dB because the signal is doubled while the noise power is doubled. However, because the power and chip area are also doubled, there is no significant improvement if the same power and chip area are used. If the same differential case is compared to the LC VCO given by Equation (8.72), the differential three-stage ring oscillator has an effective Q of about 1.5 from Equation (8.86). This linear oscillator theory predicts that the delay-type ring oscillator cannot beat the phase noise performance of the high-Q LC oscillator due to the low effective Q.

To clearly see the Q effect on the phase noise, two amplifiers with and without an inductor load are compared under identical conditions as shown in Figure 8.59. Assume that an LC oscillator and a three-stage ring oscillator are made with one gain stage, and feedback is applied using ideal phase shifters of −180° and −120°, respectively.

Then the transimpedance function for the LC oscillator given by Equation (8.71) is modified as follows:

$$\frac{v_n}{i_n}(\omega) = \frac{Z_o(\omega)}{1 + g_m Z_o(\omega) e^{-j180°}} = \frac{R}{1 + jQ\left(\dfrac{\omega}{\omega_o} - \dfrac{\omega_o}{\omega}\right) - g_m R} \approx \frac{R}{j2Q\left(\dfrac{\Delta\omega}{\omega_o}\right)}. \tag{8.87}$$

Similarly, the transimpedance function for the three-stage ring oscillator given by Equation (8.83) is also modified as follows:

$$\frac{v_n}{i_n}(\omega) = \frac{Z_o(\omega)}{1 + g_m Z_o(\omega) e^{-j120°}} = \frac{R}{1 + j\omega RC + g_m R \times e^{-j120°}} \approx \frac{R}{j\sqrt{3}\left(\dfrac{\Delta\omega}{\omega_o}\right)}. \tag{8.88}$$

From Equations (8.87) and (8.88), the effective Q of a single inverter is about 0.866, which is lower than 1.5 previously obtained by comparing differential LC and ring oscillators.

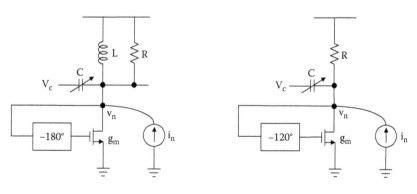

FIGURE 8.59
Oscillators with and without inductor.

8.8.4 Low Phase Noise Design for Ring-Oscillator VCO

Ring oscillator VCOs are low-cost alternatives with wider tuning range and multiphase outputs, but they exhibit poorer phase noise due to the lack of the energy-storing element. Their main design goal is not to achieve low phase noise but flexibility and simplicity. Although the phase noise analysis based on the linear model is valid for the high-Q LC oscillator, the phase noise of the ring oscillator is affected by many other factors. If the oscillation grows larger, inverters stay in the linear mode momentarily only during the low-to-high or high-to-low transients, and operate mostly in the nonlinear limiting mode. If inverters are switched on and off heavily, their load capacitances are either charged or discharged.

High-Q LC oscillators depend on the resonant tank circuit, but ring oscillators depend mostly on the inverter delay as well as on the charging and discharging times. Therefore, it is difficult to analytically obtain the oscillation frequency of the inductor-less VCO, but in general, the following relation holds true for the ring oscillator:

$$f_o \propto \frac{1}{RC} \times \frac{I_{bias}}{V_{swing}}. \tag{8.89}$$

Note that the short RC time constant helps to raise the oscillation frequency for small signal, but the high bias current and the small output swing shorten the charging and discharging times for large signal. Therefore, any one of R, C, I_{bias}, and V_{swing} values can change the oscillation frequency.

It is common to control the oscillation frequency of the ring oscillator by starving inverters with limited bias current as shown in Figure 8.60.

The current limiting is very effective because it controls both small and large signal parameters. Also as shown, the oscillation frequency can be varied if the supply voltage is regulated, but the oscillation magnitude changes, too. Other ways to vary capacitive loading and to interpolate feedback delay with the control voltage V_c are also valid as shown in Figure 8.61.

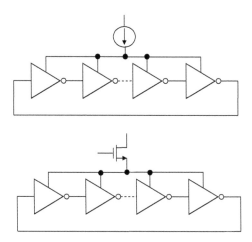

FIGURE 8.60
Current starving versus supply limiting.

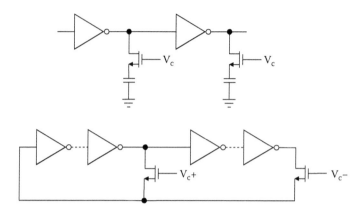

FIGURE 8.61
Variable capacitor loading and delay interpolation.

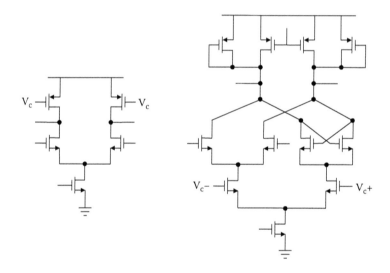

FIGURE 8.62
Differential delay cells with variable resistor loading.

The oscillation frequency changes if the load resistance is made variable. One way is to use triode-biases transistor as a load, and the other way is to use a negative resistance using local positive feedback as shown in Figure 8.62.

It is easy to make positive feedback in differential topologies. The similar negative resistance concept by positive feedback is used to make a preamplifier for comparators. In this case, the negative resistance to be subtracted from the positive diode resistance is adjusted by V_c. Although the differential delay is immune to supply noise, and flexible to give both in-phase and quadrature (I/Q) outputs, it is slower than a single-ended one due to the current limiting, and the swing is limited.

In general, jitter of the ring oscillator is inversely proportional to a square root function of power if the linear oscillator theory is valid. It also predicts that delay-type ring oscillators cannot beat LC oscillators in phase noise. However, for lower jitter or phase noise, the

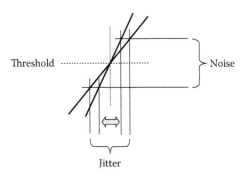

FIGURE 8.63
Noise-to-jitter transfer.

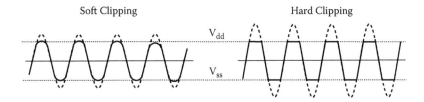

FIGURE 8.64
Soft and hard clipping.

slope of rising or falling transients can be made steeper by consuming more power and by operating them heavily in nonlinear mode. The noise-to-jitter transfer during the rising and falling transients of the ring oscillator is linear at the inverter threshold as shown in Figure 8.63.

Divided by the rising or falling slope, the RMS noise at the inverter threshold can be directly translated into the RMS jitter noise. The three-stage ring oscillator operating in linear mode has a soft-clipped output and exhibits a maximum effective Q of 1.5. However, if the inverter output is hard clipped, the transient time can be reduced further due to the steeper rising or falling slope as shown in Figure 8.64.

An ideal ring oscillator is a square-wave oscillator, and the delay element should switch fast. However, it is not possible to operate any delay elements with both delay and steeper slope because the delay is also related to the steepness of the slope.

The closest to this is to add a transition edge enhancer to the inverter as shown in Figure 8.65. The inverter is loaded with a positive-feedback latch. The bistable latch switches from one state to another with an extremely short time constant equivalent to the device unity-gain frequency of g_m/C. However, the cross-coupled latch structures are sensitive to supply and common-mode noises, and exhibit some hysteresis.

To sum up the ring-oscillator design, high power reduces noise while steeper slope reduces jitter. There should be a trade-off in the design. Steeper slope means shorter delay. Shorter delay requires more delay stages. More delay stages add more noise sources. To avoid this situation, capacitive loading can be increased as power is increased, though larger loading reduces the slope and somewhat lessens the high slope effect.

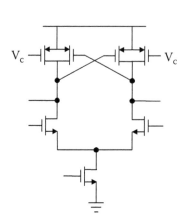

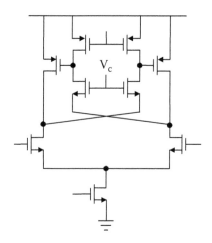

FIGURE 8.65
Hard-clipping inverter.

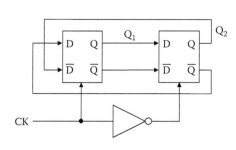

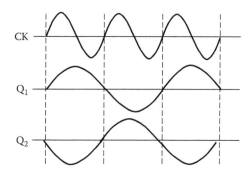

FIGURE 8.66
Divide by two using two delay flip/flops (D-F/Fs).

8.9 Prescaler

Other digital designs in PLL systems are mostly programmable dividers. There are no specific logic design issues other than the fast rise/fall time requirement. The rise/fall times of properly sized digital logic with minimum fan-outs approach the speed limit set by the device f_T. Most CMOS programmable asynchronous dividers work well in the range below hundreds of MHz. Therefore, to divide RF frequencies in the GHz range, fast dividers with pulse swallowing function such as dividing by 4/5 or 8/9 are required [5].

The RF frequency divider is called a prescaler. It is mostly made using a source-coupled logic (differential pair) and operates in class-A mode. The most common block to divide the frequency by is the delay flip/flop (D-F/F), which can be clocked either at the rising or falling edge.

Figure 8.66 shows an example of the rising-edge triggered divide-by-2 using two D-F/Fs. The D-F/F is just a clocked positive-feedback latch that latches the input at the clock edge.

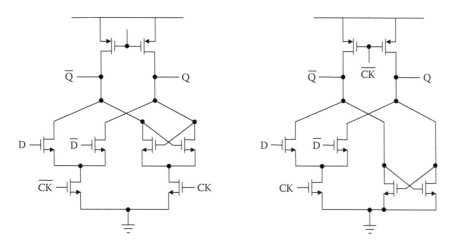

FIGURE 8.67
Examples of static and dynamic clocked D-F/Fs.

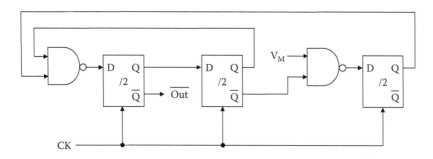

FIGURE 8.68
Divide by four and five.

At RF frequencies, the logic waveform cannot maintain the square waveform, and it looks more like a sinusoidal waveform. When clock (CK) crosses zero, Q_1 latches on the invert of Q_2, while Q_2 latches on Q_1 at the falling edge of CK. So the output waveform is the clock divided by two. The D-F/F can be made with a positive-feedback load as in the latch for the ADC comparator or a bit slicer.

Examples of D-F/F are shown in Figure 8.67. As in the comparator example, the data input D becomes a seed to decide the latch output when the latch is activated. The positive-feedback latch flips with the fast time constant of C/g_m close to $1/2\pi f_T$, and does not degrade the phase noise performance of the VCO. However, its dynamic input capacitance loads the VCO, and can disturb the sensitive VCO output. Therefore, the VCO output is always buffered to drive the prescaler.

8.9.1 Pulse Swallower

General programmable dividers are difficult to operate at RF frequencies. A simple pulse swallowing circuit is often used to skip one pulse count. It is the similar high-frequency logic like the prescaler, but the counter modulus is controlled so that the counter clock can be disabled for one clock period as shown in Figure 8.68.

The flag V_M is for modulus control. If V_M is low, two rising-edge triggered divide-by-2 circuits make a divide-by-4. If V_M goes high, the feedback is stopped during one clock pulse period, and the one pulse is swallowed. Then the counter resumes counting after five counts. Programmable counters for large numbers can be implemented using a pulse swallower placed at the front of low-frequency counters.

References

1. M. Danesh and J. Long, "Differentially driven symmetric microstrip inductors," *IEEE J. Solid-State Circuits*, vol. SC-50, pp. 332–340, January 2002.
2. F. Svelto, P. Erratico, S. Manzini, and R. Castello, "A metal oxide semiconductor varactor," *IEEE Electron Device Letters*, vol. 20, pp. 164–166, April 1999.
3. A. Hajimiri and T. Lee, "Design issues in CMOS differential *LC* oscillators," *IEEE J. Solid-State Circuits*, vol. SC-34, pp. 717–724, May 1999.
4. D. Leeson, "A simple model of feedback oscillator noise spectrum," *Proc. IEEE*, vol. 54, pp. 329–330, 1966.
5. J. Cranickx and M. Steyaert, "A 1.75-GHz/3-V dual modulus divide-by-128/129 prescaler in 0.7-µm CMOS," *IEEE J. Solid-State Circuits*, vol. SC-31, pp. 890–897, July 1996.

9

Frequency Synthesis and Clock Recovery

Phase-locked loop (PLL) finds its uses in mainly five different areas. The first area is the conventional synchronous demodulation for amplitude modulation (AM), phase modulation (PM), and frequency modulation (FM). The second area is the clock/data recovery (CDR) for digital communications, such as phase shift keying (PSK), frequency shift keying (FSK), compact disc (CD), digital video disc (DVD), and hard disk. All data receivers require CDR, which is now being implemented mostly in the digital domain. Even receivers for fiber networks will be implemented digitally. PLL for this application is a narrow-band PLL for jitter attenuation, and uses no-return-to-zero (NRZ) phase detector. The third area is RF frequency synthesis. RF signals are still analog and require very low phase noise. PLL for this is a wideband PLL with low spurious tones to reduce the voltage-controlled oscillator (VCO) phase noise. The fourth area is the clock generation and multiplication. Most digital systems need numerous synchronized clocks. The PLL design for this application is the most difficult as it requires both low VCO phase noise and low input clock jitter. Last, the fifth area is clock synchronization. Clocks and data in high-speed digital systems need to be time-aligned to eliminate clock skew without jitter accumulation, and this is often implemented using a delay locked loop (DLL).

9.1 Phase-Locked Loop Applications

Digitally implemented PLLs are ubiquitously used in modern data communications. Most demodulator functions are now rarely implemented in the analog domain. Low-frequency digital PLLs have numerical VCOs and digital PDs, but in the recent high-speed clock recovery for disk drive and networking, high-speed digital data are directly sampled and processed in parallel. Therefore, remaining analog PLL applications are now mostly limited to the high-frequency clock generation and recovery. Those two PLL systems are compared in Figure 9.1.

The most important design parameter in the PLL design is bandwidth. It mainly depends on the function PLL performs. In the synchronous clock generation, the output clock can be higher than the input or reference clock, and can also be an RF carrier. Because this application requires low phase noise, the loop should be broad-banded to suppress the VCO phase noise with high phase noise transfer function (PNTF). On the other hand, the clock recovery system should be narrow-banded to filter out the input clock jitter as it is low-pass filtered by the PLL. The PLL design strategy varies widely depending on its function. Therefore, different approaches should be taken for the PD and VCO designs.

9.1.1 General Clock Generation

PLL is to generate a local clock phase-locked to the input clock, and the low-jitter or low phase noise clock generation is the most basic function of PLL. Although every component in the PLL system contributes to jitter or phase noise, there are two major jitter sources in the PLL. One is from the input, and the other is from the VCO. If PLL is narrow-banded to filter out the former, the VCO jitter dominates. On the other hand, if broad-banded, the input jitter is not filtered. This bandwidth constraint severely limits the way clock generator systems are designed.

Figure 9.2 shows the input and output of the PLL clock generator both in time and frequency scales. The function is to generate synchronous clock multiples from the input clock. Whether PLL is narrow-banded or not, the output phase-locked to the input can be jittery, as explained. A better way to avoid the bandwidth constraint is to use a low-jitter VCO to minimize the jitter added by the PLL, while PLL is still narrow-banded to filter out the input jitter. The common notion that only broad-banding PLL helps to reduce the VCO phase noise needs to be overcome.

9.1.2 Low-Jitter Clock Generation

The low-jitter VCO can be either a voltage-controlled crystal oscillator (VCXO) or another frequency synthesizer, possibly fractional-N that is controlled by the input digital modulus.

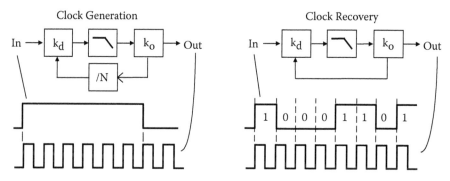

FIGURE 9.1
Clock generation and recovery.

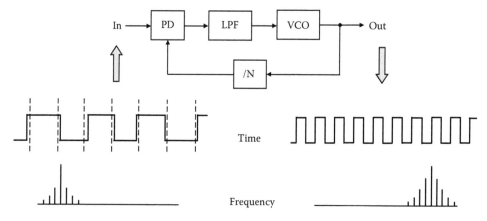

FIGURE 9.2
Clock generation.

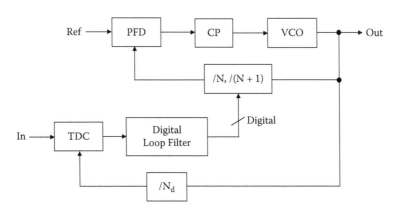

FIGURE 9.3
Clock generator with fractional-*N* synthesizer as voltage-controlled oscillator (VCO).

The former is not flexible because the crystal frequency is fixed, and its oscillation frequency range is very narrow. On the other hand, the latter can synthesize a wide range of frequencies. However, in the latter case, the PD should be a time-to-digital converter (TDC), and the loop filter should be a digital filter because the synthesizer VCO takes only the digital input modulus. This narrow-band clock generation scheme using a low-jitter frequency-synthesizer VCO is shown in Figure 9.3.

The upper portion is a wideband fractional-N frequency synthesizer, and the bottom portion is a narrow-band digital integer-N clock generator PLL. The wideband charge-pump (CP) PLL is now used as a low-jitter VCO that is digitally controllable. It can synthesize any frequencies with low jitter, and its jitter is further filtered in the narrow-band main loop filter. The TDC quantizes the time difference between the input and the VCO output, which performs as a PD. The output clock frequency is related to the input frequency by the integer divider ratio N_d in this example.

9.2 Digital PLL

In the system-on-chip (SoC) environment, the trend is that most PLL functions are implemented predominantly in the digital domain. The term *digital PLL* also refers to the PLL with a digital loop filter but which is implemented with an analog VCO. Such a digital PLL requires a TDC for digital loop filtering, and also a digital-to-analog converter (DAC) to drive the analog VCO as shown in Figure 9.4.

The VCO together with an input DAC is often called a digitally controlled oscillator (DCO). The advantage of digital loop filtering is the flexibility and accuracy that digital signal processing (DSP) provides. The digital loop filter can easily replace large loop-filter capacitors with small digital circuitry. TDC works as a phase detector and feeds the digital phase error back to the digital loop filter. However, both TDC and DAC have finite quantization noises, which should be lower than the VCO phase noise. The simplest TDC is to latch the input reference with polyphase VCO outputs, which creates a snapshot of the zero-crossing point. The location of the zero-crossing point becomes the thermometer-coded digital output like the time-domain analog-to-digital converter (ADC) discussed in Chapter 5.

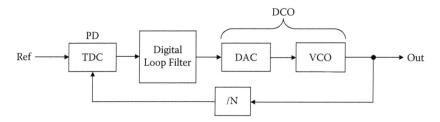

FIGURE 9.4
PLL with digital loop filter.

9.2.1 Time-to-Digital Converter

Assume that the multiphase VCO outputs are equally spaced apart by Δt in time. Then the root-mean-square (RMS) jitter power is

$$(\Delta t_{rms})^2 = \frac{(\Delta t)^2}{12}. \tag{9.1}$$

If this jitter power is spread evenly over the reference frequency band, the RMS phase noise spectral density would be

$$\frac{(\Delta\phi)^2}{\Delta f} = \frac{(2\pi f_{vco} \times \Delta t_{rms})^2}{f_{ref}}, \tag{9.2}$$

where f_{ref} and f_{vco} are the reference and VCO frequencies, respectively.

Although digital filtering gives a nice feature of convenient programmability, this quantization time step should be made very small to be effective. For example, to achieve a phase noise of −100 dBc/Hz level, the RMS jitter should be less than 3.6ps from Equation (9.2) if f_{ref} = 20 MHz and f_{vco} = 2 GHz. This RMS jitter requires that the quantization time step be smaller than 12.5ps from Equation (9.1), which corresponds to about $T_{vco}/40$. That is, the VCO period should be subdivided into 40 subintervals to generate time steps in such fine time resolution.

However, in clock recovery, even an RMS timing jitter equivalent to one-tenth of the symbol period is allowed, and jitter is not much of an issue. That is,

$$(\Delta t_{rms})^2 = \frac{(\Delta t)^2}{12} < \left(\frac{T}{10}\right)^2. \tag{9.3}$$

This condition can be easily met unless the time step Δt is excessively large and comparable to the symbol period T. However, for low phase noise clock generation, the quantized jitter spectral density should be kept lower than the VCO phase noise spectral density.

$$\frac{(\Delta\phi)^2}{\Delta f} = \frac{\left(2\pi f_{vco} \times \Delta t_{rms}\right)^2}{f_{ref}} < \frac{\left(\Delta\phi_{vco}\right)^2}{\Delta f}. \tag{9.4}$$

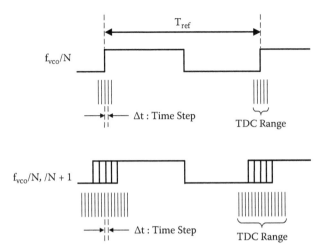

FIGURE 9.5
Time-to-digital converter (TDC) range for integer-N and fractional-N synthesizers.

In particular, considering the other quantization noise introduced by the DAC for the DCO input, this condition is not easy to meet in the very low phase noise radio frequency (RF) synthesizer design.

For integer-N frequency synthesis, the divided frequency makes periodic transitions at every reference period interval, T_{ref}, as shown at the top of Figure 9.5. Therefore, the TDC range needs only to cover the peak jitter of the divided frequency. However, in fractional-N synthesis shown at the bottom, the divider modulus changes randomly, and the jittered transition range of the divided frequency becomes much wider. It would be challenging to cover a wider jitter range with very fine time steps equivalent to a fraction of the VCO period interval. One way to cover a wider jitter range is to select multiples of the ring-oscillator VCO phase outputs at the estimated VCO transition intervals, but the complexity in switching high-frequency multiphase clocks with such a fine time step would be substantial. The same analogy can be found in the high-resolution ADC design. A wide input range should be covered with very fine resolution steps. As long as the time quantization noise can be made lower than the phase noise, this digital approach offers great flexibility and opens up the possibility of integrating all PLL components using scaled low-voltage CMOS.

9.3 Frequency Synthesis

One important application of PLL is to generate local carriers in RF systems with low phase noise and low spurious tones. RF frequency synthesis is a special case of the clock generation. The input to the frequency synthesizer is a fixed reference frequency of a crystal that exhibits extremely low phase noise. The Q of crystal is typically in the range of tens of thousands compared to the single-digit Q of on-chip spiral inductors. Therefore, the output phase noise of the frequency synthesizer comes mostly from the VCO phase noise.

The frequency synthesizer design is a trade-off between the output phase noise and reference/spurious tones. The loop bandwidth should be maximized so that the VCO phase

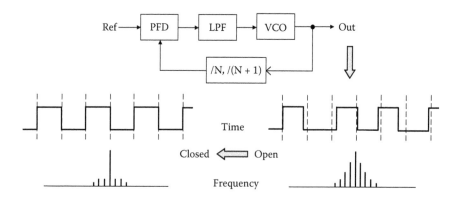

FIGURE 9.6
Frequency synthesizer.

noise can be suppressed with high PNTF given by the low-frequency loop filter gain. However, too wide of a bandwidth can risk the loop stability and also adversely affect the suppression of the out-of-band reference and spurious tones.

Figure 9.6 shows an example of the frequency synthesizer. There is no need for initial frequency acquisition if an up/down phase-frequency detector (PFD) with a charge pump is used. As shown, the in-band open-loop VCO phase noise is suppressed by the loop gain, and the low-frequency phase noise around the carrier is significantly reduced. The modulus for the feedback divider can be any integer or fractional number. If fractional, it is called fractional-N synthesizer. Depending on whether the divider is integer or fractional, PLLs presents quite different sets of design issues.

9.3.1 Integer-N versus Fractional-N Synthesizers

If the synthesizer PLL is in lock, the output frequency of the integer-N synthesizer is Nf_{ref}, and the reference frequency f_{ref} becomes a minimum frequency step to be synthesized. Because the loop bandwidth of the synthesizer should be narrower than $f_{\mathrm{ref}}/8$, if the frequency step is small, the loop bandwidth would get prohibitively narrow. Furthermore, it is difficult for narrow-band PLLs to sustain locking. Most synthesizers need wider bandwidth for agile synthesis and low phase noise.

Figure 9.7 compares synthesizable frequencies for integer-N and fractional-N cases. The vertical scale represents the frequency to synthesize marked with divider modulus. In the integer-N case, as N increases, the synthesized frequency increases by f_{ref}. In the fractional-N case, the frequency step for one integer step is much larger, and the small intermediate fractional frequency steps can be interpolated using integer numbers.

A frequency synthesizer based on the charge-pumped PLL is shown in Figure 9.8 with a divider inserted in the feedback loop. Because the reference frequency f_{ref} is the output frequency f_o divided by N, the VCO output frequency is high, but the charge pump (CP) is updated at a much slower rate of $1/f_{\mathrm{ref}}$. Therefore, the PLL loop bandwidth is scaled down according to N, and is obtained by dividing the unity loop-gain frequency by N.

$$\omega_k = \frac{IR}{2\pi}k_o \times \frac{1}{N}. \tag{9.5}$$

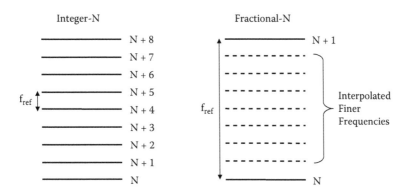

FIGURE 9.7
Integer-*N* versus fractional-*N* frequency synthesis.

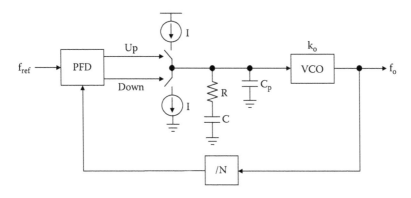

FIGURE 9.8
Frequency synthesizer with charge pump.

Also note that the phase noise is lower due to the same frequency division factor *N*. The effect of the loop bandwidth on phase noise is shown in Figure 9.9.

One most straightforward way to achieve low phase noise is to increase the loop bandwidth. The inverse of the open loop gain, which is defined as PNTF, high-pass shapes the VCO phase noise within the PLL loop bandwidth. Assume that the output frequency of $f_o = 2.441$ GHz is synthesized with a channel spacing of 1 MHz. If an integer-*N* synthesizer is used, $N = 2441$ and $f_{ref} = 1$ MHz, but the same frequency can also be synthesized with a fractional-*N* synthesizer, $N = 244.1$ and $f_{ref} = 10$ MHz. That is, fractional-*N* PLL can suppress the VCO phase noise with a 10× wider bandwidth. Note that if smaller than 1 MHz channel spacing is desired, the loop bandwidth of the integer-*N* synthesizer gets narrower accordingly.

9.3.2 Fractional Spur

There exists no fractional frequency divider. Therefore, in fractional-*N* synthesizers, the frequency divider averages many integer divider cycles over time to make an effective fractional divider ratio. A simple accumulator with an overflow can be used to control an $N/(N + 1)$ divider to interpolate any fractional values between *N* and $(N + 1)$ as in the first-order $\Delta\Sigma$ ratio modulator. For example, to divide frequency by a fractional number 244.1, it

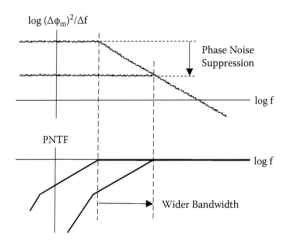

FIGURE 9.9
Phase noise versus loop bandwidth.

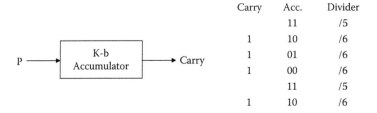

Carry	Acc.	Divider
	11	/5
1	10	/6
1	01	/6
1	00	/6
	11	/5
1	10	/6

FIGURE 9.10
Example of fractional division by 5.75.

can be divided nine times by 244, and once by 245 on average per every 10 divisions. Then, the average division ratio becomes

$$\overline{N} = \frac{(9 \times 244 + 1 \times 245)}{10} = 244.1. \tag{9.6}$$

The problem with this simple fractional division is that the same phase error pattern repeats for every 10 divisions, which creates a strong fractional spur occurring at $f_{ref}/10$.

The control for any fractional division using two adjacent integer numbers can be generated using an accumulator as shown in Figure 9.10. Assume that the fractional number to divide by is $N.F$, where N is an integer, and F is a fraction. Then dividing by $N.F$ is equivalent to dividing by N and $(N + 1)$ with an average duty of $(1 - F)$ and F, respectively. For example, if $N.F = 5.75$, $N = 5$ and $F = P/2^k = 3/4$. Therefore, $P = 3$ and $K = 2$. If the binary 11 is accumulated using 2-b counter, the carry comes out three out of four times. An effective fractional division of $N.F$ is obtained if the divider ratio is normally N, but the divider ratio jumps to $(N + 1)$ when the carry comes out.

This modulus switching is called pulse swallowing because it skips counting one pulse as discussed in Chapter 8. This simple fractional division inevitably produces a periodic phase jump due to the pulse swallowing, thereby creating a fractional spur in the PLL output. The

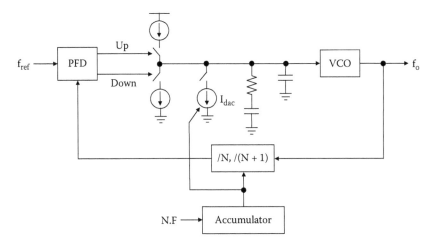

FIGURE 9.11
Digital-to-analog converter (DAC)-based spur cancellation scheme.

spur frequency is directly proportional to the frequency of pulse swallowing. In this example, swallowing one out of every four times creates a fractional spur at $f_{ref}/4$. The problem of the fractional spur is that the spur frequency can be far lower than the reference frequency, and often falls inside the PLL bandwidth. It gets far worse if the fractional number is close to the integer number N or $(N + 1)$ due to the spectral folding. The spur tones are created at

$$f_n = n \times (F\, f_{ref}),$$ (9.7)

where n is the harmonic number. For example, dividing by N.001 or N.999 would create a very low fixed fractional spur at $f_{ref}/1000$ and its multiples. Suppressing the in-band fractional spur has been one of the key design issues in fractional-N synthesizers.

9.3.3 Spur Cancellation by Digital-to-Analog Converter

An early spur cancellation scheme was to use a DAC pulse to offset the pumped charge as shown in Figure 9.11 [1]. The accumulator controls the modulus of the divider and also feeds the same divider phase information to the DAC so that the fractional spur can be offset at the CP output.

Figure 9.12 explains both the pumped charge and the DAC cancellation charge in the time domain. Note that the phase error is pulse-width-modulated (PWM), but the spur-cancellation DAC output is pulse-amplitude-modulated (PAM) with constant pulse width. The charge to be pumped into the loop filter at an interval of $1/f_{ref}$ is a gate function, and its high-frequency spectrum is shaped by the sinx/x (SINC) function depending on the pulse duration. Therefore, spur cancellation is not exact at high frequencies. However, if the DAC gain matches the CP gain, the low-frequency fractional spur can be canceled effectively.

9.3.4 Spur Shaping by ΔΣ Divider-Ratio Modulator

There are several ways to eliminate the fractional spur. Spur charge packets can be averaged over time or delayed for the synchronous charge update using analog sampled-data charge-domain filters. It is to mitigate the low-frequency spur-related fluctuation in the pumped charge, or the modulus divider can be dithered by injecting digital random white noise with mean zero so that the output spur noise spectrum can be whitened. The former analog method adds complexity to the sensitive analog loop filter, while the latter raises the overall phase noise floor.

One common way to avoid the periodic tones is to use $\Delta\Sigma$ divider ratio modulators that randomize the fractional divider ratio as shown in Figure 9.13 [2,3]. The divider-ratio modulator shapes the fractional divider modulus without adding white noise. However, representing a fractional number with a series of integer numbers still produces high-pass shaped ratio noise similar to the $\Delta\Sigma$ ADC quantization noise, which affects the phase noise performance significantly at high offset frequencies. If the divider ratio is noise shaped, the

Pumped Charge (PWM)

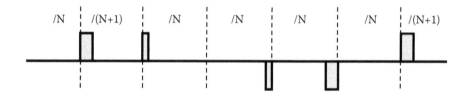

Spur-Canceling DAC Charge (PAM)

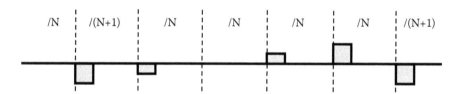

FIGURE 9.12
Pumped charge versus spur-canceling DAC charge.

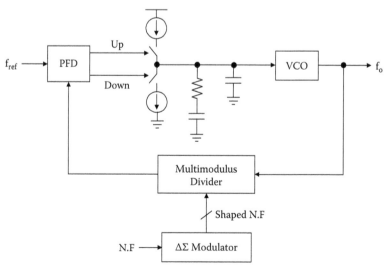

FIGURE 9.13
Multimodulus shaping by $\Delta\Sigma$ modulator.

divided VCO clock contains an average modulus plus its high-frequency shaped noise, and the clock phase jumps randomly with a minimum step of the VCO period. Therefore, the high-frequency divider noise should be filtered using out-of-band poles of the loop filter.

Figure 9.14 shows three examples of the phase jump associated with the divider modulus $(N + 1)$, N, and $(N - 1)$, respectively. Because the VCO frequency is divided, note that the time error step in the divided clock is one VCO period T, which corresponds to the phase error step of $2\pi f_{VC}T$. The minimum phase jump of one VCO period is similar to the quantization step of ADC and DAC. Fractional spurs are generated since this PWM error in the divided clock occurs periodically at an interval of T_{ref}. Depending on the number of modulus levels out of the $\Delta\Sigma$ divider ratio modulator, this phase step error is directly translated into a pumped charge with widely varying width. Therefore, in the fractional-N case, random charges with variable widths are pumped onto the loop filter as shown in Figure 9.15. However, in the integer-N case, the fixed divider makes no ratio error in the divided clock, and only the random phase error charge will be pumped.

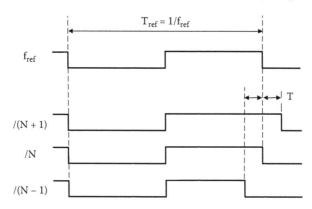

FIGURE 9.14
Integer-N versus fractional-N phase errors.

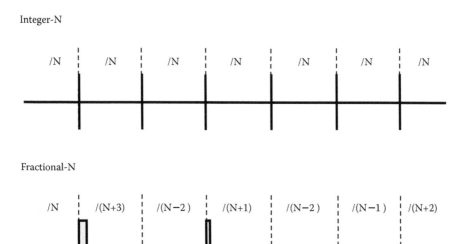

FIGURE 9.15
Integer-N versus fractional-N pumped charge.

9.3.5 Phase-Frequency Detector/Charge-Pump Nonlinearity

If high-order $\Delta\Sigma$ noise shaping is used, no low-frequency fractional spurs should show up even though the fraction is near any integer values. However, in reality, low-level fractional spurs are common because phase-frequency detector (PFD) and charge-pump (CP) are not ideal. The linearity of the pumped charge is very sensitive to any phase detector nonlinearity due to the large phase jump involved in the noise-shaping process. The PFD nonlinearity is mostly to blame for both in-band and out-of-band fractional spurs. They appear more prominently in the output spectrum proportional to the pumped charge.

Ideally, fractional spurs appear in the phase noise spectrum as sketched in Figure 9.16. The base spectrum is for the integer-N, which has no fractional spur by definition. The first-order modulator is the same as the simple accumulator and exhibits strong fractional spurs. Fractional spurs arise more prominently when the fractional number approaches the integer value. The second- and third-order shaped noises are more random and appear mostly at high frequencies. Two common sources of nonlinearity are the dead zone and the mismatch of the up/down CP currents as shown in Figure 9.17.

There are ways to suppress fractional spurs. The dead zone can be eliminated with a delay in the reset pulse of the PFD or using a class-A CP for linearity [4]. The up/down current mismatch can also be improved by feedback. Although the digital pattern depends on

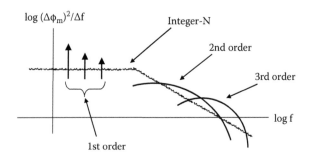

FIGURE 9.16
Fractional spur in the phase noise spectrum.

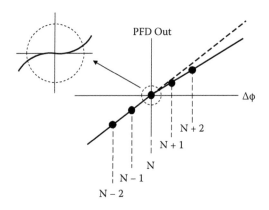

FIGURE 9.17
Phase-frequency detector (PFD) nonlinearity.

the digital $\Delta\Sigma$ ratio modulator, the large jump in the modulus values can be avoided using modulator architectures such as a single-loop high-order modulator that spreads modulus values less and keeps the modulus closely spaced and contained within the neighboring integer values. The cascaded high-order modulator tends to spread modulus values more widely than does the single-loop high-order modulator. This implies that the former injects more random charges into the loop filter than the latter [5]. One more indirect way to avoid the fractional spur is to operate the synthesizer within a narrow fraction range so that the modulus to synthesize can be tightly centered in the middle of two integers, for example, from 0.25 to 0.75. Then it is possible by covering the continuous frequency range by staggering the even and odd integer ranges. The spur will fall within a narrow frequency range from $f_{ref}/4$ to $f_{ref}/2$, and become an out-of-band tone that can be filtered by the extra loop filter pole.

9.3.6 Bandwidth Constraint of Fractional-*N* Synthesizers

The nice feature of the fractional-*N* frequency synthesizer with $\Delta\Sigma$ ratio modulator is to achieve both wide bandwidth and fine frequency resolution. However, the shaped high-frequency noise can be filtered using only out-of-band poles in the loop filter. Broad-banding helps to suppress the in-band VCO phase noise, and narrow-banding helps to suppress the out-of-band shaped $\Delta\Sigma$ noise. Therefore, a design trade-off for bandwidth is necessary once the reference frequency is given.

In typical synthesizers built around a low reference frequency in the 10 to 20 MHz range, the critical loop bandwidth is about 100 kHz or lower. For integer-*N* synthesizers, broad-banding PLL effectively suppresses the phase noise, but the charge-pump noise will limit the upper range of the PLL bandwidth to a few hundreds of kHz range. For fractional-*N* synthesizers, the upper range will be severely limited due to the drastic increase of the high-pass shaped $\Delta\Sigma$ ratio noise.

Figure 9.18 compares the loop bandwidth effect on the integrated phase noise for integer-*N* and second-order fractional-*N* frequency synthesizers. The total integrated phase noise includes all the contributions from the VCO, $\Delta\Sigma$ ratio modulator, charge pump, and reference noise. Integer-*N* synthesizers suffer mainly from the VCO noise due to the common narrow-band design. By widening the PLL loop bandwidth, a lower integrated phase noise can be achieved until it is limited by the charge pump and reference noise. However, wide bandwidth would mean high reference frequency, which, in turn, results in low frequency resolution. On the other hand, wideband fractional-*N* synthesizers circumvent this trade-off and achieve high frequency resolution, offering greater design flexibility at the system level.

If the loop bandwidth is made wide, the high-pass shaped noise becomes dominant, and the total integrated phase noise increases. Therefore, to sufficiently attenuate this high-frequency noise, the loop filter should be narrow-banded. This bandwidth constraint reduces the PLL loop gain that is needed to suppress the VCO phase noise. This translates into the similar noise and bandwidth trade-off as in the integer-*N* synthesizer case. Because the number of out-of-band poles are limited in PLL, the $\Delta\Sigma$ modulator noise spectrum is typically third-order shaped only below $f_{ref}/2\pi$, and stays flat after there. Extra loop filter poles can suppress it, but they can be placed only within a narrow range from about $4f_k$ to $f_{ref}/2$. The range further shrinks as the bandwidth is widened. Although high-order modulators randomize fractional spurs, the integrated phase noise increases drastically as the PLL is broad-banded. Furthermore, the elevated side tones make it difficult to meet the blocker specifications of the RF receiver as spurs can mix down blockers into the signal band.

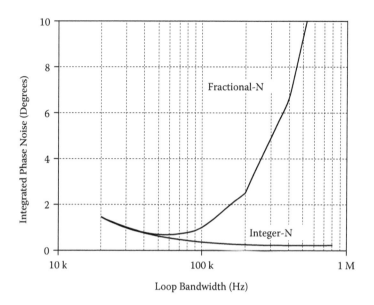

FIGURE 9.18
Integrated phase noise versus loop bandwidth.

9.4 Spur-Canceled Fractional-*N* Frequency Synthesizer

The simplest way to overcome this bandwidth constraint in the fractional-*N* synthesizer is to use higher reference frequencies, and filter out the out-of-band noise. However, most crystal frequencies in RF systems are chosen by the system requirements, and are in the low 10 to 20 MHz range. Crystal frequencies over 30 MHz are often overtone frequencies. In fact, high reference frequencies can be synthesized using another PLL. Additional PLL increases hardware complexity, and one more VCO phase noise is added to the synthesized VCO phase noise.

An alternative way is to use elaborate analog charge-domain sampled-data filters though limited by their analog imperfections. Spur averaging also achieves the same spur cancellation. It relies on the fact that the ratio error in a first-order $\Delta\Sigma$ PLL is periodic, and in a time period corresponding to the fractional frequency, the total ratio error charge injected by the CP is zero. Charge errors are accumulated on a capacitor for the time period and dumped on the loop filter at the end of each period though are limited in frequency resolution and loop bandwidth [6,7].

9.4.1 DAC-Based Spur Cancellation

The DAC-based spur cancellation schemes shown in Figure 9.11 can be generalized for the second- or third-order shaped ratio noise. The only difference is that the accumulator that is basically the first-order modulator needs to be replaced by higher-order modulators. This fractional-*N* divider ratio modulator takes a fractional word *F* and generates a sequence of time-modulated integers that is averaged to be *F*. The VCO frequency is divided by the time modulated integer values $N(n)$. Once the VCO is locked to $(N + F)f_{ref}$, the divider and reference edges would differ by $\{N(n) - F\}T$, where T is the period of the VCO output. This pulse-width modulated (PWM) ratio error is averaged over time to give the correct fractional modulus.

The PWM charge error has a fixed magnitude, but its width is variable depending on the divider ratio. To cancel this CP charge, a PWM cancellation charge is necessary. However, generating such a cancellation charge requires a precise time resolution. To avoid this, a pulse-amplitude modulated (PAM) DAC cancellation charge with a fixed minimum pulse width of T_{min} can be used instead. Then, the DAC pulse magnitude is approximately given by

$$I_{dac}(n) = \frac{I \times \left[T_{min} + \sum_{k=0}^{n} \{N(k) - F\}T \right]}{T_{min}} = I \times \left\{ 1 + \frac{\Delta T(n)}{T_{min}} \right\}, \tag{9.8}$$

where $\Delta T(n)$ is an integrated time error representing the integrated ratio error. The minimum pulse width T_{min} should be chosen considering the trade-off between the rise/fall time difference and the DAC/CP thermal noise injection. Note that the PWM charge is a zero-mean charge error for all values of F. Therefore, a fixed DAC range can be set, and a zero-mean signal can be generated for the DAC/CP gain calibration.

Any high-pass shaped $\Delta\Sigma$ ratio noise can be canceled using the spur-canceling DAC regardless of the modulator order. However, it is important to closely match the gain of the spur-canceling DAC with that of the PFD/CP. Both the PWM CP and PAM spur-cancellation charges are explained in Figure 9.19.

A DAC-based spur cancellation concept is first applied to an accumulator-type first-order fractional-N PLL [1], and later generalized for higher-order $\Delta\Sigma$ fractional-N PLLs [8–13], in order to subtract the dominant divider ratio error in real time using a DAC. If the $\Delta\Sigma$ divider ratio noise is canceled accurately, the phase noise performance of the fractional-N PLL can approach that of the integer-N PLL, irrespective of the order of the $\Delta\Sigma$ modulator

Pumped Charge (PWM)

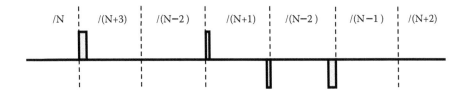

Spur-Canceling DAC Charge (PAM)

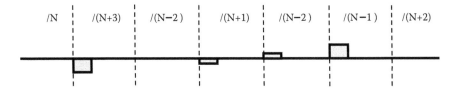

FIGURE 9.19
Pumped charge versus spur-canceling DAC charge.

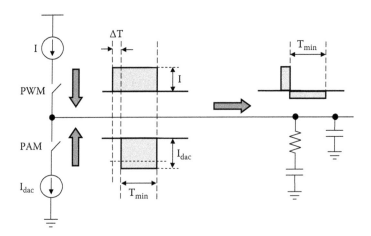

FIGURE 9.20
Pulse-width modulated (PWM) phase error versus pulse-amplitude-modulated (PAM) DAC pulse.

used. Because there is no need for the noise and bandwidth trade-off, it is possible to widen the PLL loop bandwidth independently of the frequency resolution and the high-pass shaped divider ratio noise as in any integer-N synthesizer until the phase noise is capped by the CP noise.

Figure 9.20 shows two current CP pulses injected into the loop filter. The Fourier transform of the current pulse of the PWM phase error with a minimum pulse width T_{min} is given by

$$PWM(\omega) = \frac{I}{j\omega}\left\{1 - e^{-j\omega(T_{min} + \Delta T_{rms})}\right\}. \tag{9.9}$$

where ΔT_{rms} is the average time error equivalent to the phase jump that leads to the fractional spur. The spur-canceling current pulse of the PAM DAC with a fixed pulse width T_{min} cancels this PWM phase error, and its Fourier transform is similarly given by

$$PAM(\omega) = -\frac{I}{j\omega}\left(1 + \frac{\Delta T_{rms}}{T_{min}}\right)\left\{e^{-j\omega(\Delta T_{rms})} - e^{-j\omega(T_{min} + \Delta T_{rms})}\right\}. \tag{9.10}$$

Note that the PAM pulse is shifted in time by ΔT_{rms}. The pulse width is T_{min}, and the height is $(1 + \Delta T_{rms}/T_{min})$.

Although the total PWM error charge is equal in average to the total PAM spur-cancellation charge, they do not match in both time and frequency domains and, therefore, leave some residual voltage at the loop filter for the duration of the PWM and PAM pulses. Both PWM and PAM errors in the frequency domain are well matched at low frequencies. In particular, at DC, they can be perfectly matched.

$$PWM(0) = -PAM(0). \tag{9.11}$$

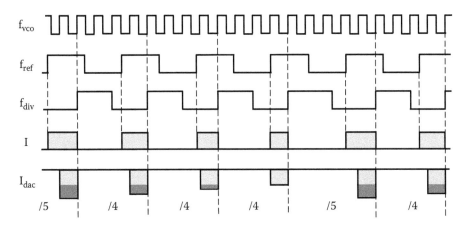

FIGURE 9.21
Dividing by 4.25 using class-A charge pump (CP).

However, they are not matched precisely at all frequencies, and the residual spur will be suppressed to a less degree. Such residual spurs remain as uncanceled divider ratio noise at high frequencies.

Consider a fractional division by 4.25 using just an accumulator as a modulus controller that produces /4, /4, /4, and /5 periodically as shown in Figure 9.21.

Note that figures like this are usually drawn for explanation purposes assuming that the VCO is in lock, and its frequency does not change. However, the fact is that spurious tones are generated because it changes. The edges of the reference clock and the divided clock are separated by the minimum pulse width of $(T_{min} + \Delta T)$. The CP current pulse is PWM, but the DAC current pulse is PAM as dark shaded. The rising edge of the reference sets the up pulse, and the rising edge of the divided clock reset it. In actual implementations, the short pulse T_{min} can be generated with multiple VCO periods.

9.4.2 Adaptive DAC Gain Calibration

Figure 9.22 sketches the spur spectrum before and after cancellation. Because the gain error exists where the maximum $\Delta\Sigma$ shaped ratio noise is present, the DAC pulse width should be minimized to reduce the frequency-dependent gain mismatch. This high-frequency mismatch becomes important if a much wider PLL bandwidth is desired, and the digital compensation is necessary for the correction of the sinx/x (SINC) function [10]. This DAC/CP gain mismatch problem is rather fundamental in all DAC-based spur cancellation schemes, and their effectiveness depends on how accurately the cancellation works. The DAC/CP matching accuracy can be enhanced further by making the DAC gain adjustment more adaptive applying feedback.

An example of an adaptive DAC-based spur-cancellation system is shown in Figure 9.23 [12]. A zero-forcing servo feedback concept is applied to calibrate the DAC gain mismatch in real time. The sign-sign least-mean-square (LMS) algorithm correlates the actual accumulated divider ratio error at the loop filter with its sign polarity, and to trim the DAC gain adaptively based on the sign polarity of the gain mismatch error. The adaptive process goes on until the DAC/CP gains are closely matched within the step size used for update.

The polarity of the actual spur energy is detected because the accumulated ratio error at the loop filter is directly correlated with the sign of the integrated phase error. A 1b $\Delta\Sigma$ modulator converts this correlation into a 1b stream, which is then fed into the digital

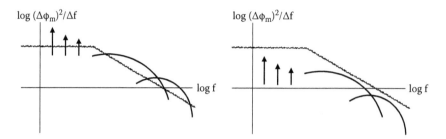

FIGURE 9.22
Fractional spur spectrum before and after cancellation.

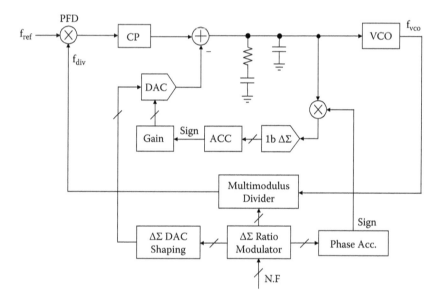

FIGURE 9.23
Spur cancellation with adaptive DAC gain.

block for low-pass filtering. An up/down counter can control the gain of the DAC, and this feedback adaptively corrects the DAC and CP gain mismatch based on the actual perturbations occurring at the loop filter. The ratio modulator generates the control word for the frequency divider. The difference between the control word and the fractional word is integrated to get the total phase error of the CP. This integrated phase error is then passed through the $\Delta\Sigma$ modulator that truncates the phase error for the DAC control, and high-pass shapes the resulting quantization noise.

9.4.3 Minimum DAC Pulse Width

The PFD/CP related nonlinearity is one of the well-known sources of fractional spurs, which are more prominent at the integer boundaries of the fractional words. Consider only the static current gain mismatch between the up/down pulses. The average time error ΔT_{rms} for fractional modulus is equivalent to the phase error of $2\pi(\Delta T_{rms}/2T)$. If the mismatch between the up/down currents is ΔI, the phase noise can be further scaled by $\Delta I/2I$, and the phase noise power of the fundamental spur f_{spur} can be approximated as

$$\left(\Delta\phi_{rms}\right)^2 \approx \left(2\pi \times \frac{\Delta T_{rms}}{2T}\right)^2 \times \left(\frac{\Delta I}{2I}\right)^2 \times \left|H(f_{spur})\right|^2, \tag{9.12}$$

where the phase noise is shaped by the closed-loop transfer function $H(f)$. Note that ΔT_{rms} also depends on the divider ratio modulator architecture and should be minimized for low-spurious frequency synthesis.

The DAC/CP gain mismatch not only results from the DAC/CP static current mismatch but also from the rise/fall time difference of the DAC/CP current pulses. If the CP is operated in a class-A mode away from the nonlinear zero-crossing point, a fixed timing offset is introduced in the CP up/down pulses by sinking or sourcing a fixed offset current for a time period of T_{min} so that only either the up or down pulse modulation can be used to maintain the steady locked state. The nonlinearity resulting from the rise/fall time mismatch is reduced if T_{min} is much longer than the rise/fall time of up/down pulses. However, it cannot be set to be very long because it will increase the CP phase-noise contribution, which is given by

$$\frac{\left(\Delta\phi_{rms}\right)^2}{\Delta f} \approx \left\{2\pi\overline{(N+F)} \times \left(\frac{2T_{min} + \Delta T_{rms}}{T_{ref}}\right)\right\}^2 \times \frac{i_{cp}^2}{I^2} \times \frac{\left|H(f)\right|^2}{\Delta f}, \tag{9.13}$$

where $i_{\mathrm{cp}}^2/\Delta f$ is the charge-pump noise, and T_{ref} is the reference period. The mismatch effect is severe if T_{min} is short. The shorter T_{min} is, the higher are the rise/fall related nonlinearity and the phase noise.

9.4.4 Quantization Noise of the Spur-Canceling DAC

The word width of the DAC is also an important design factor. For perfect cancellation of the $\Delta\Sigma$ divider ratio noise, the DAC resolution on the order of the frequency resolution is needed. For example, with a 20 MHz reference clock, a 20b DAC is needed for the frequency resolution of

$$\frac{f_{ref}}{2^N} = \frac{20MHz}{2^{20}} \approx 19Hz. \tag{9.14}$$

However, a 20b DAC is not practical, and using a low-resolution 7 ~ 8b DAC for 20b frequency resolution creates periodic quantization noise errors, which depend on both the word width and the fractional word.

For finite N-bit DAC resolution, such perturbations remain as uncanceled divider ratio noise, and residual fractional spurs appear due to the quantization error at

$$f_n = n \times (2^N F f_{ref}). \tag{9.15}$$

For the fractional value of F close to 0 or 1 due to spectral folding, the DAC quantization noise is more severe because the harmonics fall inside the PLL bandwidth. The more bits the DAC has, the higher the spur tones will be, and the lower the DAC quantization noise power will be. To alleviate the problem of F-related spurs, the DAC output can be shaped using a higher-order $\Delta\Sigma$ modulator (third-order would suffice.). This modulator takes the full 20b resolution phase error, and generates a truncated 7 ~ 8b DAC output. It is to high-pass shape any quantization noise generated by the truncation process. The $\Delta\Sigma$ modulator moves this DAC quantization error out of the loop bandwidth so that it can be sufficiently filtered by the loop filter.

The spur-canceling DAC nonlinearity also has the same effect as the PFD/CP nonlinearity on the PLL output phase noise. It results mainly from the mismatches among the DAC elements, and can be greatly reduced by using a thermometer-coded DAC or by using dynamic element mismatch shaping techniques [13].

9.4.5 Sign-Sign LMS Algorithm

For spur cancellation, the actual charge error at the loop filter is correlated with the sign of $\Delta T(n)$ to obtain the correlation energy, which is then used to calibrate the gain of the DAC current. Assume that the charge error between the PWM phase error and PAM spur-cancellation DAC output is

$$\Delta Q(n) = -\alpha(n) \times I \times \left\{ T_{\min} + \Delta T(n) \right\}, \tag{9.16}$$

where α is the mismatch ratio between the two. After neglecting uncorrelated products, low-pass filtering of the charge error correlated by the sign of the integrated phase error yields

$$LPF\left[\Delta Q(n) \times \operatorname{sgn}\left\{ \Delta T(n) \right\} \right] = -\alpha(n) \times I \times LPF\left\{ \left| \Delta T(n) \right| \right\}. \tag{9.17}$$

Hence, the correlation with the sign sequence gives us the power of the uncanceled residual $\Delta\Sigma$ divider ratio noise. The sign of its DC value can be used to update the gain of the DAC current. A sign-sign LMS DAC gain correction equation can be derived from the above parameters as

$$G_{dac}(j+1) = G_{dac}(j) + \mu \times \operatorname{sgn}\left\{ -\alpha(j) \right\}, \tag{9.18}$$

where G_{dac} is the DAC gain coefficient, and μ is the step size. This algorithm is stable and converges well. The value of μ should be small so that the gain correction can suppress a DAC/CP gain mismatch down to the desired resolution. Based on the DC energy of the correlation, the DAC gain coefficients are updated.

9.4.6 $\Delta\Sigma$ Divider Ratio Modulator

The fractional ratio modulator can be implemented using any high-order modulator architectures, but the order should be less than three because PLL cannot suppress any higher-order shaped noises. The $\Delta\Sigma$ modulus generator contributes extra phase noise. One is the quantization step of one VCO period, and the other is the frequency resolution limited by the number of bits. The phase noise equivalent to one VCO period jump is estimated as

$$\Delta\phi_{rms} = 2\pi f_{VCO} \times \left(\frac{T}{\sqrt{12}}\right) = \frac{2\pi}{\sqrt{12}}, \tag{9.19}$$

which is suppressed further by the third-order noise transfer function. Depending on the order of the modulator, this phase noise affects the output phase noise differently. If the third-order modulator is used, the in-band phase noise contribution is negligible, but out-of-band shaped noise would limit the phase noise performance. The out-of-band pole and the spur cancellation using DAC are to suppress this shaped fractional spur.

Similarly, the frequency resolution due to the truncation in the N-bit modulator amounts to the phase noise of

$$\Delta\phi_{rms} = 2\pi \times \left(\frac{f_{ref}}{2^N}\right) \times \left(\frac{T}{\sqrt{12}}\right) = \frac{2\pi}{2^N \times \sqrt{12} \times (N+F)}. \tag{9.20}$$

As it is divided by 2^N, this phase noise contribution by digital truncation is negligible. It is mainly set low just for low-frequency resolution.

Figure 9.24 shows a cascaded $\Delta\Sigma$ divider ratio modulator widely used as a fractional modulus generator. It is easy to implement without any overflow stability concern. It produces widely spread digital outputs that help randomization. In the example, it is processed with 23 bits that will give lower than 10 Hz resolution with a typical low 10s of MHz reference clock.

A single-loop third-order example scaled for stability is also shown in Figure 9.25 [5]. It is known to exhibit much lower spread in the digital outputs than the cascaded type, which

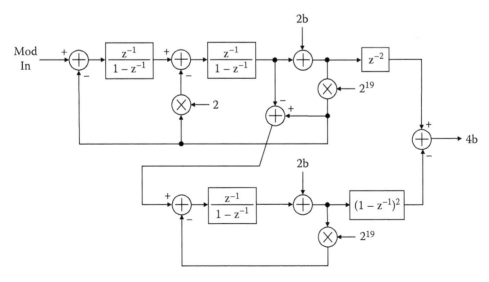

FIGURE 9.24
Cascaded 2-1 $\Delta\Sigma$ divider ratio modulator.

directly leads to lower phase jump and lower spurious tones. It interpolates fractional numbers mostly with two adjacent integer divider ratios.

9.4.7 Calibration of VCO Frequency

The VCO input can be well isolated from the digital noise, but with a high VCO gain of ~100 MHz/V, only a 10 μV noise coupled to the VCO input can produce a frequency noise of 1 kHz. Therefore, in most PLL designs for low phase noise, the VCO gain is intentionally set to a range of low tens of MHz/V to prevent any noise coupling through the loop filter. Then the operational frequency range of the PLL is severely limited. Furthermore, in low-voltage operations, it is difficult to set the VCO control range to be at the middle of the supply rail. As a result, it is necessary to auto-tune the VCO free-running frequency.

Figure 9.26 shows a digitally controlled VCO with a coarse tuning scheme that constantly monitors the loop filter voltage. It adaptively centers the input control range to be at the mid supply, and compares it to the high (V_H) and low (V_L) threshold of the control range. After averaging the up and down pulses over a long period, the fixed varactors are switched so that the input control voltage can remain in the mid range. Thereby, a wide tuning range can be achieved with a low VCO gain, and the high VCO gain requirement over process, supply voltage, and temperature becomes irrelevant.

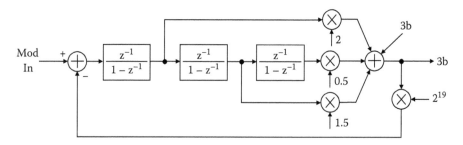

FIGURE 9.25
Single-loop third-order $\Delta\Sigma$ divider ratio modulator.

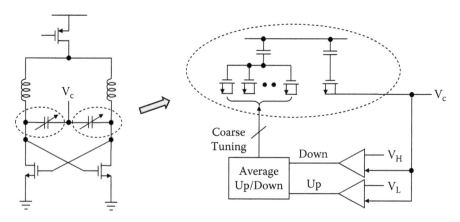

FIGURE 9.26
Servo loop to center the VCO control range.

9.5 Data Symbols

All digital communications receivers need to recover the timing information first from the received data before data are sliced for bit decisions. PLL is a key component used for clock and data recovery (CDR). The CDR is a special case of the general PLL clock generation as explained in Figure 9.27.

It performs two functions. One is to align the clock rising and falling edges with the data so that bit decisions can be made at the falling and rising edges, respectively. The recovered clock frequency is the same as the symbol rate $1/T$. The input is typically the no-return-to-zero (NRZ) jittery input data. In this application, jitter is not a serious issue because the jitter as large as one-tenth of the symbol period is often tolerable. The other function is to filter the jitter of the incoming data. PLL should be narrow-banded for that.

Digital communication is to transmit data symbols directly over broadband channels or over narrow band-limited channels, and to receive them. Two common data symbols used in digital communications are impulse and binary pulses as shown in Figure 9.28.

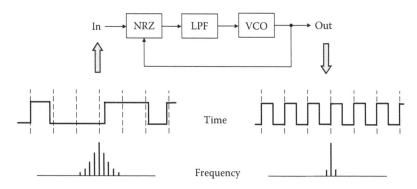

FIGURE 9.27
No-return-to-zero (NRZ) clock recovery.

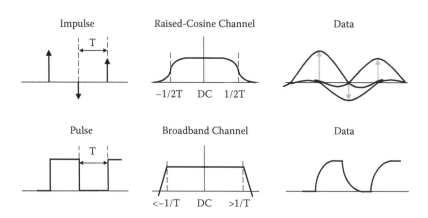

FIGURE 9.28
Channel and shaping functions for two data symbols.

9.5.1 Impulse Symbol

The impulse is a mathematical symbol for modern digital modems such as Ethernet 100 Base-T, 1000 Base-T, quadrature phase shift keying (QPSK), and quadrature amplitude modulation (QAM). I/Q carriers are directly AM-modulated by data symbols for band-limited channels. The synchronous demodulation removes the carrier to recover the impulse data. The similar impulse symbol is also used in hard disks and magnetic tapes. The impulse symbol is spread over time when band limited, and can cause both pre- and postcursor intersymbol interference (ISI). The ISI is an overflow of the data symbol over adjacent symbol periods.

When the impulse data sequence passes through an ideal brick-wall filter with the channel bandwidth of $1/2T$ as shown in Figure 9.29, the impulse symbol becomes the sinx/x (SINC) function.

$$s(t) = \sum_{n=-\infty}^{\infty} s_n(t) = \sum_{n=-\infty}^{\infty} A_n \frac{\sin\{\pi(t-nT)/T\}}{\pi(t-nT)/T}. \tag{9.21}$$

This baseband signal is band limited if $s_n(t)$ is. In QAM, the complex impulse symbol is used for A_n as follows:

$$A_n = a_n e^{j\phi_n} = a + jb, \tag{9.22}$$

where a and b make a two-dimensional vector. If this baseband is modulated up to the carrier ω_c, the QAM band-pass signal is defined as

$$s(t) = \sum_{n=-\infty}^{\infty} \text{Re}\{s_n(t)e^{j\omega_c t}\}. \tag{9.23}$$

The channel bandwidth limits the number of symbols that can pass through without ISI within a given amount of time. Note that this ideal brick-wall filter does not cause ISI, though the channel is band limited. It is because the sinx/x (SINC) function is nulled at every sampling interval T. However, the brick-wall filter is not feasible, and the closest real channel filter is a raised-cosine filter as sketched in Figure 9.28. The filter response has a smooth transition across the channel band edge. The impulse responses of the filter family are modified to allow an excess bandwidth as follows:

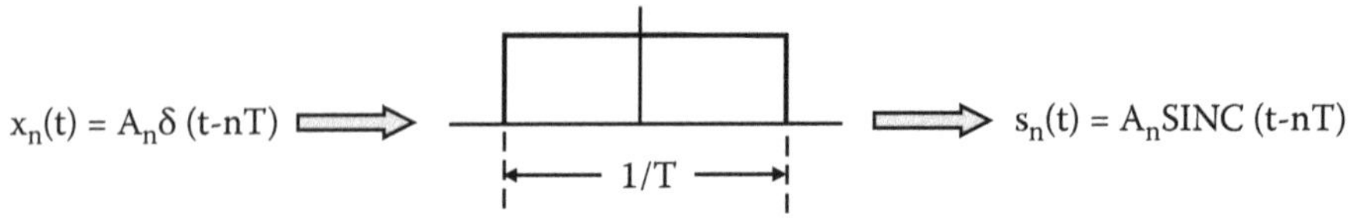

$$x_n(t) = A_n\delta\,(t\text{-}nT) \Longrightarrow \quad \overline{\phantom{\rule{1cm}{0pt}}} \quad \Longrightarrow \quad s_n(t) = A_n\text{SINC}\,(t\text{-}nT)$$
$$\longleftarrow 1/T \longrightarrow$$

FIGURE 9.29
Band-limited impulse symbol.

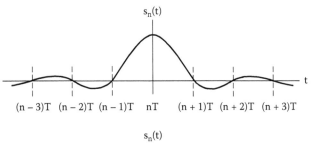

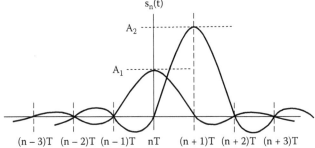

FIGURE 9.30
Intersymbol interference (ISI)-free impulse symbols.

$$g(t) = \frac{\sin\left(\dfrac{\pi t}{T}\right)}{\left(\dfrac{\pi t}{T}\right)} \times \frac{\cos\left(\dfrac{\alpha \pi t}{T}\right)}{1 - 4\left(\dfrac{\alpha \pi t}{T}\right)^2}, \quad 0 < \alpha < 1. \tag{9.24}$$

Once raised-cosine shaped, the channel frequency response extends beyond $1/2T$ by the factor of α. If $\alpha = 0$, it becomes the brick-wall filter.

Figure 9.30 shows the raised-cosine shaped impulse response, and this family of two adjacent impulse symbols with magnitudes of A_1 and A_2 at nT and $(n + 1)T$. Because there are nulls at every T, there exists no ISI. However, the exact timing is required to recover data correctly. This synchronous demodulation involves very sophisticated digital signal processing in modern data receivers, and requires that the channel be equalized first to recover the original shaped symbol. In most digital communication and hard disk, decision-directed adaptive equalizers shape the total channel response to be raised-cosine shaped, and the timing recovery is done digitally at the symbol rate.

An example of the symbol timing recovery used in hard disk and magnetic tapes is sketched in Figure 9.31. From three data A, B, and C sampled at the symbol rate, the early or late timing error can be derived. When the impulse peak is sampled at the right time, both pre- and postcursors are symmetric, and early and late conditions can be estimated digitally by comparing $(B - A)$ and $(B - C)$. This phase error is used to adjust the sampling clock or digitally shift the data in time so that PLL can be implemented digitally.

9.5.2 Binary Pulse Symbol

The plain binary pulse is another symbol commonly used in baseband digital communications such as wire-line or optical networks. Unlike the impulse symbol that requires exact

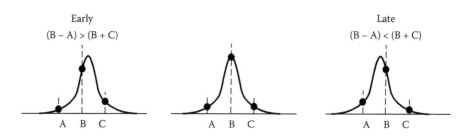

FIGURE 9.31
Clock recovery in the digital domain.

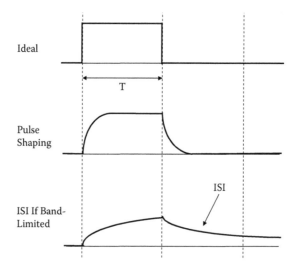

FIGURE 9.32
Pulse-shaped binary pulse symbol and ISI.

equalization, the binary pulse just needs broad bandwidth. The pulse shaping for binary pulse symbols is shown in Figure 9.32.

Although ideal rectangular pulses are transmitted, the receiver side can only receive low-pass filtered waveforms. If band-limited, data cannot complete the full logic transitions during the symbol period T. Because they overflow into the subsequent symbol periods, incomplete symbol transitions cause ISI that affects the decisions of the subsequent symbols.

NRZ binary pulses have no timing information built in. The Manchester coding used in the old Ethernet 10 Base-T includes the clock transition in every symbol, but the data rate is reduced to half. Due to the randomness of the data, the PD for CDR should be triggered by NRZ data so that the PD can work only when data make transitions. Because they detect no frequency error, all NRZ data systems require initial frequency acquisition during the preamble period. When NRZ binary symbols are transmitted over band-limited channels, the DC baseline wanders. With DC wandering, a bit slicer with a fixed threshold cannot detect the symbol correctly as shown in Figure 9.33. Note that unlike the impulse-shaped symbol, the pulse-shaped binary symbol exhibits no precursor ISI.

Each symbol can have a long tail affecting many subsequent symbols. The identical situation occurs when driving heavily capacitive bus lines in large chips.

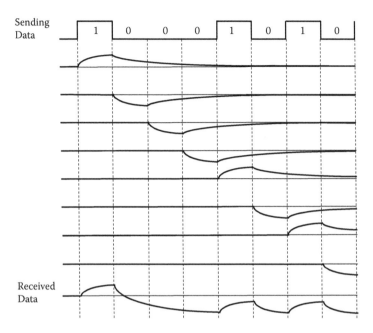

FIGURE 9.33
NRZ pulses in band-limited channel.

9.6 Data Channel Equalization

The received data should be properly equalized before bit slicing. The channel equalization is a must in all data communications as data are altered by the channel characteristic. Impulse symbols need exact raised-cosine equalization, and binary symbols need broad bandwidth. The critical raised-cosine equalization is mostly performed in the digital domain using finite impulse response (FIR) filters. The tap weights of the FIR filter are adaptively updated using decision-directed LMS algorithms. Therefore, such equalizers are called decision-directed equalizers. The exact pulse shape is of paramount interest in modern disk drive read channels because even the partial responses of an impulse are used to encode more data per symbol.

9.6.1 Linear Equalizer

Binary pulse symbols do not need exact equalization. Even in old disk drives, the symbol was slimmed by raising high frequencies, and the peak detector was used to detect the isolated impulse symbols. The pulse slimming effect can be achieved by linear equalizers, which raise high-frequency bands. The problem with the linear equalizer is that high-frequency noise is also amplified, and it is not that effective in terms of the signal-to-noise ratio in the bit decision. In the decision-directed equalizers, the amount of equalization is adaptively set, and the detection algorithms such as maximum likelihood further improve the receiver performance.

The linear equalizer system applied to the NRZ data is sketched in Figure 9.34. It raises the high-frequency portion of the data to make up for the channel attenuation and restores original pulse shape. Since the high-frequency boost also raises the noise level, it is of no use to boost the high-frequency band too much.

9.6.2 Decision-Feedback Equalizer

An alternative way to restore the high-frequency portion of the symbol is to use decision-feedback equalizer (DFE) as shown in Figure 9.35. The DFE is a feedback system that subtracts the estimated prior symbols affecting the current symbol before bit slicing. It is only effective for postcursor ISI as in the binary pulse symbol. As conceptually sketched, if all prior symbol ISI is assumed subtracted, only the new symbol makes a transition during the symbol interval. This is equivalent, in effect, to adding the missing high-frequency portion of the data. Because it is additive rather than multiplicative as in the linear equalizer, the signal-to-noise ratio is not degraded in bit slicing. An example of typical implementation of the DFE is shown in Figure 9.36.

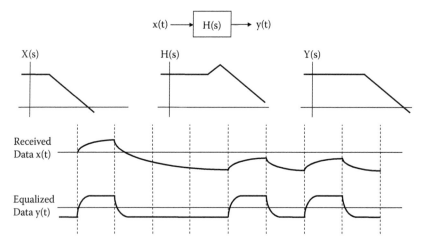

FIGURE 9.34
Linear equalizer for no-return-to-zero (NRZ) data.

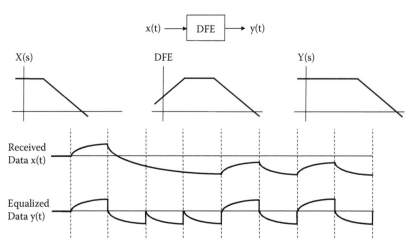

FIGURE 9.35
Decision-feedback equalizer for NRZ data.

Assume that the channel attenuates the pulse symbol of the magnitude 1 by a_1, a_2, and a_3, respectively, at the next three symbol periods. Then before the current symbol decision is made, the previous three decisions in the delay line can by multiplied by the same attenuation factors, and can be subtracted so that the decision can be ISI-free [14]. The common problem of the DFE is that it relies on the fact that all previous decisions are correct. However, if decisions are wrong, it is possible to get the burst of errors. This error propagation is a serious disadvantage, but in the modern error-correcting digital system environment, the advantage of using DFE far outweighs the burst error propagation problem.

9.6.3 Zero versus Cosine Equalizers

Linear equalizers insert zero at the channel band edge for high-frequency boost. There are many ways to insert zeros in the transfer function. Two examples are shown in Figure 9.37.

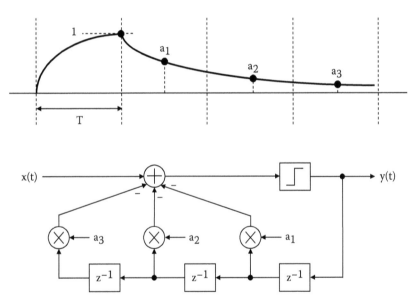

FIGURE 9.36
Implementation of decision-feedback equalizer (DFE).

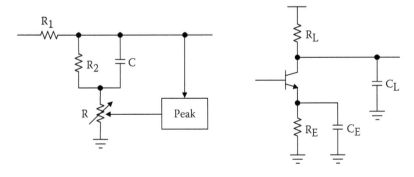

FIGURE 9.37
Two zero equalizers.

The one on the left is a standard automatic line build-out circuit that tracks the channel pole with a zero. The peak detector senses the attenuation of the magnitude, and adjusts the zero location accordingly. The transfer function is

$$H(s) = \frac{R}{R_1} \times \frac{s + \dfrac{R + R_2}{RR_2C}}{s + \dfrac{R_1 + R_2}{R_1R_2C}}. \tag{9.25}$$

The one on the right is an emitter peaking circuit that also introduces a zero as follows:

$$H(s) \approx -\frac{R_L}{R_E} \times \frac{1 + sR_EC_E}{1 + sR_LC_L}. \tag{9.26}$$

Both create negative real zeros, which compensate for the channel loss, but their group delay varies. The magnitude error after equalization affects the vertical eye opening, and the phase error affects the horizontal eye opening. For example, zeros for pulse slimming certain data symbols such as optical disc or disk drive are to boost the high-frequency portion, but should not introduce any irregular group delay. For that purpose, symmetric zeros are necessary. The symmetric zeros are obtained by feeding signal forward across two integrators as shown in Figure 9.38.

The top one is an integrator followed by a low-pass filter, and the bottom one is a resonator. Their normalized transfer functions are as follows:

$$H(s) = \frac{1 - ks^2}{s(1 + s)}. \quad H(s) = \frac{1 - ks^2}{1 + s^2}. \tag{9.27}$$

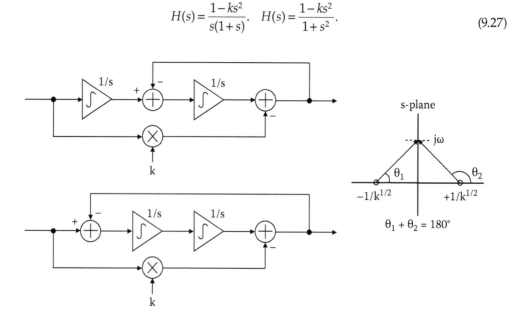

FIGURE 9.38
Symmetric zero equalizers.

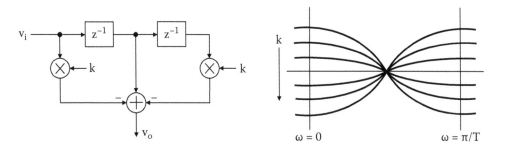

FIGURE 9.39
Cosine equalizer.

Both have symmetric real zeros at $\pm 1/k^{1/2}$, and the phase shifts due to the zeros are summed to be constant 180° as shown.

Another equal group delay filter named cosine equalizer can be implemented in a sampled-data form as shown in Figure 9.39.

It is a three-tap delay line filter, with a frequency response of

$$H(j\omega) = e^{-j\omega T}(1 - 2k\cos\omega T). \tag{9.28}$$

As shown, the tilt of the gain response can be adjusted with no group delay variation because it is an FIR filter.

9.7 Clock and Data Recovery

Once the incoming data waveform is equalized, the clock should be recovered for bit slicing. Impulse symbol data systems rely on digital processing both for equalization and symbol detection, but some NRZ systems still use analog PLL.

Figure 9.40 shows a common format for the data packet used in most digital communications. The length of the preamble depends on the standard from the shortest four symbols of 1010 for Bluetooth and the fairly long 192 μs preamble for wireless local area network (WLAN) standards. The CDR PLL should recover timing during the preamble period, and keep the timing to receive the data load. NRZ PDs work only when data make transitions. When data stay high or low, the PLL is in open-loop condition, and the VCO is not updated. If the data load has too long sequence of 1s or 0s, the VCO may drift away due to the lack of timing information, and the CDR PLL can lose phase locking. To avoid this situation, most standards use run-length-limited (RLL) coding, and data should make transitions after a limited number of long 1 or 0 sequences so that the CDR PLL can sustain phase locking.

Figure 9.41 shows the NRZ timing recovery. CDR is to align the rising edge of the local clock with the zero-crossings of the incoming data so that the falling edge can be used for bit slicing. During the preamble period, CDR acquires frequency and phase locking, and PLL should be broad-banded using PFD. However, during the data load period, PLL just sustains locking, and NRZ PD works only when data make transitions as marked with stars. In the tracking mode, PLL should be narrow-banded. Although the PLL can be operated with both rising and falling edges of the data, it would be jittery. If only either rising or falling edge is used, PLL is slower though it is less jittery.

9.7.1 CDR with Band-Pass Filter

The simplest CDR has been implemented using a high-Q band-pass filter as shown in Figure 9.42. The band-limited data stream has no clock component as shown. A nonlinear device such as a squarer or a rectifier is needed to broaden the spectrum to create the clock component. High-Q band-pass filter recovers the exact phase clock. This simple CDR scheme has been used in most repeater applications like Sonet.

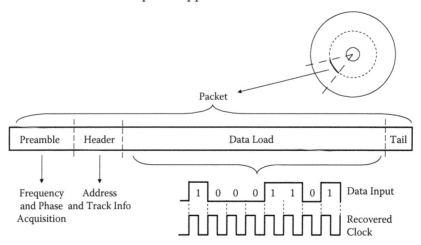

FIGURE 9.40
Data packet for digital communications.

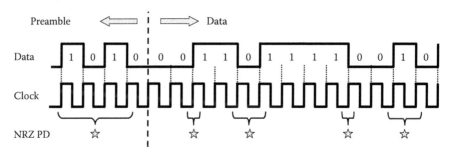

FIGURE 9.41
NRZ timing recovery.

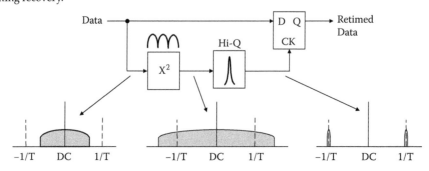

FIGURE 9.42
Clock and data recovery (CDR) using band-pass filter.

9.7.2 Oversampling Digital CDR

An oversampling polyphase sampler/latch explained in Figure 9.43 can be used as a PD for CDR [15]. Polyphase VCO outputs are obtained from ring-type VCOs. The polyphase clocks slice the incoming data and produce snapshots of the data showing where the zero-crossings are. The digital loop filter can be used to control the center of the window aligned to the 0-1 or 1-0 transitions relative to the middle of the window. Because the phase error is in the digital domain, it is a digital PLL, and the PD is time-to-digital converter (TDC). It is also called a hybrid analog/digital CDR. The finite time step contributes to the phase noise, which is not serious in CDR, but in frequency synthesis, the quantized phase noise should be lower than the VCO phase noise after phase locked.

9.7.3 Delay-Locked Loop for CDR

Delay-locked loop (DLL) is often used for CDR to adjust clock and data skews as shown in Figure 9.44. The clock is set, and the data are aligned with the clock. Because there is no VCO used, it makes the first-order loop, and there is no stability issue. Furthermore, no clock jitter is accumulated because the clock is fixed. Note that the variable wideband group delay for the signal is difficult to implement because the group delay varies as in the gain-bandwidth trade-off. It is also possible to delay the

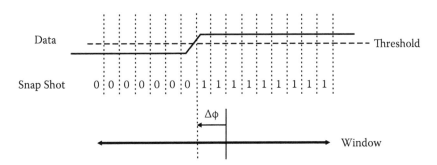

FIGURE 9.43
Digital snapshot of the data transition.

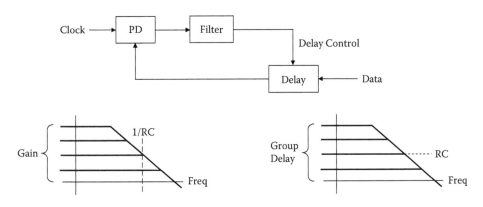

FIGURE 9.44
Delayed-locked loop.

clock to be aligned with the incoming data. Then, the delay element becomes the same complicated design as the VCO. It is the clock synchronization scheme widely used in high-definition multimedia interface (HDMI) and random access memory (RAM) bus at the board level.

9.7.4 PLL for CDR

The common CDR is based on PLL as shown in Figure 9.45. As discussed, it has dual loops for frequency acquisition and phase tracking, though they do not work simultaneously. The frequency acquisition loop is a wideband loop using PFD for fast initial acquisition during the preamble period, and the phase tracking loop is a narrow-band loop to sustain phase locking. The NRZ PD works only at data transitions.

Figure 9.46 shows the PLL bandwidth requirement for NRZ CDR during the tracking mode. The NRZ data cover a broad range from the lowest f_{min} when the data pattern repeats the long sequence of 1s and 0s to f_{max} when the data repeats 1010... at the maximum rate. Therefore, the PLL bandwidth should be set lower than $f_{min}/8$. During the acquisition mode, the PLL bandwidth can be wider because the preamble pattern repeats at high rates.

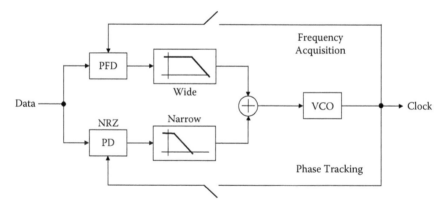

FIGURE 9.45
PLL-based timing recovery.

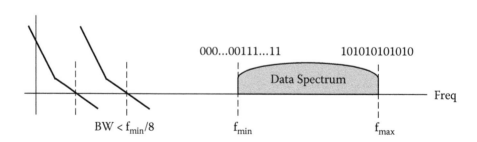

FIGURE 9.46
PLL bandwidth for NRZ clock and data recovery (CDR).

9.8 NRZ Phase Detector

For frequency synthesis, the state-machine up/down PFD is predominantly used with the CP. However, for CDR, there are many variations of the data-triggered NRZ PD, and even very coarse bang-bang-type PD based on early/late sampling is also used.

The most common NRZ PD is the Hogge-type PD as shown in Figure 9.47 [16]. Consider the logic is falling-edge triggered. The first edge-triggered D flip/flop makes the up pulse proportional to the phase difference of the data transition point from the clock falling edge, while the down pulse is fixed to be half the clock period. Therefore, if the logic swing is V_{DD}, this PD will have a gain of

$$k_d = \frac{V_{DD}}{4\pi}.\qquad(9.29)$$

Another common NRZ PD is the Alexander-type PD [17]. Its operation is explained in Figure 9.48. This simple early-late- or bang-bang-type PD is useful for very high-speed CDR when elaborate PD is not possible. It is based on the simple logic like the oversampling digital PD. Assume that the data are sampled three consecutive times across the zero-crossing points to get three decisions for early E, middle M, and late L, respectively. Depending on the following logic decisions, early or late phase error can be detected.

$$\text{Early}: \quad E = M \neq L.$$
$$\text{Late}: \quad E \neq M = L.\qquad(9.30)$$

This is the same nonlinear feedback as the 1b $\Delta\Sigma$ modulator, and the feedback quantization error is only high or low as sketched. Of course, sampling more points is equivalent

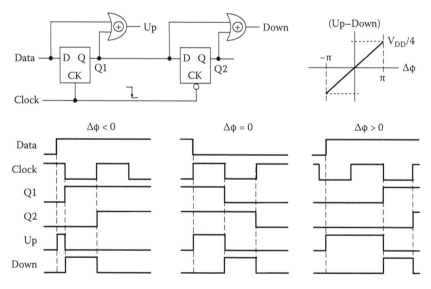

FIGURE 9.47
Hogge NRZ phase detector (PD).

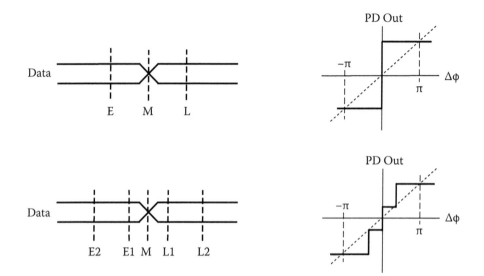

FIGURE 9.48
Early-late- or bang-bang-type PD.

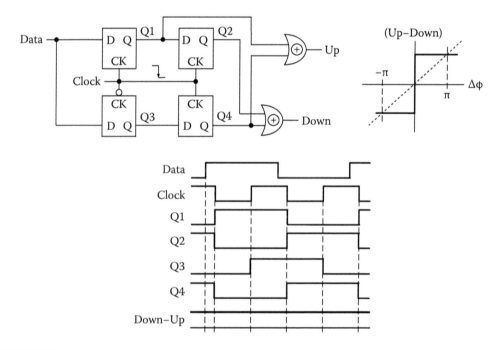

FIGURE 9.49
Alexander-type PD.

to the multibit quantizer case of the $\Delta\Sigma$ modulator. Note that the quantization step would be smaller than in the 1b case. This quantized feedback would be noisy and nonlinear but acceptable in CDR.

Figure 9.49 shows an implementation of the Alexander-type PD. It produces two high or low output levels as follows:

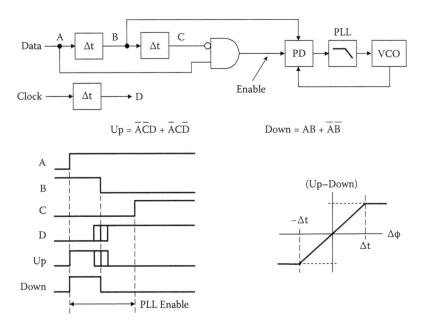

$$Up = \overline{A}\overline{C}D + \overline{A}C\overline{D} \qquad\qquad Down = AB + \overline{A}\overline{B}$$

FIGURE 9.50
Data-triggered NRZ PD.

$$Up: \quad Q_2 = Q_4.$$
$$Down: \quad Q_1 = Q_4. \tag{9.31}$$

Figure 9.50 also shows the Hogge-type data-triggered PD using simple logic gates [14]. The PLL enable signal can be generated every time the incoming data make transitions, and the up and down PD outputs can be generated using the simple logic as shown. The up pulse outputs the phase difference between the data edge and the clock only during the PLL enable period. The down pulse is always half of the enable period.

References

1. G. Gillette, "Digiphase synthesizer," *Proc. 23rd Annual Freq. Cont. Symp.*, pp. 201–210, 1969.
2. B. Miller and R. Conley, "A multiple modulator fractional divider," *IEEE Trans. Instrum. Meas.*, vol. 40, pp. 578–583, June 1991.
3. T. Riley, M. Copeland, and T. Kwasniewski, "Delta-sigma modulation in fractional-N frequency synthesis," *IEEE J. Solid-State Circuits*, vol. SC-28, pp. 553–559, May 1993.
4. I. Bietti, E. Ternporitil, G. Albasini, and R. Castello, "An UMTS SD fractional synthesizer with 200kHz bandwidth and –128dBc/Hz @1MHz using spurs compensation and linearization techniques," *IEEE Custom Integrated Circuits Conf.*, pp. 463–466, 2003.
5. W. G. Rhee, A. Ali, and B. S. Song, "A 1.1-GHz CMOS fractional-N frequency synthesizer with a 3b 3rd-order delta-sigma modulator," *IEEE J. Solid-State Circuits*, vol. SC-35 pp. 1453–1460, October 2000.

6. Y. Koo, H. Huh, Y. Cho, J. Lee, J. Park, K. Lee, D. Jeong, and W. Kim, "A fully integrated frequency synthesizer with charge-averaging charge pump and dual path loop filter for PCS and cellular CDMA wireless systems," *IEEE J. Solid-State Circuits*, vol. SC-37, pp. 536–542, May 2002.

7. S. Pellerano, S. Levantino, C. Samori, and A. L. Lacaita, "A dual-band frequency synthesizer for 802.11a/b/g with fractional-spur averaging technique," *Dig. Tech. Papers, IEEE Int. Solid-State Circuits Conf.*, pp. 104–105, February 2005.

8. W. Rhee and A. Ali, "An on-chip compensation technique in fractional-N frequency synthesis," *Proc. IEEE Int. Sym. Circuits and Systems*, pp. 363–366, 1999.

9. S. Meninger and M. Perrott, "A fractional-N frequency synthesizer architecture utilizing a mismatch compensated PFD/DAC structure for reduced quantization-induced phase noise," *IEEE Trans. Circuits and Systems II*, pp. 839–849, November 2003.

10. S. Pamarti, L. Jansson, and I. Galton, "A wideband 2.4GHz delta-sigma fractional-N PLL with 1-Mb/s in-loop modulation," *IEEE J. Solid-State Circuits*, vol. SC-39, pp. 49–62, January 2004.

11. Y. Dufour, "Method and apparatus for performing fractional division charge compensation in a frequency synthesizer," U.S. patent 6 130 561, 2000.

12. M. Gupta and B. S. Song, "A 1.8GHz spur-cancelled fractional-N frequency synthesizer with LMS-based DAC gain calibration," *IEEE J. Solid-State Circuits*, vol. SC-41, pp. 2842–2851, December 2006.

13. A. Swaminathan, K. Wang, and I. Galton, "A wide-bandwidth 2.4 GHz ISM-band fractional-N PLL with adaptive phase-noise cancellation," *IEEE J. Solid-State Circuits*, vol. SC-42, pp. 2639–2650, December 2007.

14. B. S. Song and D. C. Soo, "NRZ timing recovery for band-limited channels," *IEEE J. Solid-State Circuits*, vol. SC-32, pp. 514–520, April 1997.

15. B. Kim, D. N. Helman, and P. R. Gray, "A 30-MHz hybrid analog/digital clock recovery circuit in 2-μm CMOS," *IEEE J. Solid-State Circuits*, vol. SC-25, pp. 1385–1394, December 1990.

16. C. R. Hogge, Jr., "A self correcting clock recovery circuit," *IEEE Trans. Electron Devices*, vol. ED-32, pp. 2704–2706, December 1985.

17. J. D. H. Alexander, "Clock recovery from random binary data," *Elect. Lett.* vol. 11, pp. 541–542, November 1975.

Index